Lectures on
Surface Science

Lectures on Surface Science

Proceedings of the Fourth Latin-American Symposium
Caracas, Venezuela, July 14–18

Editors: G. R. Castro and M. Cardona

With 204 Figures

Springer-Verlag Berlin Heidelberg New York
London Paris Tokyo

Professor Dr. German R. Castro
Universidad Central de Venezuela, Apartado 21201,
Caracas 1021-A, Venezuela

Prof. Dr., Dres. h. c. Manuel Cardona
Max-Planck-Institut für Festkörperforschung, Heisenbergstraße 1
D-7000 Stuttgart 80, Fed. Rep. of Germany

ISBN 3-540-17318-8 Springer-Verlag Berlin Heidelberg New York
ISBN 0-387-17318-8 Springer-Verlag New York Berlin Heidelberg

Library of Congress Cataloging-in-Publication Data. Lectures on surface science. Papers presented at the Fourth Latin-American Symposium on Surface Physics. 1. Surfaces (Physics) – Congresses. 2. Surface chemistry – Congresses. I. Castro, G.R. (German R.), 1950–. II. Cardona, Manuel, 1934–. III. Latin-American Symposium on Surface Physics (4th: 1986: Caracas, Venezuela)
QC173.4.S94L43 1987 530.4'2 87-319

Printed in Germany

Offset printing: Weihert-Druck GmbH, D-6100 Darmstadt
Bookbinding: J. Schäffer GmbH & Co. KG, D-6718 Grünstadt
2153/3150-543210

Preface

The Fourth Latin-American Symposium on Surface Physics (SLAFS 4) was held from July 14 to 18, 1986, at the Unversidad Central de Venezuela, Caracas, Venezuela, as part of a program of the Centro Latinoamericano de Fisica (CLAF) aimed at improving the scientific relations and cooperation among Latin-American researchers in the field of surface physics. This symposium, which brought together about 60 scientists, included 12 invited talks given by internationally recognized experts covering a wide range of surface physics topics and showing the most recent developments of the theory and spectroscopies involved. This proceedings volume contains the text of the 12 invited lectures plus a selection of the Latin-American contributed papers. The topics covered during the conference were: thin films, chemisorption theory, and surface spectroscopies such as high resolution electron energy loss spectroscopy (HREELS), inverse photoemission, Auger deexcitation, SEXAFS, LEED, Raman spectroscopy (SERS) and deuteron solid state FT-nuclear magnetic resonance spectroscopy. There were also some contributions related to the application of surface analysis to the chemical and semiconductor industries. In addition to the talks there were two poster exhibitions, which showed mainly Latin-American contributions, making it possible to discuss extensively the work presented.

Finally, we wish to acknowledge the assistance given by the local organizing committee, especially the diligent cooperation of Luis Cortina, Iván Escalona, Helena Isérn, Miguel Martin, Claudio Mendoza and Ivar Pettersson in developing the symposium program. We are also indebted to the international organizing committee for cooperation in editing the contributions presented at the conference. We acknowledge financial support provided by UNESCO/CLAF and the local institutions, especially by the Universidad Central de Venezuela, IBM de Venezuela, Fundación Polar and Petróleos de Venezuela.

Caracas, September 1986

G.R. Castro
M. Cardona

Contents

Part II Theory of Clean Surfaces and Chemisorption

Part III Surface Spectroscopies

Part IV **Structure and Characterization of Surfaces**

Part I

Thin Films and Superlattices

Folded, Confined, Interface, and Surface Vibrational Modes in Semiconductor Superlattices

M. Cardona

Max-Planck-Institut für Festkörperforschung, Heisenbergstrasse 1, D-7000 Stuttgart 80, Fed. Rep. of Germany

The vibrational modes of semiconductor superlattices are related to those of their bulk constituents. The reduction in the translational symmetry, however, makes possible the observation by optical spectroscopy of modes which are symmetry forbidden in the bulk materials. Interesting but subtle concepts, such as those of Brillouin zone folding and mode confinement, arise. Also, electrostatic interface phonons exist in ideal infinite superlattices. In practical superlattices with a finite number of periods, surface vibrational modes also occur. Most of the past experiments have been performed by means of Raman spectroscopy on GaAs-$Al_xGa_{1-x}As$ superlattices. The properties of these modes, their symmetries and relationship to the modes of the bulk components are discussed. Vibrational spectroscopy can also be applied to the characterization of superlattices (e.g. strained superlattices).

1. INTRODUCTION

Semiconductor superlattices have become one of the most fruitful objects of physical investigation since they were proposed in the pioneering work of Esaki and Tsu /1/. In this work, vibrational spectroscopy has played a prominent role since 1977 /2,3/. In Refs. 2 and 3 GaAs-AlAs small period superlattices were prepared and their vibrational modes investigated by means of Raman and ir spectroscopy. The authors found vibrational modes related to the $\vec{q}=0$ Raman phonons of the bulk materials and also modes at lower frequencies which they attributed to the "folded" dispersion relations of acoustic phonons. The latter, however, were more recently shown to be partly spurious, due to scattering from air at the sample surface /3/. Nevertheless the concept of Brillouin zone folding has been very useful in all subsequent work to interpret vibrational spectra of superlattices.

In the present paper we shall emphasize the difference between the concepts of zone folded modes, applicable to acoustic-like phonons, and confined modes applicable to the phonons derived from the optical branches of the bulk materials. This difference does not come clearly through in most of the pertinent literature /5/.

Much of the subsequent work has been performed on GaAs-$Ga_{1-x}Al_xAs$ superlattices (in contrast to the original GaAs-AlAs work of Refs. 2-3). While these superlattices for x<1 are easier to grow and more stable than those for x=1 they have the disadvantage that the vibrational structure of the bulk random alloy constituent ($Ga_{1-x}Al_xAs$) has not yet been satisfactorily described theoretically. Even if such description were available it would be rather complicated and difficult to transfer to the treatment of the superlattices /4/. Hence best understanding of the basic problem of vibrations of these types of superlattices is obtained by performing measurements on GaAs-AlAs superlattices /5/. We shall confine ourselves here mostly to this type of materials.

While the original work of Ref. 2 included infrared studies, it has become obvious in later work that Raman spectroscopy is more suitable for such investigations. The penetration depth of infrared radiation is usually $\gtrsim 1\mu m$ and thus rather thick superlattices must be made if reasonable signals are to be obtained. Transmission through the substrate also complicates the problem. Raman spectroscopy, however, only samples the penetration depth of the laser (or the scattered radiation) which, in the case of visible lasers (e.g. Ar^+) is typically between 200 Å and $1\mu m$. This gives a very convenient superlattice thickness range to work with. Raman spectroscopy also offers the possibility of tuning the laser radiation (or the scattered one) to any of the interband gaps of the material, a fact which leads to rather rich resonant phenomena. This, and also changes in the polarization of the laser light and analysis of the polarization of the scattered beam, plus the possibility of varying the scattering wave vector, has given Raman spectroscopy a considerable lead over the other spectroscopic techniques when it comes to investigating superlattices. Very recently, data obtained with high-resolution electron energy loss spectroscopy (HREELS) have also become available /6/. They are particularly suitable to investigate excitations localized in the immediate vicinity of the surfaces ($\lesssim$40 Å).

The GaAs-AlAs superlattices are easy to prepare by molecular beam epitaxy (MBE) techniques on GaAs substrates /1/. They have the interesting property of a very small lattice constant mismatch between GaAs (a_0 = 5.653 Å) and AlAs (a_0 = 5.660 Å). This favors their growth and the absence of mismatch dislocations. Recently, considerable interest has been devoted to other types of superlattices in which the a_0 mismatch is larger, such as GaSb-AlSb (0.65% mismatch) /7/, or Si-$Ge_{1-x}Si_x$ (up to 4% mismatch) /8/ and GaAs-$In_xGa_{1-x}As$ /9/. If the superlattice periods are thin, the lattice constants equalize themselves through the establishment of extensive and compressive strains in both components, respectively. As the periods grow, it becomes energetically more favorable to release the strain through the build-up of misfit dislocations /10/. Raman scattering is an ideal tool to investigate these phenomena: strains shift and split the Raman phonons of the bulk material. These shifts and splittings can be used to evaluate the stress once their dependence on stress is known /11/.

Superlattices made of two amorphous components (e.g. Si-$Si_{1-x}N_x$) have also received considerable attention on account of ease of preparation and interesting optoelectronic properties /12/. Their acoustic phonons also exhibit "folding" effects /13/, a fact which confirms the quasicontinuous, elastic nature of these vibrations, independent of the periodicity of the microscopic crystal lattice. It is interesting to point out that the elastic continuum theory of acoustic phonons in superlattices was worked out by Rytov as early as in 1955, in connection with the propagation of seismic waves through stratified media /14/. The information to be gained from the optic phonons of amorphous superlattices, however, is rather limited, due to the broad nature of their Raman spectra /15/.

Recently, considerable interest has been generated by the so-called quasicrystals which contain several sets of basis vectors which are incommensurate with one another. The one-dimensional case, which can be realized experimentally in the form of quasiperiodic superlattices, represents the simplest prototype of such crystals. Particularly interesting are the so-called Fibonacci superlattices, based on the repetitional algorithm which bears that name /16/. They have been recently built on the basis of GaAs and AlAs and their vibrational properties investigated with Raman spectroscopy /16/.

This article discusses recent experimental and theoretical developments in the field of vibrations in superlattices, with emphasis on the GaAs-AlAs system. Since a number of review articles on this subject has already appeared /17-20/, we shall mainly concentrate on a few basic concepts to which the author has recently devoted his attention.

2. INTRODUCTION TO RAMAN AND BRILLOUIN SCATTERING IN BULK CRYSTALS

2.1 Selection Rules and Coupling Constants

The principles of light scattering in solids are thoroughly discussed in the series "Light Scattering in Solids" /20/. The mechanism of light scattering by phonons is easy to understand in simple terms. Semiconductors have a very high electronic dielectric constant (GaAs: ε=12) and thus a large electric susceptibility or polarizability, connected with the small interband gap. The phonons modulate this susceptibility χ at their frequency ω_P by altering periodically the atomic positions around their equilibrium. Thus we have components of χ modulated like $\chi_P \sim \exp(\pm i\omega_P t)$. When multiplying this by the incident laser field, of frequency ω_L, a polarization (dipole moment per unit volume) ~ $\exp[-i(\omega_L \pm \omega_P)t]$ results. This acts as a radiating dipole and thus scattered radiation at the frequency $\omega_L \pm \omega_P$ results. The (+) sign corresponds to the Stokes (antistokes) component. The scattering efficiency (scattered intensity per unit incident intensity, unit solid angle, and unit path in the sample) contains as a factor $1+n_B(\omega_P)$ in the Stokes and $n_B(\omega_P)$ in the antistokes case, where n_B is the Bose-Einstein phonon occupation number. It also contains as the coupling constant the square of the so-called Raman susceptibility, i.e., the derivative of $\chi(\omega_L)$ with respect to the phonon coordinate displacement. $\chi(\omega_L)$ contains singularities at the so-called interband gap energies or critical points. Its derivative will be even more singular at these points. Thus resonance effects in Raman scattering by phonons arise.

In a crystalline sample there is an important selection rule which obtains from wavevector or crystal momentum conservation:

$$\vec{k}_S = \vec{k}_L \pm \vec{q}, \tag{1}$$

where $k_{S,L}$ are the wavevectors of the scattered and incident light, respectively, and q that of the phonon. The (+) sign applies to Stokes (antistokes) scattering. The order of magnitude of q at a general point of the Brillouin zone (BZ) is $2\pi/a_0$. The magnitudes of $\vec{k}_S$ and $\vec{k}_L$ are $2\pi n/\lambda_{S,L}$ where n is the refractive index of the medium and λ the wavelength in vacuum (~5000 Å). Hence k_S and k_L are nearly negligible compared with the dimensions of the Brillouin zone and only phonons with $\vec{q}$ very close to the center of the BZ (q=0, Γ-point). In the case of acoustic phonons with the corresponding small value of q, $\omega_P = qv_P$ (v_P is the speed of sound): The phenomenon is called Brillouin scattering.

The argument given above for the scattering efficiency must be slightly modified in the case of acoustic phonons with q≈0. Translational invariance arguments indicate that for phonons with q=0 (all atoms move by the same amount u, pure translation) χ is not modified by u and thus the "Raman susceptibility" vanishes. A non-vanishing differential susceptibility is obtained for $q \neq 0$, however. In this case $du/dz = \varepsilon = iqu$ (z is the propagation direction and ε a component of the strain tensor). The strain ε multiplied by the corresponding component of the strain-optical constants (p_{ij}) gives a time-dependent component of the susceptibility (the Brillouin susceptibility) which is responsible for the Brillouin scattering. This Brillouin susceptibility exhibits resonance phenomena near electronic transitions, somewhat similar to those found in Raman scattering. Examples of Brillouin and Raman resonances and their detailed theory for bulk semiconductors can be found in Ref. 21.

The scattering efficiency is usually written as:

$$\frac{\partial^2 S}{\partial \Omega \partial \ell} = e_S \cdot \overleftrightarrow{R} \cdot e_L \; ^2, \tag{2}$$

where Ω is the solid angle, ℓ the path in the sample, $e_{S,L}$ the unit vectors of the scattered (S) and the incident (L) electric fields and $\overleftrightarrow{R}$ the Raman (or Brillouin) tensor (second rank) for the phonons under consideration. The tensor $\overleftrightarrow{R}$ must have the same symmetry, i.e., belong to the same row of the same irreducible representation, as the corresponding phonon. It is a symmetric tensor (like χ), except possibly very close to a resonance. Phonons which belong to representations other than those of a (symmetric) second rank tensor are said to be Raman-(dipole) forbidden. We should mention that for phonons which are simultaneously ir and Raman active (this is only possible in the absence of inversion symmetry) the longitudinal component of these phonons may be seen in dipole forbidden configurations close to resonance. It is a quadrupole effect which can be represented by a completely symmetric tensor $\overleftrightarrow{R}$ in Eq. (2). This tensor does not have the symmetry of the corresponding q=0 phonons hence the finite q vector is of the essence for their Raman coupling: The scattering efficiency is proportional to q^2 for small q.

2.2 Comparison with Inelastic Neutron Scattering

The $q \approx 0$ selection rule is common to all optical spectroscopies, in particular to ir absorption spectroscopy. Besides, in centrosymmetric materials there is an exclusion selection rule: Raman active phonons must be even, ir-active ones odd with respect to the inversion. ir and Raman activity are thus mutually exclusive.

It is of interest to compare Raman spectroscopy with its inelastic neutron scattering counterpart. The wavelengths of thermal neutrons lie typically around 5 Å, hence Eq. (1) enables one in this case to reach all values of q in the BZ. The neutron technique thus seems to be far superior to that of Raman spectroscopy since it yields information about all phonons, not only those near q=0. Inelastic neutron scattering has indeed been used to determine the full phonon dispersion relations of many crystals /22/. Nevertheless a number of reasons make Raman spectroscopy preferable for many investigations of phonons. For one thing, the cross-section for neutron scattering by phonons is rather small and large samples ($\lesssim 10$ cm^3) are needed (useless for superlattices!). Also, the frequency accuracy and resolution of phonon spectrometers is a factor of 10 to 100 times smaller than that of Raman spectrometers (in the latter accuracies and resolutions better than 0.1 cm^{-1} can be obtained /23/). With Raman spectroscopy scanning microprobing of a sample surface is possible (commercial Raman microprobes are available), obviously not with neutron spectroscopy. Last but not least, neutron reactors are expensive, cumbersome, and questionable in ecological terms.

2.3 Ways of Circumventing the $\vec{k}$-Conservation Selection Rule

Equation (1) is a consequence of the stringent translational symmetry of the perfect crystal. Several techniques are used to circumvent it. Tetrahedral semiconductors can be prepared in either crystalline or amorphous form. The latter possesses the same tetrahedral short-range order as the crystalline form but no long-range order at all (this is the commonly accepted interpretation. Ourmazd et al. have recently suggested that amorphous semiconductors may be actually composed of very small crystallites /24/. This may not alter our present argument). As a result, instead of the sharp optical phonons at q=0 the Raman (and also ir) spectra of amorphous tetrahedral semiconductors contain four broad bands which correspond to the TA, LA, LO, and TO phonon branches (see Fig. 1). These spectra represent approximately (after removing the dispersion of coupling constants and Bose-Einstein factors) the density of phonon states vs. ω_p.

Another way of observing phonons away from q=0 with optical techniques is by measuring the rather weak second-order spectra in which two phonons are generated. One can say that one phonon destroys the translational invariance while the other samples the thus distorted crystal (this is similar to the situation in a superlattice to be discussed below). All q's can be reached since only the sum of the two-phonon wavevectors must be nearly zero ($q_1+q_2 \approx 0$). If both phonons are of the same frequency (scattering by overtones) one obtains a Raman spectrum which represents the density of one-phonon states with the energy scale multiplied by two (Fig. 2).

Another way of partially destroying the translational symmetry is through the random addition of substitutional or interstitial impurities. Particularly interesting in connection with superlattices is the case of the mixed crystals such as $Al_{1-x}Ga_xAs$ /25,26/. Figure 3 shows the appearance of disorder activated TA, LA, LO, and TO bands (DATA, DALA...) in $Ga_{0.8}Al_{0.2}As$. These bands represent the corresponding broadened density of phonon states of GaAs.

A more "civilized" way of destroying or reducing the translational invariance is by the formation of a one-dimensional superlattice. This lifts most of the original translational symmetry operations and leaves only a few which correspond to the period of the superlattice. A GaAs-AlAs superlattice with a period of n_1 layers of GaAs (layer thickness a_1) and n_2 (001) layers of AlAs (thickness a_2) is shown schematically in Fig. 4. The new period becomes $(n_1a_1 + n_2a_2) = d$ instead of $a_1(a_2)$ of the bulk materials. The reduced BZ of the superlattice now extends from q=0 to $q = \pm\pi/d$ instead of the much larger boundaries π/a_1 and π/a_2 of the bulk materials (see Fig. 5). The dispersion relations of the bulk crystals must now be mapped onto the BZ of the superlattice (Fig. 5). While the bulk crystal has only one set of optical phonons (one LO and two TO phonons each set) at q=0 the superlattice has $2(n_1+n_2)-1$ sets. All of these q=0 phonons may, in principle, be Raman active. In this manner one can see by Raman spectroscopy more phonons than in the corresponding bulk crystals.

We should remark that extensive use of the concept of increased period and BZ-folding has been made for natural crystal, i.e., for the various polytypes of SiC /27/. The assumption was made that zone folding leaves unaltered the dispersion relations since it does not change the nearest neighbor structure. Thus, under this assumption, Raman measurements for several SiC polytypes (3C, 4H, 6H, 15R, 21R) should yield accurate dispersion relations for zincblende-type SiC (3C). An analysis of the intensi-

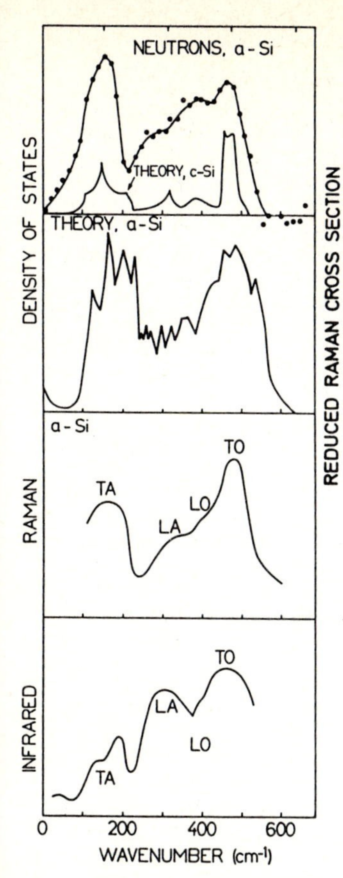

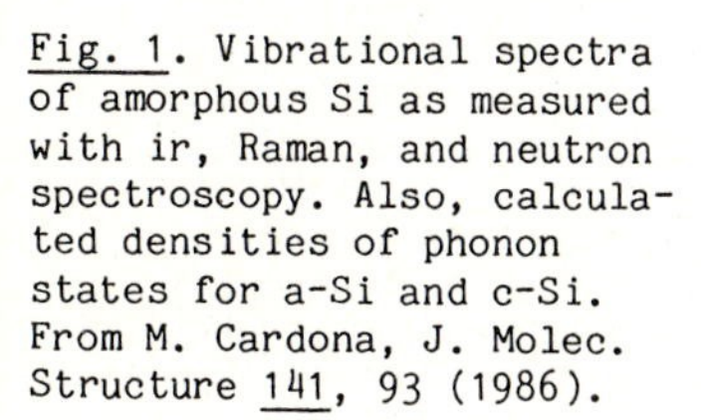
Fig. 1. Vibrational spectra of amorphous Si as measured with ir, Raman, and neutron spectroscopy. Also, calculated densities of phonon states for a-Si and c-Si. From M. Cardona, J. Molec. Structure 141, 93 (1986).

Fig. 2. Two-phonon spectrum of germanium as compared with the density of overtone states (from B.A. Weinstein and M. Cardona, Phys. Rev. 7, 2542 (1973).

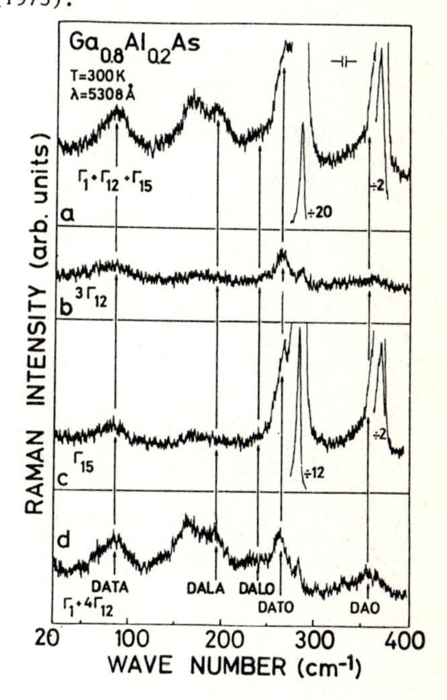

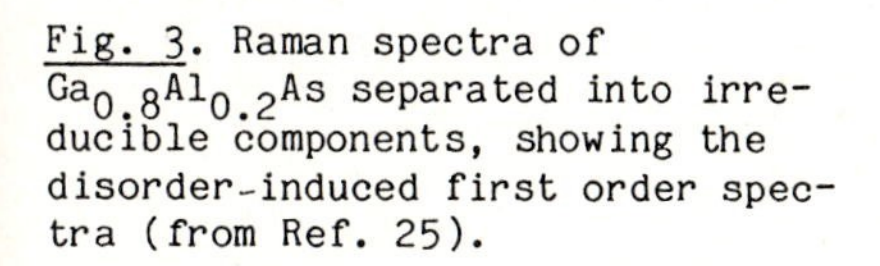
Fig. 3. Raman spectra of $Ga_{0.8}Al_{0.2}As$ separated into irreducible components, showing the disorder-induced first order spectra (from Ref. 25).

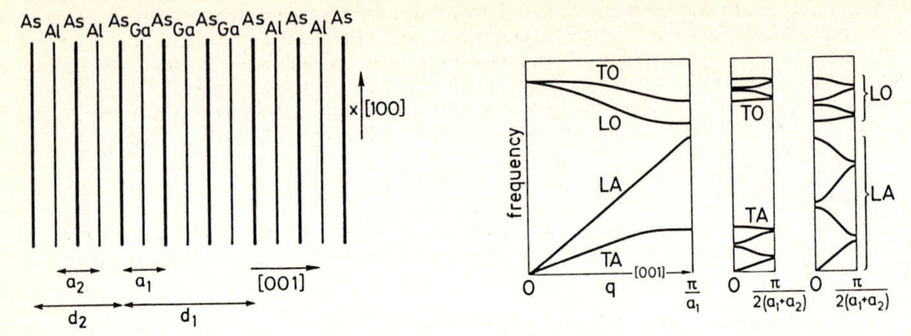

Fig. 4. Sketch of a GaAs-AlAs superlattice (n_1,n_2) with $n_1=3$ and $n_2=2$.

Fig. 5. Schematic representation of the folding of the Brillouin zone for the superlattice with $n_1=2$, $n_2=2$ (2,2). Note that for the optical modes to be of the "folded" type, as shown here, the difference in the cation masses would have to be much smaller than the masses of Al and Ga.

ties of the folded modes has been published recently /28/. Vibrations in a polytype of germanium (4H) and their zone-folding effects have been investigated recently /28a/. Also, similar investigations of a layered tetrahedral crystal whose unit cell involves two molecules of GaGeTe and has a trigonal, superlattice-like structure, have been performed /28b/.

3. DISPERSION RELATIONS:FOLDING AND PHONON CONFINEMENT

Let us consider the superlattice of Fig. 4 or any similar superlattice. The difference in the lattice dynamical parameters (atomic mass, restoring force constants, speeds of sound, densities) of both constituents is not large, otherwise they would not form alloys nor grow epitaxially on each other. In fact, the main difference lies in the atomic mass of Al (27 atomic mass units) vs. Ga (70 amu). As far as the acoustic phonons are concerned, the atomic mass enters in their frequency as the density and thus has to be averaged with the mass of As (75 amu): the densities of GaAs and AlAs only differ by 30% (the corresponding frequencies by 15%). This difference can be treated by perturbation theory. Let us consider a fictitious bulk crystal with the average parameters of GaAs and AlAs, in particular the average density. The superlattice of Fig. 4 can be obtained by applying a square wave perturbation to these parameters /29/. Sinusoidal superlattices have also been treated in Ref. 29.

There is a strong formal analogy between the lattice dynamical problem of acoustic phonons in a superlattice and that of electrons in a crystal lattice. Let us first consider free electrons in a crystal with a given structure but vanishingly small potential. The electrons have a parabolic dispersion $E = (\hbar^2/2m)q^2$ which must be folded into the reduced Brillouin zone of the crystal:

$$E = \frac{\hbar^2}{2m}(\pm\vec{q}+\vec{G})^2, \qquad (3)$$

where $\vec{G}$ are reciprocal lattice vectors /30/. Switching on a weak potential (nearly free electron or weak binding approximation WBA) leaves most of the "energy bands" of Eq. (3) nearly unchanged, except for the appearance of splittings at the center and the edge of the reduced BZ of the superlattice (the so-called minizone). Similarly, for acoustic phonons in the bulk crystal one has a linear dispersion relation ($\omega = v_p q$ which must be "folded" into the minizone of the superlattice in order to treat the corresponding perturbation (square wave modulation of parameters). For small perturbations one obtains splittings of the folded dispersion relations at the center and edges of the one-dimensional minizone (Fig. 5), i.e., at

$$q = \frac{\pi}{n_1 a_1 + n_2 a_2} m = \frac{\pi}{d} m, \quad m = 1,2,... \tag{4}$$

of the bulk crystal. A small deviation of the linearity of the dispersion relation, as expected for large q (bulk), should not significantly change this argument.

The "folding" of the BZ onto a minizone will substantially alter the optical (Raman) selection rule based on $\vec{q}$ conservation (Eq. (1)): while in the bulk crystal only acoustic phonons of very small frequency (up to $\omega = v_P (4\pi n/\lambda_L)$ for back scattering) are allowed (Brillouin scattering), many acoustic phonons of the bulk become "optic" phonons of the superlattice and for $q \approx 0$ (reduced minizone) they can be allowed in optical transitions. The new selection rules thus become in the backscattering configuration

$$q = \pm(4\pi n/\lambda) + \frac{2\pi}{n_1 a_1 + n_2 a_2} m, \quad m=0,\ 1,\ 2... \tag{5}$$

The argument just given is based on the continuum, elastic limit of the acoustic vibrations, valid for q (bulk) $\ll 2\pi/a_0$ (a_0 = bulk lattice constant $\approx a_1 + a_2$). Hence it should not matter whether the material is amorphous or crystalline. The dispersion relation of the acoustic modes in the superlattice should depend only on macroscopic parameters well defined in both cases, such as the density and the speed of sound. As an example we show in Fig. 6 the first folded LA doublet, at frequencies:

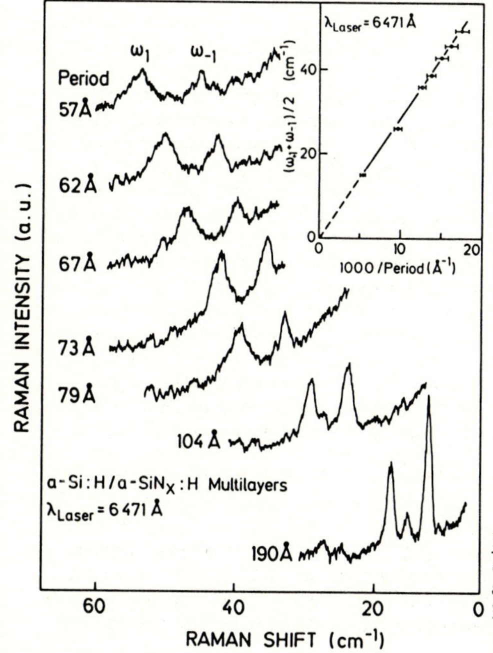

Fig. 6. First doublet in the Raman spectrum of the folded LA phonons of amorphous $Si\text{-}SiN_{1.3}$ superlattices. From Ref. 13.

$$\omega = v_p(\pm 4\pi n/\lambda + \frac{2\pi}{d}) \qquad (6)$$

obtained for longitudinal acoustic phonons in a series of amorphous superlattices with compositions Si-Si $N_{1.3}$ and several periods d /13/. The doublet observed reproduces well the trend given in Eq. (6).

The perturbative treatment just described is valid if the amplitude of the perturbation is small compared with the width of the dispersion band under consideration. This is certainly the case for the acoustic phonon branches of the bulk materials which enter into typical superlattices. It is no longer true for the corresponding optical branches. In the case of GaAs-AlAs, for instance, the TO(q=0) optical frequency of bulk GaAs is 267 cm^{-1} (at 296 K) while that of AlAs is 361 cm^{-1} (the large difference is mainly due to the difference in the reduced atomic masses, the frequencies being inversely proportional to the square root of those masses). The widths of the TO bands are <30 cm^{-1}, i.e., much less than the difference in the TO(q=0) frequencies. Hence averaging both frequencies and treating the fluctuation as a perturbation is not very useful. It is more meaningful to treat the vibrations as localized on either the GaAs slabs, with no vibration of the AlAs, or on the AlAs slabs, with no participation of GaAs. In order to eliminate completely the vibration of one type of slabs one must go to very high order of perturbation theory and hence the perturbative type of description is not very useful. The procedure of localizing or confining the phonons is similar to the tight binding or LCAO method of calculating energy band structures of materials with two atoms per unit cell. When the difference between the energies of corresponding electronic states in the two atoms is larger than the overlap integrals this method, and not the WBM, gives the better description since the electrons are nearly localized in one type of atom or the other. Thus we have introduced the somewhat orthogonal concepts of zone folding, appropriate to acoustic phonons, and confinement, appropriate to flat optical branches. In the former case, and for $\lambda_L \gg d$, acoustic phonons with:

$$q = \frac{2\pi}{d}m, \quad m = 0, \pm 1, \pm 2, \qquad (7)$$

are activated by the superlattice perturbation. In the case of confined TO phonons we may assume that, approximately, the vibrational amplitude vanishes at the interfaces. Thus we observe vibrational frequencies nearly equal to those of each bulk material with

$$q_1 = \frac{\pi}{d_1}m, \quad q_2 = \frac{\pi}{d_2}m; \quad m = \pm 1, \pm 2. \qquad (8)$$

Note that for $d_1 = d_2$ Eqs. (8) and (7) coincide, and thus confusion between the clearly distinct concepts of folding and confinement may arise. Unfortunately this confusion is enhanced in much of the pertinent literature.

As an example of the excitation of TO phonons with q (bulk) $\neq$ 0 in superlattices we show in Fig. 7 the GaAs-like TO phonons seen for a GaAs-AlAs superlattice with $n_1 = n_2 = 7$ /5/. They correspond to Eq. (8) with m = 1, 2,...5. In Fig. 8 we display the frequencies of these phonons vs. q, and compare them with the dispersion relation of bulk GaAs. The agreement is excellent. We should point out, however, that the bulk dispersion relations which were obtained from neutron scattering had ±5 cm^{-1} error bars. Hence the superlattice data of Fig. 8 may yield a better representation of the bulk dispersion relations than the neutron data. In order to establish this point, however, one should clarify through theoretical analysis what are the possible deviations of the bulk phonon frequencies introduced by

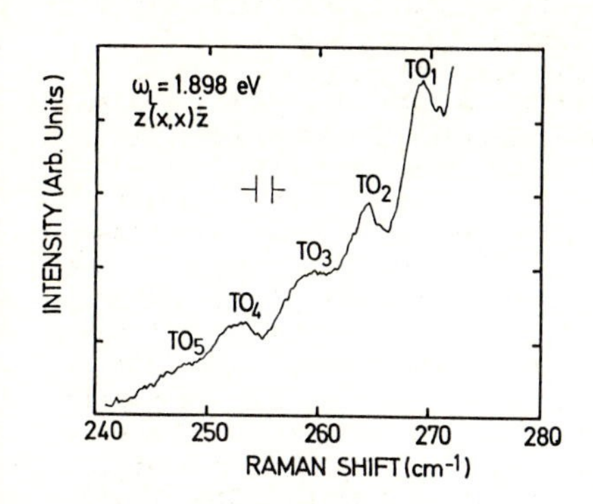

Fig. 7. TO-phonons confined in GaAs observed in a (7,7) GaAs-AlAs superlattice. The observation is only possible very close to an excitonic resonance (see Fig. 20). From Ref. 5.

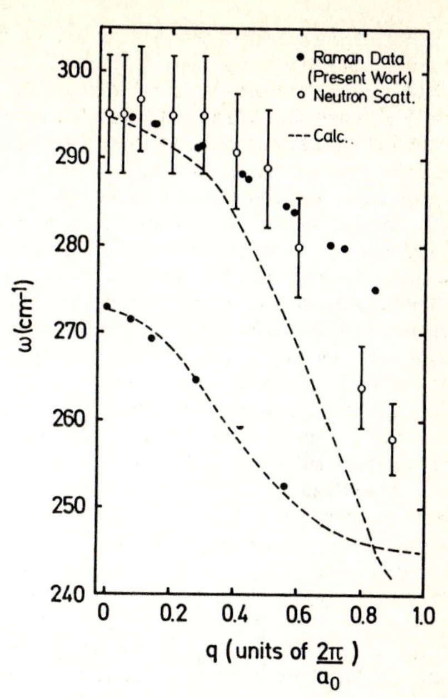

Fig. 8. Dispersion relations of TO- and LO-phonons obtained by plotting superlattice data according to Eqs. (8). Also neutron scattering data and calculated dispersion relations for the bulk. From Ref. 5.

the presence of the superlattice. We should point out, for instance, that theoretical analysis suggests that n_1 and n_2 may have to be replaced by n_1+1 and n_2+1 in Eqs. (4) /31/.

Before closing this section we would like to point out the analogy between the concept of confinement and that of Anderson localization /32/. The latter occurs when the perturbation at a random site is larger than the band width, a condition similar to that for confinement. In the latter case, however, the perturbing potential is regularly distributed and, although localization in one set of wells occurs, the localized states must be symmetrized according to Bloch's theorem and extended states result again. The phenomena nevertheless are rather similar, even hard to distinguish for a finite solid.

4. ACOUSTIC PHONONS: ZONE FOLDING

The zone folding of acoustic phonons can be easily treated on the basis of the elastic continuum model /14/ or the linear chain model /29,33/. The former is exact in the limit of small q's. The latter applies to arbitrary q's but it is only of easy applicability in the case of nearest neighbor interactions. The linear chain applies to 3d crystals provided one replaces interatomic by interplanar force constants /34/. We should also remark that the folded acoustic phonons of superlattices are rather similar to the so-called LAM (longitudinal acoustic modes) found in crystallized folded polymer chains /35/.

Folded acoustic modes in superlattices are usually observed in the backscattering configuration. Hence the scattering q is along the superlattice axis ($q = q_z$, $q_x = q_y = 0$). We thus limit ourselves to this case. The atomic displacements $\vec{u}$ for either longitudinal or transverse modes fulfill Laplace's equation /14/, hence they can be written as sums of sines and cosines of $ZQ_{1,2}$ where the $Q_{1,2}$'s are related to the vibrational frequency ω_p through $\omega_p = Q_{1,2}v_{1,2}$ ($v_{1,2}$ is the appropriate speed of sound in medium 1 or 2). After imposing the pertinent boundary conditions at the interface (continuity of $\vec{u}$ and of the components of the stress tensor which correspond to forces perpendicular to the interface) and using Bloch's theorem to relate u(q) in one period to u(q) in the next (through the phase factor e^{iqd}) we find the following secular equation which determines q vs. ω_p /14/:

$$\cos(qd) = \cos\frac{\omega_p d_1}{v_1}\cos\frac{\omega_p d_2}{v_2} - \alpha\sin\frac{\omega_p d_1}{v_1}\sin\frac{\omega_p d_2}{v_2}, \tag{9}$$

where

$$\alpha = 1 + \delta = \frac{1}{2}\left(\frac{\rho_2 v_2}{\rho_1 v_1} + \frac{\rho_1 v_1}{\rho_2 v_2}\right). \tag{10}$$

Equations (9) and (10) are valid for both longitudinal and transverse phonons provided one uses the appropriate values of $v_{1,2}$. Equation (9) has a form which is formally similar to that found for other propagation phenomena in stratified media like, for instance, optical interface modes /36/.

As already mentioned, $\rho_2 v_2$ is usually not very different from $\rho_1 v_1$, hence $\delta \ll 1$. Equation (9) can thus be rewritten as:

$$\cos(qd) = \cos\left[\omega_p\left(\frac{d_1}{v_1} + \frac{d_2}{v_2}\right)\right] - \delta\sin\frac{\omega_p d_1}{v_1}\sin\frac{\omega_p d_2}{v_2}. \tag{11}$$

For q not too close to the center or edge of the BZ (minizone) of the superlattice ($q \neq \pi m/d$) and δ small we may approximately set $\delta=0$ in Eq. (11). We thus find:

$$\omega_p(q) = \left(\frac{2\pi}{d}m \pm q\right)\langle v^{-1}\rangle^{-1}, \tag{12}$$

where <> represents the average over the two superlattice components, i.e.:

$$\langle v^{-1}\rangle = \frac{1}{d}\left(\frac{d_1}{v_1} + \frac{d_2}{v_2}\right). \tag{13}$$

Equation (12) reproduces the result obtained in Sect. 3, i.e., the folded dispersion relation of a bulk material with an <u>averaged</u> speed of sound. The error committed in Eq. (13) by neglecting δ is quadratic in δ provided $q \neq 0, \pi/d$. For $q = 0$ or π/d we find a correction linear in δ which produces a splitting of the two-fold degenerate frequencies given by Eq. (12) for q=0, πd: This splitting is:

$$\Delta\omega_p = \frac{\delta}{d\langle v^{-1}\rangle}\sin\frac{m\pi d_1}{d v_1\langle v^{-1}\rangle}. \tag{14}$$

The splitting $\Delta\omega_p$ is not easy to see under backscattering conditions since q is sufficiently large to make Eq. (12) an excellent approximation. Under forward scattering conditions, however, q becomes negligible compared with $2\pi/d$ and we can see a splitting of the Raman-active modes of Eq. (12). The corresponding measurement for a GaAs-AlAs superlattice with d = 37 Å and $d_1/d = 0.23$ (d_1 = GaAs thickness) is shown in Fig. 9. The observed split-

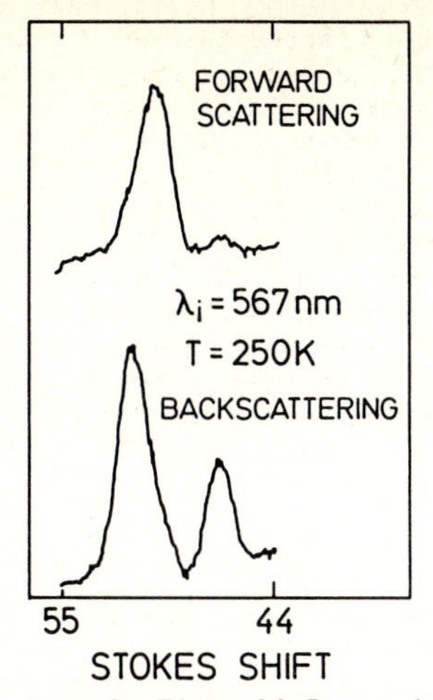

Fig. 9. First LA Raman doublet of a GaAs-AlAs superlattice with d = 37 Å and $d_1/d = 0.23$ as observed in forward and backscattering. The forward scattering spectrum displays the even-odd selection rule discussed in the text. From B. Jusserand et al. Phys. Rev. B 33, 2897 (1986).

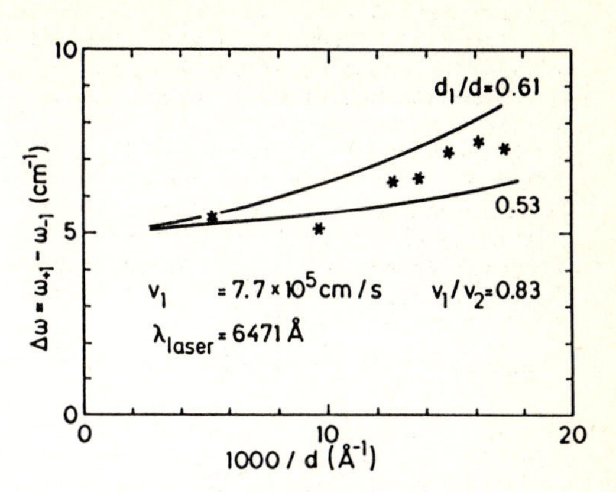

Fig. 10. Dependence of the m=1 LA doublet splitting of the superlattice of Fig. 6 on the period d. The points are experimental while the solid curves have been calculated with Rytov's equation (Eq. (9)). From Ref. 13.

ting $\Delta\omega_p$ = 2.5 cm^{-1} agrees well with the predictions of Eq. (14) for this system. We note that the upper component of the forward scattering doublet seen in Fig. 9 is strong while the lower component is very weak. In backscattering the strengths of both components are similar. This is due to the fact that for q=0 the upper component u is odd (~sin qz) with respect to the twofold rotation axes along x or y passing through the center of one layer while the lower component is even (~cos qz). The scattered intensity is obtained by averaging over a period the strain ($\partial u/\partial z$) times the appropriate photoelastic constant. The lower component, for which u is even, yields an odd strain which averages to zero if q is exactly zero. The upper component, odd in u, is even in the strain and averages to yield a strong signal. Away from q=0 the odd and even modes mix and both signals become nearly equal.

Another more indirect way of determining the splittings $\Delta\omega_p$ at the zone center (minigaps) has been suggested in Ref. 13. In the region of validity of Eq. (12), in which we can take δ=0 and thus the minigaps do not affect $\omega_p(q)$, we find a splitting of $\omega_p(q)$ which is independent of d. The backscattering measurements for amorphous SiN_x superlattices shown in Fig. 10 indicate that this is not the case when d becomes very small. This corresponds to the increase in $\Delta\omega_p \propto d^{-1}$ according to Eq. (14). The lines calculated with Eq. (11) for two values of d_1 encompass the experimental results.

The existence of minigaps at the BZ boundaries of GaAs-$Ga_xAl_{1-x}As$ superlattices has been demonstrated in a very elegant experiment involving the transmission of monochromatic phonons produced by a superconductivity tunnel junction /37/. Acoustic phonons in superlattices with a period composed of three different materials ($GaAs-Al_{0.5}Ga_{0.5}As-AlAs$) have been recently investigated by Nakayama et al. /9/.

5. OPTICAL PHONONS: CONFINEMENT

5.1 Boundary Conditions

We have already mentioned in Sect. 3, and also in connection with Figs. 7 and 8, that the optical phonons of GaAs-AlAs superlattices are confined either to the GaAs or the AlAs slabs. For TO phonons it is usually assumed that the appropriate boundary condition is u=0 at the interface. However, some penetration of the "forbidden" layers occurs. This penetration can be easily estimated by considering a parabolic expansion of the dispersion relation of the bulk material:

$$\omega_{TO}(q) = \omega_{TO}(0) - B_{TO}q^2. \tag{15}$$

From the dispersion relation of Fig. 8 we find for TO phonons in GaAs $B_{TO} = 88\ cm^{-1}\times Å^2$. For AlAs B_{TO} is estimated to be around $60\ cm^{-1}\times Å^2$. For q real Eq. (15) yields the standard down-dispersing TO bands. For q pure imaginary ($q = i\tilde{q}$, $\tilde{q}$ real) it represents up-dispersing bands which, however, decay in space like $\exp(-\tilde{q}z)$. The penetration depth of the AlAs-like mode into GaAs can be estimated with Eq. (15) from the difference between the ω_{TO}'s of GaAs (267 cm^{-1}) and AlAs (361 cm^{-1}). We write:

$$361-267 = 94\ cm^{-1} = B_{TO}\tilde{q}^2. \tag{16}$$

For the AlAs-like mode Eq. (16) yields a penetration depth into GaAs $\tilde{q}^{-1} \approx 1$ Å, not quite negligible when compared with $a \approx 2.7$ Å. A similar estimate can be made for the LO phonons ($\tilde{q}^{-1} \approx 0.9$ Å according to Ref. 38). For the GaAs-like mode the estimate is somewhat more difficult. The TO bands of AlAs now bend downwards, towards the ω_{TO} of GaAs. The reason why the GaAs modes do not penetrate into AlAs is because the ω_{TO} of AlAs never reaches ω_{TO} (q=0) of GaAs. Since it gets close, however, the corresponding penetration depth would be expected, in the spirit of Eq. (15), to be larger than that for the AlAs-mode into GaAs. An estimate in Ref. 38 gives a penetration depth of 2.8 Å for the GaAs-like mode into AlAs. Since $d_1 \approx d_2 = 2.7$ Å, a slight correction to Eq. (8), in the sense of increasing n_2 from the nominal value to about n_1+1, may be in order (see Eq. (19)).

The above discussion has a macroscopic character. When we talk about penetration depths of the order of a fraction of $d_{1,2}$, however, one should look at the microscopic details of u, i.e., at the vibrational eigenmodes. Such discussion has been presented in the three Phys. Rev. Letters Comments of Ref. 31. In the comment by Jusserand and Paquet a simple, linear chain calculation of the Kronig-Penney-type is performed /18,39/. Within this model the full dispersion relation is (note the formal analogy to Eq. (9)):

$$\begin{aligned} \cos(qd) &= \cos(d_1 . q_1)\cos(d_2 q_2) - \alpha \sin(d_1 q_1)\sin(d_2 q_2) \\ \alpha &= \frac{1-\cos(q_1 a)(\cos q_2 a)}{\sin(q_1 a)\sin(q_2 a)} \\ d &= d_1+d_2 = (n_1+n_2)a, \end{aligned} \tag{17}$$

where $a \approx a_1, a_2$ is the common monolayer thickness and q_1 and q_2 are related to the frequency ω_p through the corresponding bulk dispersion relations. Note that, in the case of optical modes, for q_1 real q_2 will be pure imaginary and viceversa. Taking k_1 real and $q_2 = i\tilde{q}$ with $\tilde{q} >> a^{-1}$ (see discussion of Eq. (16)) Eq. (17) becomes:

$$\sin(n_1+1)q_1 a = 0, \tag{18}$$

i.e.,

$$q_1 = \frac{\pi}{(n_1+1)a} m, \quad m = \pm 1, \pm 2, \ldots \tag{19}$$

instead of Eq. (8). For large $d_1 = n_1 a >> a$ both equations give similar results but considerable derivations may result for small d_1.

The physical interpretation of Eqs. (19) and its relationship to Eq. (8) is quite simple. Equation (8) corresponds to setting the vibrational amplitude equal to zero at the beginning and the end of a period (i.e., from As to As, see Fig. 11). Equation (18), however, implies that u first vanishes at the first Al atoms outside of the GaAs layer (Fig. 11). This is physically reasonable since Al is much lighter than either Ga or As and thus will not be able to follow the GaAs vibration. The As atoms tied to Ga, however, should follow.

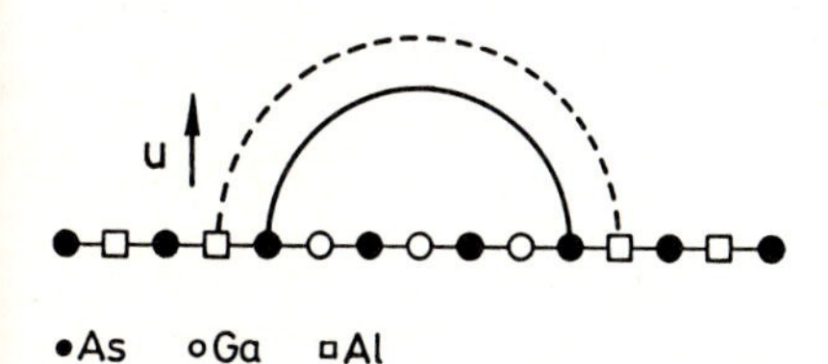

Fig. 11. Schematic diagram of the vibrational amplitudes of optical phonons of GaAs under the assumption u=0 at the interface (solid line) and at the first Al atom outside the interface in a (3,2) GaAs-AlAs superlattice.

A microscopic investigation of this boundary condition effect was performed by Molinari et al./31/. They carried out a calculation of the dispersion relations of GaAs-AlAs superlattice with $n_1 = n_2 = 7$. This calculation, which can be regarded as "ab initio", was based on planar force constants obtained from a first principles pseudopotential calculation performed for bulk GaAs. Figure 12 displays the frequencies found in this calculation plotted according to Eq. (19) vs. q. The agreement with the bulk dispersion relation calculated with the same force constants is excellent, better than that obtained using Eq. (8) (shown in Fig. 1 of the paper by Molinari et al. /31/). Figure 13 illustrates the cancellation of u for Al which is obtained in the "ab initio" calculation provided that m is small (m=1 in Fig. 13). However, for larger values of m (e.g. m=7 in Fig. 13) considerably more leakage into AlAs and Eq. (19) may lead to poor results. Figure 12, however, indicates that the approximation is excellent for all m's for the calculated eigenvalues although it is not for the eigenvectors.

In the zincblende structure the q=0 optical phonons are both Raman and infrared active. Because of the infrared activity, i.e., the existence of an electric dipole moment associated with the displacement u, the Raman phonons are split into an LO singlet and a TO doublet (except for forward scattering where polariton effects are observed /21/). The electrostatic fields associated with these phonons are subject also to boundary conditions. For the LO phonons neglecting retardation the electrostatic field $\vec{E}$ is related to a potential ϕ: $\vec{E} = -\vec{\text{grad}}\, \phi$. Sometimes in the literature these modes are treated as consisting only of electrostatic fields and thus the boundary condition $\phi=0$ at the interface is imposed /40,41/. Since

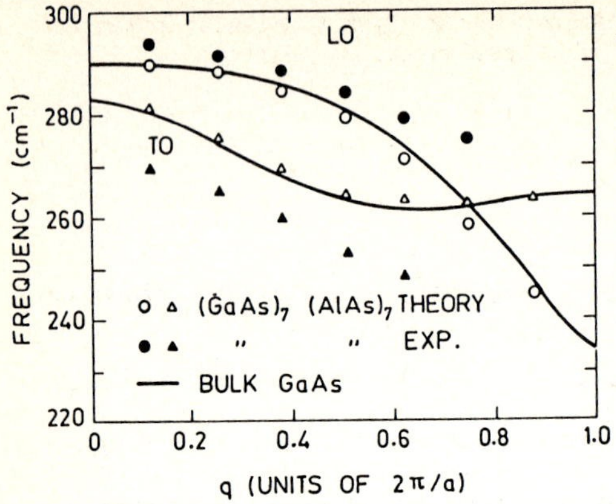

Fig. 12. The solid lines represent the bulk dispersion relation of GaAs. The hollow points the calculated dispersion relation of a (7x7) GaAs-AlAs superlattice. The full points are experimental for this superlattice, plotted according to Eq. (19). From Ref. 31.

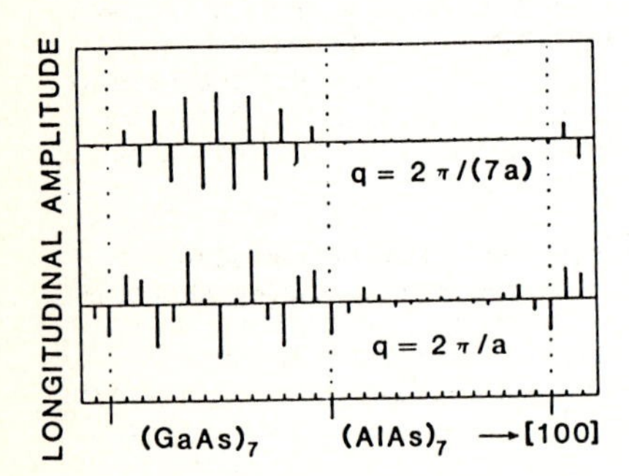

Fig. 13. LO-phonon amplitude in the superlattice of Fig. 12 for m=1 and m=7. From Ref. 31.

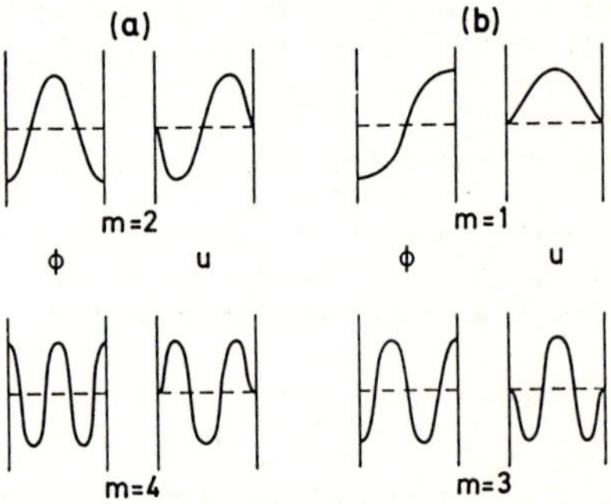

Fig. 14. Schematic diagrams of the electrostatic potential ϕ and the mechanical displacement u for confined LO phonons. (a) corresponds to Eq. (23a) and (b) to Eq. (23b).

the polarization P, and thus the depolarizing field E, are proportional to u, the potential ϕ is out of phase with u. A zero of ϕ corresponds to a maximum in u and viceversa. It is, however, unjustifiable to impose boundary conditions on ϕ and not on u. Hence, the matter has to be looked at more closely. We shall find that the boundary condition u=0 at the interface, neglecting the boundary condition on ϕ, yields a justifiable approximation. The use of incorrect boundary conditions leads to incorrect treatment of phenomena in which the details of the wavefunction are important, such as the electron-phonon interaction /41/ (and thus the Raman scattering, see below).

In the absence of free charges, and neglecting retardation effects (i.e., assuming that the speed of light is infinite) the electric field and the electric displacement $\vec{E}$ and $\vec{D}$ must fulfill the electrostatic equations:

$$\vec{\mathrm{curl}}\ \vec{E} = 0 \quad \mathrm{div}\ \vec{D} = 0 \quad \vec{D} = \varepsilon\vec{E}, \tag{20}$$

where ε is the dielectric constant which includes electron and phonon contributions. Following Eqs. (20) we may derive $\vec{E}$ from a scalar potential χ or $\vec{D}$ from a vector potential $\vec{V}$ /17/. These equations lead to:

$$\varepsilon\nabla^2\phi = 0 \tag{21a}$$

or to:

$$\varepsilon^{-1}\nabla \times \nabla \times \vec{V} = 0, \tag{21b}$$

where ϕ and $\bar{V}$ are a scalar and a vector potential, respectively. Equation (21a) can be fulfilled by making either:

$$\varepsilon(\omega) = 0 \tag{22a}$$

or

$$\nabla^2\phi = 0. \tag{22b}$$

Equation (22a) leads to the confined longitudinal modes of either component while Eq. (22b) leads to the interface modes to be discussed in the next section /36,42-44/. The solution of Eq. (21b) obtained for $\varepsilon^{-1} = 0$ (i.e., $\varepsilon=\infty$) yields the confined TO modes. We shall now examine the details of these TO and LO confined modes and the boundary conditions at the interface. Since the treatment is macroscopic we shall not make any allowance for the difference between Eqs. (8) and (19).

In the case of LO modes of slab 1 ($\varepsilon^{(1)}=0$) the potential in slab 1 can be expanded as a sum of the following Fourier components, taking the center of slab 1 as origin:

$$\phi_1(x,z) = \phi_0 e^{ikx}\cos qz \tag{23a}$$

$$\phi_1(x,z) = \phi_0 e^{ikx}\sin qz \tag{23b}$$

For simplicity, we have only included in Eqs. (23) the transverse dependence on x. On the neighboring slabs 2 we must have no dependence of ϕ on z so as to fulfill the continuity equation for D_z (remember that $\varepsilon^{(1)}= 0$ at the LO frequency of medium 1). The boundary condition for E_x yields for ϕ_2:

$$\phi_2(\pm d_1/2) = \phi_0 e^{ikx} \cos \frac{qd_1}{2} \tag{24a}$$

$$\phi_2(\pm d_1/2) = \pm\phi_0 e^{ikx} \sin \frac{qd_1}{2}. \tag{24b}$$

Equations (24a,b) correspond to Eq. (23a,b), respectively. We now impose the condition $u_{x,z} = 0$ at the interfaces. Equation (23a) yields:

$$ik\phi_0 e^{ikx} \cos \frac{qd}{2} = 0 \tag{25'}$$

$$q\phi_0 e^{ikx} \sin \frac{qd}{2} = 0. \tag{25''}$$

It is obviously impossible to fulfill Eqs. (25) simultaneously except in the case k=0. In this case we find:

$$q = \frac{\pi}{d}m,\ m = 2,\ 4,\ 6. \tag{26}$$

We note that for optically excited phonons even if $k\neq0$, k will be much smaller than π/d in typical superlattices ($k \lesssim \pi/1000$ Å^{-1}; $q \sim 2\pi/100$ Å^{-1}). Thus the boundary condition of Eq. (25') is much stronger than that of Eq. (25"). We may neglect the latter which will have to be fulfilled through small admixture of other modes around the interface /17/. The same argument applies to Eq. (23b). We find in this case:

$$q = \frac{\pi}{d}m, \quad m = 1, 3, 5... \tag{27}$$

The potentials $\phi(z)$ and the displacements $u(z)$ corresponding to Eqs. (23a,b) are shown schematically in Fig. 14 for the four first values of m. We note that the boundary condition is indeed u=0 and that ϕ must be a maximum at the boundary.

5.2 Raman Selection Rules

Conventional GaAs-AlAs superlattices are grown with the z-axis along [001]. Raman back-scattering on the [001] face of the bulk crystals is only allowed for LO phonons which vibrate perpendicular to the surface. According to the discussion in 5.1 phonons vibrating along [001] are also LO-like provided $k \ll 2\pi/d$. TO-like phonons (Eq. (21b)) must be polarized along x or y for $k \ll 2\pi/d$. The point group of this infinite superlattice is D_{2d} (symmetry operations: improper four-fold rotation along z, two-fold rotation perpendicular to z, two perpendicular reflection planes containing z). For this symmetry the vibrations along x,y are two-fold degenerate and belong to the E representation (Table 2.1 of Ref. 21). They have Raman tensors of the form:

$$\begin{matrix} 0 & 0 & 1 \\ 0 & 0 & 0 \\ 1 & 0 & 0 \end{matrix} \quad , \quad \begin{matrix} 0 & 0 & 0 \\ 0 & 0 & 1 \\ 0 & 1 & 0 \end{matrix} \quad . \tag{28}$$

It is not possible to couple to these TO-like phonons in backscattering, i.e., with e_L and e_S along x and/or y. Nevertheless these phonons are clearly seen, though weakly, in Fig. 7. This was possible only under extremely resonant conditions, i.e., with the laser frequency very close to that of the first excitonic transitions of the superlattice. The mechanism for the breakdown of the dipole selection rule is not known in this case.

The TO phonons of a GaAs-AlGaAs superlattice were observed in an allowed configuration (90° scattering) by Zucker et al./45/. The laser was incident perpendicular to the (001) face and the scattered light excited sidewise along (010). The superlattice had been clad with AlGaAs so as to channel the scattered light (waveguide) along the y direction. Under these conditions, it is easy to see that one couples to the tensors of Eq. (28). One can also couple to the LO-like mode whose Raman tensor has the form:

$$\begin{pmatrix} 0 & 1 & 0 \\ 1 & 0 & 0 \\ 0 & 0 & 0 \end{pmatrix} \tag{29}$$

by taking $e_L \parallel [010]$, $e_S \parallel [100]$.

The LO phonons of the [001] superlattice can have, within the D_{2d} point group, two types of symmetries: B_2 which corresponds to Eq. (29) and thus leads to the same dipole selection rules as in bulk material, and A_1, i.e., invariant under all symmetry operations and dipole forbidden in the bulk. It is easy to see that the (a) modes of Fig. 14 have A_1 symmetry while the (b) modes have B_2 symmetry. The A_1 modes are forbidden in the bulk. Their Raman tensor for the superlattice is diagonal with $R_{xx} = R_{yy}$ (Table 2.1 of Ref. 21): Hence they should appear for parallel $\hat{e}_L$ and $\hat{e}_S$ polarizations.

While the A_1 modes are dipole forbidden in the bulk, we mentioned in Sect. 2 that for parallel polarizations quadrupole allowed, dipole forbidden modes could be seen near resonance for $\hat{e}_L \| \hat{e}_S$. Their strength should be proportional to the scattering wavevector square. In the superlattice for the LO modes the scattering wavevector must be replaced by Eq. (27) and hence the modes become dipole allowed. By analogy to the bulk case one may expect, however, that these A_1 modes should resonate stronger than the B_2 modes.

We note that using for $\omega_{LO}(q)$ the standard dispersion relations of bulk material the highest frequency LO phonon should be a B_2 mode (m=1, Eq. (27)), followed by A_1 (m=2), followed by B_2 (m=3), etc., in alternating sequence. This pattern was observed in Ref. 5. We illustrate it in Fig. 15. In the upper part of this figure one can see in the B_2 scattering configuration $[z(x,y)\bar{z}]$ the highest LO mode for m=1 (LO_1). In the A_1 configuration $[z(x,x)\bar{z}]$ we see the m=2 (LO_2) mode which peaks below LO_1, as expected. We also see m=3, 5 modes. These observations were only possible away from the strong resonance of the lowest interband exciton. Close to resonance the A_1 modes become very strong and appear even in the B_2 scattering configuration, probably as a result of an unspecified breakdown in the selection rule. We should keep in mind that the A_1 phonons must couple to the incident and scattered fields through their electrostatic potential ϕ (Fröhlich interaction) while B_2 couples through u (deformation potential interaction) (see Fig. 14). The former is known to be more strongly resonant than the latter.

The phenomenon just described was also observed in Ref. 46 (see Fig. 4 of this reference). Its origin, however, was not identified as ϕ-coupling vs. u-coupling. This was done in Ref. 5.

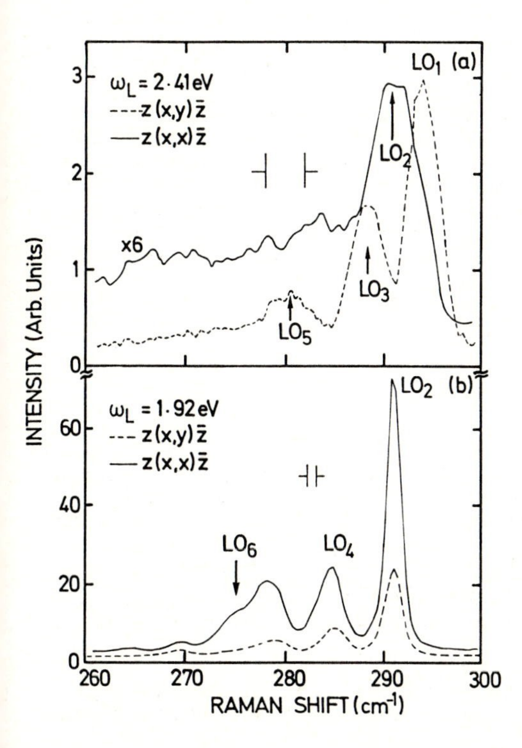

Fig. 15. Upper half: non-resonant scattering in a (7,7) GaAs-AlAs superlattice showing the difference between A_1 and B_2 phonons. Lower half: resonant scattering showing that only A_1 phonons are seen. From Ref. 5.

5.3 Dispersion Relations of LO-Phonons

Figure 8 shows the frequencies of the various LO peaks obtained in the Raman spectra of several superlattices plotted in the bulk BZ with the use of Eq. (8). The use of Eq. (19) does not substantially alter this plot for the superlattices under consideration. We note that the superlattice points fall above the neutron data, although the discrepancy only surpasses the error flags near the BZ edge. More accurate neutron measurements are possible with present day equipment but they have not yet been performed. They are needed to ascertain any possible true superlattice effects. Such effects, however, have not been obtained in the computational work of Molinari et al. (Fig. 12), except for the trivial $n_1 \to n_1+1$ replacement (Eq. (19)).

Nakayama et al./47/ have performed Raman measurements on very accurately prepared small period GaAs-AlAs superlattices (n_1, n_2 = (1,1), (2,2), (3,3), (4,4)). The frequencies of the m=1 LO peaks are plotted vs. $n_1 = n_2$ in Fig. 16 and compared with linear chain calculations. Agreement between theory and experiment is good except for n=1 where the calculation falls ~30 cm^{-1} below the experiment for the GaAs-like mode (6 cm^{-1} for the AlAs-like mode). This type of discrepancy is similar to that shown in Fig. 8 for large q. Most of the calculations do not include the long-range Coulomb forces or include them in an approximate way /48/. The effect of the thickness of the "barrier" layer on the m=1 confined modes has been investigated in Ref. 38 for the GaAs-like modes and with $n_1 \geq 3$ no effect is seen all the way down to $n_2=1$, thus giving a measure of how good the confinement is. The same thing applies to the AlAs-like m=1 modes. For n_1 = 1, 2 an increase of the phonon frequency is seen as n_2 decreases to 1. This increase amounts to 8 cm^{-1} for $n_1=1$. For these very thin GaAs layers the confinement is thus not very good.

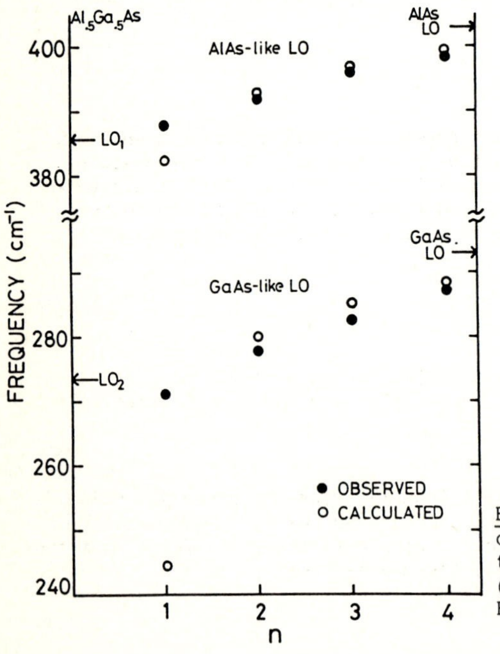

Fig. 16. Observed and calculated frequencies of the m=1 AlAs-like and the GaAs-like LO modes in GaAs-AlAs (n_1,n_2) superlattices versus $n_1=n_2$. From Ref. 47.

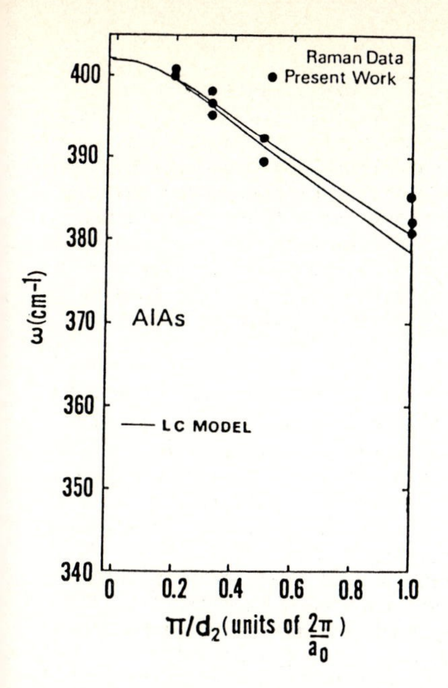

Fig. 17. Dispersion relation of AlAs obtained from Raman measurements for GaAs-AlAs superlattices and from linear chain calculation (Ref. 38). The various points shown for a given π/d_2 correspond to different GaAs thicknesses n_1a_1, ω higher the lower n_1.

We show in Fig. 17 the dispersion relation of AlAs obtained from confined phonon data for n_2 = 1-5 and m=1 and several values of n_1 (1-5): Th lower n_1, the higher the frequency for a given n_2. This plot was obtained with Eq. (8) and thus may have to be considerably altered if Eq. (19) is used (n_2+1 instead of n_2). The clarification of the correspondence of these data with those of the bulk material is rather important since no neutron data are available for AlAs.

6. INTERFACE AND SURFACE MODES

As mentioned in Sect. 5.1 interface modes, propagating along z and x,y, are found for an infinite superlattice from the solution of Laplace's equation (Eq. (22b)) for ϕ. It is of interest to treat first the problem of a single interface between GaAs and AlAs. Let us use the following decaying solutions of Eq. (22b) on both sides of the interface:

$$\begin{aligned} \phi &= \phi_1 e^{ikx} e^{qz} \ (z<0) \\ \phi &= \phi_2 e^{ikx} e^{-qz} (z>0). \end{aligned} \qquad k=q \tag{30}$$

Applying the condition of continuity of E_x and D_z we find the frequency ω_I and the dispersion relation of these interface modes by solving:

$$\varepsilon^{(1)}(\omega_I) = -\varepsilon^{(2)}(\omega_I). \tag{31}$$

We note that there are two ω_I's, one GaAs-like and the other AlAs-like. These ω_I's are independent of k and lie between ω_{TO} and ω_{LO}. If the electronic contributions to the dielectric constants are equal for both materials ($\varepsilon_\infty^{(1)} = \varepsilon_\infty^{(2)}$) the interface frequency becomes:

$$\omega_I = \left(\frac{\omega_{LO}^2+\omega_{TO}^2}{2}\right)^{1/2} \tag{32}$$

For an interface of a semiconductor to air:

$$\omega_I = \left(\frac{\varepsilon_\infty \omega_L^2 + \omega_T^2}{\varepsilon_\infty + 1}\right)^{1/2}. \tag{33}$$

Since $\varepsilon_\infty \approx 10$, Eq. (33) yields for ω_I a value rather close to ω_L. For the interface modes of the superlattices we must apply the E_x, D_z continuity conditions at two interfaces and the Bloch condition for the propagation along z. We obtain the secular equation /36,42,43/:

$$\cos qd = \cosh kd_1 \cosh kd_2 + \frac{\eta^2+1}{2\eta} \sinh k_x d_1 \sinh k_x d_2 \tag{34}$$

with

$$\eta(\omega) = \frac{\varepsilon^{(1)}(\omega_I)}{\varepsilon^{(2)}(\omega_I)}. \tag{35}$$

Note that this equation is formally similar to Eqs. (9) and (17). In the limit $q \to 0$ Eq. (34) splits into the two following equations:

$$\tanh \frac{kd_1}{2} \cosh \frac{kd_2}{2} = \begin{cases} -\eta & (36a) \\ -\eta^{-1}. & (36b) \end{cases}$$

η tends rapidly to ∞ near $\omega_{TO}^{(1)}$ and $\omega_{LO}^{(2)}$ while η^{-1} does the same thing near $\omega_{TO}^{(2)}$ and $\omega_{LO}^{(1)}$. In these regions solutions of Eq. (36) are found. Thus four ω_I modes appear for each k and q, near the ω_{LO} and ω_{TO} modes of both constituents, respectively.

In the limit $k \to 0$ Eqs. (36) take the simple form:

$$\varepsilon_1 d_2 + \varepsilon_2 d_1 = 0 \tag{37a}$$

$$\varepsilon_1 d_1 + \varepsilon_2 d_2 = 0. \tag{37b}$$

Or, using the average symbol $\langle\rangle$:

$$\langle \varepsilon^{-1} \rangle = 0 \tag{38a}$$

$$\langle \varepsilon \rangle = 0. \tag{38b}$$

Equation (38a) corresponds to TO modes of a material with the average dielectric constant $\langle \varepsilon^{-1} \rangle^{-1}$ while Eq. (38b) represents LO modes of a material with the average $\langle \varepsilon \rangle$. In the case $d_1 = d_2$ Eqs. (38a,b) yield the same solution, namely that of Eq. (31)! It may seem puzzling that modes which decay around the interfaces may become like bulk modes of an average material. This is so because we are in the limit $q \to 0$ and thus the modes actually do not decay. The treatment in Ref. 6 is equivalent to Eqs. (37a,b). We note that for $d_1 = d_2$ the two Eqs. (36a,b) lead to the same result. Correspondingly, no gap appears in the calculated dispersion relations vs. k: a gap appears when $d_1 \neq d_2$ (see Fig. 18).

Another interesting limit of Eq. (34) is that for $q = \pi/d$ (edge of the minizone): Eqs. (36a,b) are obtained with the minus signs replaced by plusses. In the limit $k \to 0$ the solutions are $\omega_I = \omega_{TO}^{(1)}, \omega_{LO}^{(1)}, \omega_{TO}^{(2)}, \omega_{TO}^{(2)}$. The spectra of ω_I extend for the (1)-like vibrations from $\omega_{TO}^{(1)}$ to $\omega_{LO}^{(1)}$ and for the (2)-like ones from $\omega_{LO}^{(2)}$ to $\omega_{TO}^{(2)}$. For $k \to \infty$ all ω_I's converge at the value given by:

$$\frac{\eta^2+1}{2\eta} = -1, \text{ i.e., } \omega_I^2 = \frac{\omega_{LO}^2 + \omega_{TO}^2}{2}. \tag{39}$$

The interface modes for k=0 and kd=π have a definite parity with respect to the 2-fold rotation for q=0, πd. The parity is indicated in Fig.

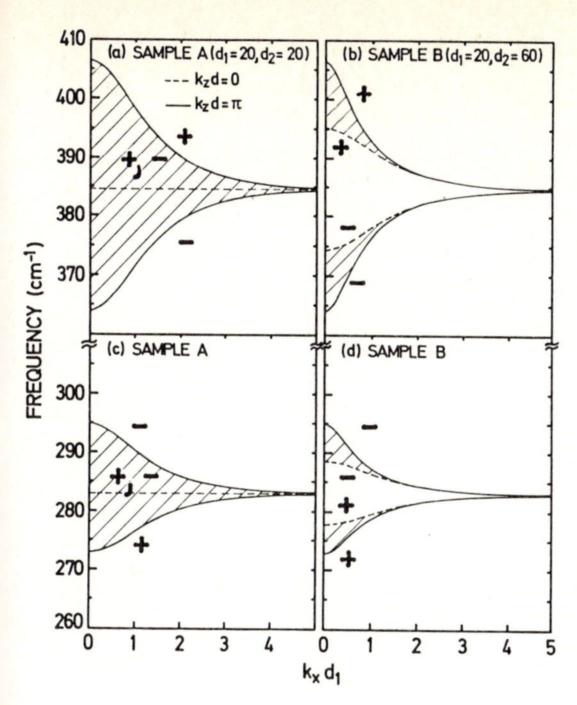

Fig. 18. Dispersion of the interface-modes of GaAs-AlAs superlattices with $d_1=d_2$ (left) and with $d_1 < d_2$ (right). From Ref. 36.

18: It reverses for a given band in going from material 1 to 2. Also, both modes q=0 and q=π/d of a given branch (LO or TO) have the same parity for the material with the thicker layers, opposite parities for the thinner layers. Like in the case of the minigap of LA phonons (Fig. 9) the parity determines the strength of the coupling: modes in which ϕ is odd should couple very little if the coupling occurs via ϕ. Figure 19 shows that this is indeed the case.

Propagating interface modes exist in an infinite crystal. For practical, finite crystals, one must impose boundary conditions at the outside surface and standing waves appear. Standing and surface waves have been recently observed for plasmons in GaAs-AlGaAs superlattices by light scattering /49/. Similar phenomena should be observable for the interface phonons. A recent publication reports the observation of surface phonons at the air-superlattice interface of GaAs-AlGaAs superlattices by means of high-resolution electron energy loss spectroscopy /6/.

7. RESONANCE EFFECTS

We have mentioned resonance effects many times in previous sections and here we shall only repeat a few additional points relevant to our discussion. Resonance effects for phonon scattering have been extensively investigated experimentally /5,36,50/. The details of their theoretical underpinning, however, are not yet well understood /51/. Resonances occur when either ω_L or ω_L equals the frequency ω_{ex} of an interband exciton (ingoing and outgoing resonances), usually also localized in the GaAs layers. This exciton must be composed of an electron and a hole with the same longitudinal quantization number (some number of nodes of the wave function),

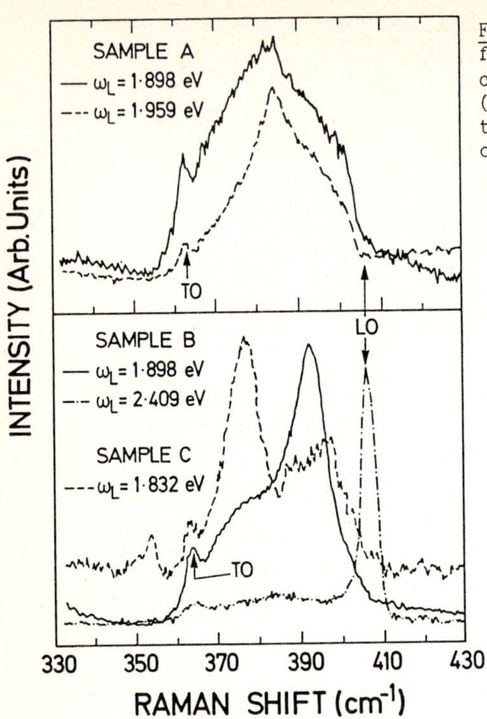

Fig. 19. Raman scattering by interface modes of GaAs-AlAs superlattices reported in Ref. 36. Sample A (B) is that whose dispersion relation is given in the left (right) of Fig. 18.

otherwise the oscillator strength is too weak to be seen. The first resonance (m=1 exciton) is much stronger than the subsequent ones (Fig. 20). It is usually observed that the "outgoing resonance" for $\omega_S = \omega_{ex}$ is stronger than the ingoing one ($\omega_L = \omega_{ex}$) /5,36/. This has been interpreted as being related to impurity-induced fourth order scattering processes /5,36/ which also occur for "forbidden" scattering by LO-phonons in bulk samples /52/.

In a recent paper /53/ Miller et al. reported the observation in a GaAs-GaAlAs superlattice, with appropriately chosen layer thicknesses, of a double resonance in which both ω_L and ω_S equal excitonic energies. This becomes possible because of the splitting of the light hole and heavy hole excitons: ω_S can resonate with the heavy hole and ω_L with the light hole exciton. Very large scattering efficiencies are observed. Also, the polarization selection rules are very interesting for such superlattices: the normal modes for the incident and scattered light being circularly polarized.

Resonance phenomena are also observed for interface modes /36/ as indicated in Fig. 19. Although these modes are either GaAs, or AlAs-like, their ϕ's and u's extend into both media and thus should be resonant at both GaAs and AlAs excitons. Since this is not the case for the confined modes (they resonate at electronic excitations of the material where they are confined) one may use the fact that they resonate at electronic gaps of both materials as a signature of the interface modes.

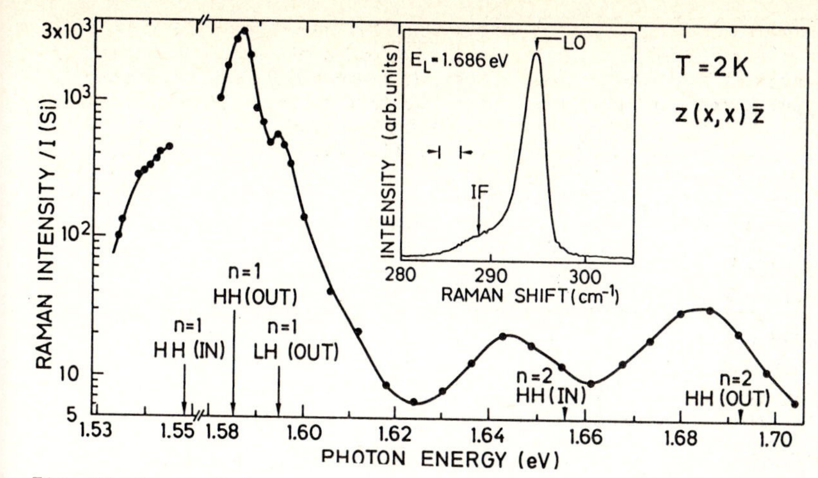

Fig. 20. Resonant Raman profile for scattering by LO-phonons in a GaAs-$Ga_{0.75}Al_{0.25}As$ superlattice (n_1=37, n_2=45). Ingoing and outgoing resonances are observed. HH (LH) labels resonances with transitions originating in localized heavy (light) hole bands. The insert indicates the observed LO-phonon lineshape and an interface mode (IF). From A.K. Sood et al., unpublished.

I would like to thank K. Christensen for an expert and speedy typing of this manuscript.

8. LITERATURE

1. L. Esaki and R. Tsu: IBM J. Res. Develop. 14, 61 (1970).
2. J.L. Merz, A.S. Baker, Jr., and A.C. Gossard: Appl. Phys. Letters 31, 117 (1977); A.S. Barker, Jr., J.L. Merz, and A.C. Gossard: Phys. Rev. B 17, 3181 (1978).
3. R. Merlin, C. Colvard, M.V. Klein, H. Morkoc, A.Y. Cho, and A.C. Gossard: Appl. Phys. Letters 36, 43 (1980).
4. B. Jusserand, D. Paquet, and K. Kunc: In "The Physics of Semiconductors", ed. by J. Chadi and W.A. Harrison (Springer, New York, 1985), p. 1165.
5. A.K. Sood, J. Menéndez, and M. Cardona: Phys. Rev. Letters 54, 2111 (1985).
6. Ph. Lambin, J.P. Vigneron, A.A. Lucas, P.A. Thiry, M. Liehr, J.J. Pireaux, R. Candano, and T.J. Knech: Phys. Rev. Letters 56, 1842 (1986).
7. B. Jusserand, P. Voisin, M. Voos, L.L. Chang, E.E. Méndez, and L. Esaki: Appl. Phys. Letters 46, 678 (1985).
8. F. Cerdeira, A. Pinczuk, J.C. Bean, B. Batlogg, and B.A. Wilson: Appl. Phys. Letters 45, 1138 (1984).
9. M. Nakayama, K. Kubota, H. Kato, and N. Sano: Solid State Commun. 51, 343 (1984).
10. A.T. Fiory, J.C. Bean, L.C. Feldman, and J.K. Robinson: J. Appl. Phys. 56, 1227 (1984).
11. A.K. Sood, E. Anastassakis, and M. Cardona: phys. stat. sol. (b) 129, 101 (1985) and references therein; P. Wickboldt, M. Cardona, and R. Sauer: to be published in Phys. Rev.

12. M. Hundhausen, L. Ley, and R. Carius: Phys. Rev. Letters 53, 1598 (1984).
13. P. Santos, M. Hundhausen, and L. Ley: Phys. Rev. B 33, 1516 (1986).
14. M. Rytov: Soviet Physics, Acoustics 2, 67 (1956).
15. N. Maley and J.S. Lannin: Phys. Rev. B 31, 5577 (1985).
16. R. Merlin, K. Bajema, R. Clarke, F.Y. Juang, and P.K. Bhattacharya: Phys. Rev. Letters 55, 1768 (1985).
17. M.V. Klein: IEEE J. Quantum Electronics.
18. B. Jusserand and D. Paquet: In "Semiconductor Heterojunctions and Superlattices", ed. by G. Allan and G. Bastard, N. Boccara, M. Lanoo, and M. Voos (Springer, Heidelberg, 1986) p. 108.
19. See chapter on Superlattices in Phonon Physics, ed. by J. Kollar, N. Kroo, N. Menyhard, and T. Siklos (World Scientific Publ. Singapore, 1985), p. 505.
20. Light Scattering in Solids I to IV, edited by M. Cardona and G. Güntherodt (Springer, Heidelberg, 1975-1984). Volume V of this series, devoted to superlattices, is in preparation.
21. M. Cardona: In Ref. 20, Vol. II, p. 58.
22. H. Bilz and W. Kress: Phonon Dispersion Relations in Insulators (Springer, Heidelberg, 1979).
23. We measure phonon frequencies or energies usually in cm^{-1}. 1 cm^{-1} = 30 GHz frequency = 1.88×10^{11} rad/sec angular frequency = 1.24×10^{-4} eV energy = 1.99×10^{-16} erg energy = 1.44 K temperature.
24. A. Ourmazd, J.C. Bean, and J.C. Phillips: Phys. Rev. Letters 55, 1599 (1985).
25. N. Saint-Cricq, R. Carlos, J.B. Renucci, A. Zwick, and M.A. Renucci: Solid State Commun. 39, 1137 (1981).
26. T. Kamijoh, A. Hashimoto, N. Watanabe, and M. Sakuba: Phys. Rev., in press.
27. D.W. Feldman, J.H. Parker, Jr., W.J. Choyke, and L. Patrick: Phys. Rev. 173, 787 (1968).
28. S. Nakashima, H. Katahama, Y. Nakakura, and A. Mitsuishi: Phys. Rev. 33, 5721 (1986).
28a 4H Ge E. Lopez-Cruz and M. Cardona: Solid State Commun. 45, 787 (1982).
28b GaGeTe E. Lopez-Cruz, E. Martinez, and M. Cardona: Phys. Rev. B 29, 5774 (1984).
29. C. Colvard, T.A. Gant, M.V. Klein, R. Merlin, R. Fischer, H. Morkoc, and A.C. Gossard: Phys. Rev. B 31, 2080 (1985).
30. See, for instance, G. Burns: Solid State Physics (Academic Press, New York, 1985).
31. A.K. Sood, J. Menéndez, M. Cardona, and K. Ploog: Phys. Rev. Letters 56, 1753 (1986); B. Jusserand and D. Paquet: Ibid. 56, 1752 (1986); E. Molinari, A. Fasolino, and K. Kunc: Ibid. 56, 1751 (1986).
32. See, for instance, Andersen Localization, ed. by Y. Nagaoka and H. Fukuyama (Springer, Heidelberg, 1982).
33. J. Sapriel, J.C. Michel, J.C. Toledano, R. Vacher, J. Kervavec, and A. Regreny: Phys. Rev. B 29, 2007 (1983); J. Sapriel, B. Djafari, . Rouhani, and L. Dobrzynski: Surface Science 126, 197 (1983).
34. M. Cardona, K. Kunc, and R.M. Martin: Solid State Commun. 44, 1205 (1982).
35. H.G. Olf, A. Peterlin, and W.I. Peticolas: J. Polym. Science 12, 359 (1974).
36. A.K. Sood, J. Menéndez, and M. Cardona: Phys. Rev. Letters 54, 2115 (1985).
37. V. Narayanamurti, H.L. Störmer, M.A. Chin, A.C. Gossard, and W. Wiegmann: Phys. Rev. Letters 43, 2012 (1979).
38. A. Ishibashi, M. Itabashi, Y. Mori, K. Kaneko, S. Kawado, and N. Watanabe: Phys. Rev. 33, 2887 (1986).

39. B. Jusserand, D. Paquet, and A. Regreny: Phys. Rev. B 30, 6245 (1984).
40. R. Fuchs and K.L. Kliewer: Phys. Rev. 140, A2076 (1975).
41. R. Lassnig: Phys. Rev. B 30, 7132 (1984).
42. E.P. Pokatilov and S.I. Beril: phys. stat. sol. (b) 110, K75 (1982) and 118, 567 (1983).
43. R.E. Camley and D.L. Mills: Phys. Rev. B 29, 1695 (1984).
44. F. Bechstedt and R. Enderlein: phys. stat. sol. (b) 131, 53 (1985).
45. J.E. Zucker, A. Pinczuk, D.S. Chemla, A. Gossard, and W. Wiegmann: Phys. Rev. Letters 53, 1280 (1984).
46. M. Nakayama, K. Kubota, T. Kanata, H. Kato, S. Chika, and N. Sano: Jap. J. Appl. Phys. 24, 1331 (1985).
47. M. Nakayama, K. Kubota, H. Kato, S. Chika, and N. Sano: Solid State Commun. 53, 493 (1985).
48. S.K. Yip and Y.C. Chang: Phys. Rev. 30, 7037 (1984); G. Kanellis: Solid State Commun. 58, 93 (1986).
49. A. Pinczuk, M.G. Lamont, and A.C. Gossard: Phys. Rev. Letters 56, 2092 (1986); G. Fasol, N. Mestres, H.P. Hughes, A. Fischer, and K. Ploog: Phys. Rev. Letters 56, 2517 (1986).
50. J.E. Zucker, A. Pinczuk, D.S. Chemla, A. Gossard, and W. Wiegmann: Phys. Rev. Letters 51, 1293 (1983).
51. See, for instance, P. Manuel, G.A. Sai-Halasz, L.L. Chang, C.A. Chang, and L. Esaki: Phys. Rev. Letters 37, 1701 (1976).
52. W. Kauschke and M. Cardona: Phys. Rev. B 33, 5473 (1986).
53. R.C. Miller, D.A. Kleinman, and A.C. Gossard: Solid State Commun., in press.

Effect of Plasma Waves on the Dispersion Relation of Conductor-Insulator Superlattices

M. del Castillo-Mussot[1], *W.L. Mochán*[1], *and R.G. Barrera*[2,a]

[1]Instituto de Física, Universidad Nacional Autónoma de México, A.P. 20-364, 01000 México, D.F., México
[2]Centro de Investigación y de Estudios Avanzados del IPN, Departamento de Física, A.P. 14-740, 07000 México, D.F., México

We obtain the dispersion relation of the electromagnetic normal modes of a conductor-insulator superlattice, taking into account the presence of plasma waves, spatial dispersion and retardation. We find resonant features corresponding to the propagation of guided plasma waves in the conducting layers coupled by transverse fields in the insulating layers.

Systems qualitatively different from the extensively studied semiconductor superlattices/1/ can be obtained by increasing the density of carriers, either by heavy doping/2,3/ of a semiconducting superlattice or by fabricating a metallic superlattice/4/. Then, either one or both kinds of layers comprising the heterostructure may behave as a 3D conductor. The normal modes of these systems can have a rich structure; in this paper we investigate those electromagnetic modes arising from the coupling between plasmons and transverse electromagnetic waves in superlattices.

Let us consider a superlattice consisting of alternating insulating and conducting layers of width a and b, respectively, stacked along the Z direction, and on which p-polarized and longitudinal waves propagate. All waves have the same frequency ω, and all their wave vectors have a common component Q parallel to the interface. We do not consider s-polarized waves since they do not couple to plasmons. Making use of Bloch's theorem, we can relate the amplitudes of the transverse and longitudinal waves in the insulator and conductor layers by imposing at the interfaces both electromagnetic boundary conditions (continuity of the components of the electric and magnetic fields parallel to the interfaces) and boundary conditions of non-electromagnetic origin or additional boundary conditions (ABCs). Given the Coulomb and statistical repulsion between the conduction electrons in the conductor it is reasonable to demand that there be no singularities in their charge density. This, together with Gauss' law implies that E_z should be continuous at the conductor's surface. For simplicity we adopt this ABC, and we ignore the discontinuity of E_z due to the accumulation at the surface of bound charges. Notice that in the model calculation that follows there are no bound charges. We will take them into account elsewhere.

Now we proceed to present the results of our calculations performed for a model superlattice, in which the conducting layers have a Drude transverse dielectric function and a hydrodynamic longitudinal response

$$\varepsilon(\vec{q},\omega) = 1 - \frac{\omega_p^2}{\omega^2 + i\omega/\tau - \beta^2 (Q^2 + \ell^2)}, \tag{1}$$

where ω is the frequency, ω_p is the plasma frequency, τ is the electronic relaxation time and $\beta^2 = 3v_F^2/5$, with v_F the Fermi velocity of the conductor. Q and ℓ are, respectively, the components of the wave vector $\vec{q}$ parallel and perpendicular to the interfaces. We chose vacuum as an insulator so that $\varepsilon_I = 1$, and we took $\tau = 100/\omega_p$ and $v_F = 0.01c$, where c is the speed of light in vacuum.

In Fig. 1 we show the dispersion relation ω vs. p, for a superlattice, where p is the 1D Bloch's wave vector. We chose $a=b=0.1\lambda_p$ and a fixed value of $Q = 1.5/\lambda_p$, where we defined $\lambda_p \equiv c/\omega_p$.

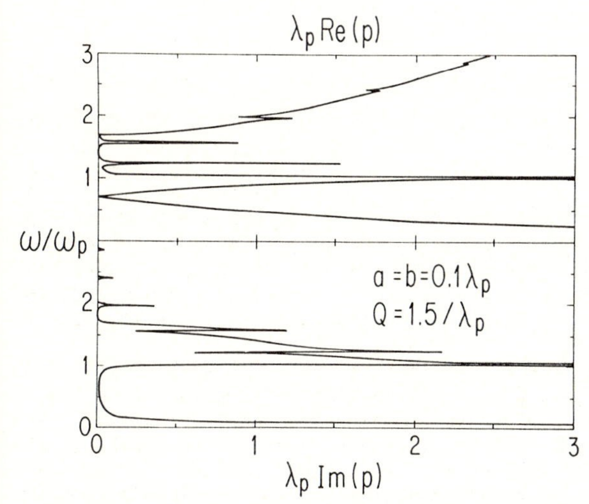

1. Dispersion relation ω (frequency) vs. p (Bloch's wave vector) of a superlattice with $a=b=0.1\lambda_p$ and with $Q=1.5/\lambda_p$. The real part of p is shown in the upper panel and its imaginary part in the lower panel.

In order to understand the main features of Fig. 1, it is useful to compare it with Fig. 2, where we plotted the dispersion relation for the same system but in the local limit ie. without plasma waves. We also plotted in Fig. 2 the dispersion relation of the longitudinal waves $\omega = \omega_n$ assuming they do not interact with transverse waves at the conductor surfaces and assuming an infinite relaxation time τ, since they can not exist with real frequencies if τ is finite. At frequencies near the surface plasma frequencies of an isolated conductor-vacuum interface, $\omega_s = \omega_p/\sqrt{2}$, two propagating normal modes can be seen: one whose frequency diminishes and another whose frequency increases with increasing wavevector p. Since these modes have $\omega < Qc$, they are made up of evanescent

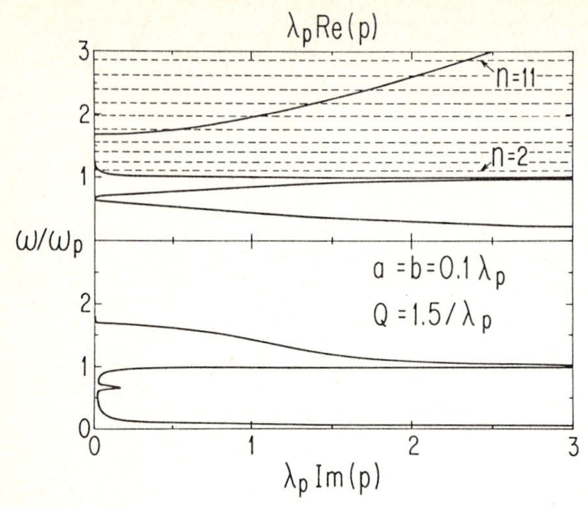

2. Dispersion relation ω vs. p of a superlattice (with a,b and Q as in Fig.1) in the local limit. Also shown with dashed lines is the dispersion relation of guided plasma waves in the absence of damping and of longitudinal-transverse interaction.

waves in both the conducting and the insulating layers. Actually, they consist of surface plasmons localized on each interface but interacting among themselves through the tails of their evanescent fields, thus giving rise to bulk modes of the whole superlattice. Recall that for an isolated thin conductor film there are two surface plasmons; a symmetric one with equal charges and an antisymmetric one with opposite charges on the two surfaces of the film. When many conducting films are brought near each other to form a superlattice, each of these two kinds of surface plasmons gives rise to a bulk band. These two bands have been studied recently in Refs. 5 and 6 within the non-retarded limit. An analogous coupling of two dimensional (2D) plasmons in layered 2D electron gas systems gives rise to 3D electromagnetic modes/7/ which have been observed recently/8/. Notice, however, the difference between 2D and surface plasmons: the former are localized at 2D electron gases while the latter are localized at the interface between two 3D systems with different dielectric response. Going up in frequency, it can be seen that the dispersion relations shown in Figs. 1 and 2 are very similar to each other except for a series of peaks in the non-local calculation at the frequencies $\omega = \omega_n$ for odd values of n. Notice that there are no visible effects of the longitudinal waves for even values of n. The relative importance of these odd and even numbered frequencies (which correspond to plasmons whose wavelength fits a half-integer and an integer number of times in the conductor's width, respectively) in the dispersion relation is discussed in Ref. 9.

Although we obtained our present results by assuming a specific ABC, namely, continuity of the normal component of the electric field, our approach can be easily adapted to other ABC's. However, we do not expect our qualitative results to be modified. Our formulas can also be used for conductor-conductor superlattices at frequencies below the plasma frequency of the more dense conductor, provided we ignore the effects of spatial dispersion in it. These effects are negligible within the hydrodynamic model when the decay distance of the plasmon is small.

[a] On leave from Instituto de Física, UNAM, México.

1. See, for example, L. Esaki, J. de Phys. (París) Coloq. 45 C5-3 (1984).
2. G. Eliasson, G.F. Giuliani and J.J. Quinn, Phys. Rev. B33, 1405 (1986).
3. H. Köstlin, R. Jost and W. Lems, Phys. Status Solidi A 29, 87 (1975); I. Hanberg, C.G. Granqvist, K. F. Berggren, B.E. Sernelius and L. Ergström, Vacuum 35, 207 (1985); F. Demichelis, E. Minetti-Mezzeti, V. Smurro, A. Tagliaferro and E. Tresso, J. Phys. D 18, 1825 (1985).
4. Z. A. Zheng, C.M. Falco and J.B. Ketterson, Appl. Phys. Lett. 38, 424 (1981); I.K. Schiller and C.M. Falco, Surf. Sci. 113, 443 (1982).
5. R. E. Camley and D.L. Mills, Phys. Rev. B 29, 1695 (1984).
6. G.F. Giuliani, J.J. Quinn and R.F. Wallis, J. of Phys. (Paris) Coloq. 45 C5-285 (1984).
7. J.K. Jain and P.B. Allen, Phys. Rev. Lett. 54, 947 (1985); Phys. Rev. B 32, 997 (1985).
8. A. Pinczuk, M.G. Lamont and A.C. Gossard, Phys. Rev. Lett. 56, 2092 (1986), G. Fasol, N. Mestres, H.P. Hughes, A. Fischer and K. Ploog, Phys. Rev. Lett. 56, 2517 (1986).
9. W.L. Mochán, Marcelo del Castillo-Mussot and Rubén G. Barrera, submitted to Phys. Rev. B.

SGFM Applied to the Calculation of Surface Band Structure of Vanadium

R. Baquero[1], *V.R. Velasco*[2], *and F. García-Moliner*[2]

[1]Departamento de Física, Instituto de Ciencias, Universidad Autónoma de Puebla, A.P. J-48 Puebla, Puebla, México
[2]Instituto de Física del Estado Sólido (C.S.I.C.), Serrano 123, E-28006 Madrid, Spain

The surface Green function matching (SGFM) method has been developed recently /1/ to deal with a great variety of problems in a unified way. The method was first developed for continuum systems /2/. The recent advances for discrete structures can deal with surfaces, interfaces, quantum wells, superlattices, intercalated layered compounds, and other systems. Several applications of this formalism are being carried out. In the present note we will describe how the formalism applies to the calculation of the electronic surface band structure of vanadium which is a quite interesting transition metal with very active magnetic properties at the surface, in particular at the (100) surface. It is straightforward, on the basis of the calculation presented here, to obtain the magnetic moment on the surface, for example, through the method followed by G. ALLAN /3/ or the surface paramagnon density which should be particularly enhanced at this surface as compared to the bulk.

The SGFM method allows the calculation of the density of states on the atomic layers following the surface into the bulk,and so one can find how deep the enhanced paramagnon distribution can enter the crystal without any ad hoc hypothesis. Interesting results might come out from tunneling experiments normal-vacuum-superconductor, in the sense that the enhanced surface magnetism could manifest itself in the I-V characteristics from which the Eliashberg function is obtained. Paramagnons in the bulk are held responsible for a high decrease of the superconducting critical temperature of vanadium/4/.

In this paper we devote ourselves to the description of the practical application of the SGFM method to the electronic band structure of the (100) surface of vanadium, based on the bulk band structure of YASUI et al. /5/ and compare to the results of GREMPEL and YING /6/.

The SGFM method considers in general the problem of the matching of wave functions at an interface (or a surface, which is a particular case) in the language of Green functions. It is assumed that the Green functions of the two bulk media, say a and b, are known. The Hamiltonian of the whole system is expressed as

$$H^S = P_a H_a^S P_a + P_b H_b^S P_b + P_a H^X P_b + P_b H^X P_a \quad , \qquad (1)$$

where P is the projector operator into the respective medium, H^X is the Hamiltonian representing the interaction between the two media and $H^S_{a,b}$ is the Hamiltonian describing the atoms in medium a and b respectively.

For a free clean non-reconstructed surface described in the spirit of a tight-binding Slater-Koster analysis including first and second nearest neighbours and parametrizing the Hamiltonian with five d-like wave functions in each atom site, the Hamiltonian can be expressed as a (10 X 10) matrix.

These matrices can be parametrized with seven parameters all together: the intra-site matrix element, three first-nearest neighbours matrix elements (σ, Π and δ-like) and three more for the corresponding second nearest neighbours.

A tight-binding Slater-Koster fit analysis for bulk vanadium using the band structure calculation of YASUI et al. /5/ was reported by GREMPEL and YING /6/. A good fit was obtained with five non-zero tight-binding parameters. We have used these in our calculations.

The SGFM analysis allows us to calculate the surface-projected Green function inverse G_S^{-1} by the following simple expression

$$G_S^{-1}(E,K) = (E\,I - P_0\,H^S\,P_0) - P_0\,H^S\,P_1\,T; \qquad (2)$$

here all the terms represent (10X10) matrices, I is the unit matrix and H^S the surface Hamiltonian. The projector operator P now acts always in the single vanadium medium and the index defines the principal layer (two atomic layers in this case) into which the Hamiltonian is to be projected. T is the transfer matrix which we have calculated following the algorithm of LOPEZ - SANCHO et al. /7/. To find the density of states we first integrate $G_S(E,K)$ in the surface Brillouin zone by using the method developed by CUNNINGHAM /8/ and then use the usual formula:

$$N(E) = -\mathrm{Im}\ \mathrm{tr}\ G_S(E)/\Pi. \qquad (3)$$

Our result is shown in Fig. 1 where we present the density of states for the vanadium (100) surface. We have repeated the calculation using the decimation technique /9/ and have obtained exactly the same result. The upper curve is the result of GREMPEL and YING /6/. The relative magnitude is not meaningful in this figure. The main feature of this result is the high increase in the surface density of states near the Fermi level. The three main peaks of the density of states in this region are common to both

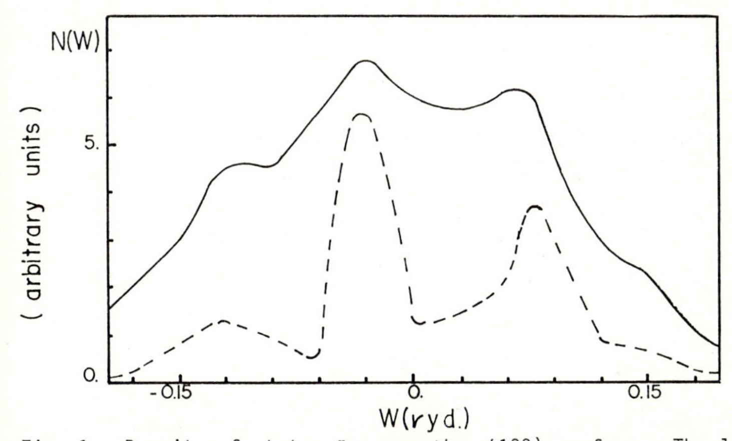

Fig. 1. Density of states for vanadium (100) surface. The lower curve has been obtained by using the transfer matrix and the decimation techniques. These two results are undistinguishable. The upper curve is from /6/. It has been corrected to take into account charge neutrality at the surface. Our result as presented here is not. The big increase in the density of states at the Fermi level is evident.

results although our result is quite sharper. This might be due to the fact that the result from GREMPEL and YING /6/ has been corrected to take into account charge neutrality while ours has not.

In conclusion, we have applied the SGFM method to the calculation of the electronic band structure at the (100) surface of vanadium. The expected sharp increase of the surface density of states at the Fermi level is evident from our calculation although a certain smoothing is to be expected if we correct the result to take into account charge neutrality. The SGFM method as developed to include discrete structures constitutes a simple unambiguous way for calculations like the one we present here.

Acknowledgements The authors acknowledge the hospitality of the Internationa Centre for Theoretical Physics (Trieste) where part of this work was done.

References

1. F. García-Moliner and V.R. Velasco, Prog. Surf. Sci. (1986) in press
2. F. García-Moliner and J. Rubio, Proc. Roy. Soc. London, A324, 257 (1971); F. García-Moliner and F. Flores, Introduction to the theory of solid surfaces, Cambridge University Press, (Cambridge, U.K., 1979)
3. G. Allan, Phys. Rev. B19, 4774 (1979)
4. J.M. Daams, J.P. Carbotte, M. Ashrat and R. Baquero, J. Low Temp. Phys. 55 1 (1984)
5. M. Yasui, E. Hayashi and M. Shimizu, J. Phys. Soc. Jpn. 29, 1446 (1970)
6. D.R. Grempel and S.C. Ying, Phys. Rev. Lett. 45, 1018 (1980)
7. M.P. Lopez-Sancho, J.M. Lopez-Sancho and J. Rubio, J. Phys. F14, 1205 (198
8. S.L. Cunningham, Phys. Rev. B10, 4988 (1974)
9. F. Guinea, C. Tejedor, F. Flores and E. Louis, Phys. Rev. B28, 4397 (1983)

Microstructure and Interfaces in Cu-Ni Multilayered Thin Films

P. Orozco D.

Departamento de Fisica, Universidad Nacional de Colombia, Ciudad Universitaria, Bogotá, Colombia

1. Introduction

The study of epitaxially grown multilayered materials has received renewed attention because of their atypical properties, HILLIARD /1/, and their potential applications in several fields, LYNN et al./2/, UNDERWOOD et al./3/. The multilayers are composite materials, made of alternating layers of different chemical composition. The most interesting repeat distance, the modulation wavelength, ranges between 1 and 10 nm. Since the size and the atomic number are different for the two components, one expects to have not only the composition variation across the interfaces but also a strain modulation.

This contribution describes experiments carried out using Cu-Ni multilayers. The difference in lattice parameter between Cu and Ni is 2.5%, sufficiently small to expect coherent interfaces, i. e. atoms in perfect register, for short modulation wavelengths. Roughly, one could expect to find coherent interfaces between the layers when the strain energy per unit volume in a given thickness is smaller than the specific energy required to create an incoherent interface. As the layer thickness grows the coherency breaks up. For Cu-Ni a coherency limit below 10 nm could be expected, HILLIARD /1/. This fact has to be considered when discussing the nature of interfaces in multilayered materials. There are several approaches to the subject in the literature, but mostly related to single thin layers on thick substrates, MATTEWS /4/.

However, there is little analysis in the literature dealing with the detailed growth mechanisms and microstructure of metallic multilayered materials. Knowledge of these features may give additional clues to enable explanation of some peculiar physical properties as: supermodulus effect, HENEIN et al./5/, anomalous electrical resistivity, OROZCO /6/, anomalous diffusivity, HENEIN et al./7/ etc. shown by metallic superlattices.

2. Experimental

The Cu-Ni multilayers were produced by alternate deposition of the materials onto a heated substrate, under high vacuum conditions. The substrate consisted of an 80 nm thick Cu layer evaporated onto a freshly cleaved muscovite mica substrate held at 400 °C. The Cu backing layer was found to improve epitaxial growth of the multilayered material.

During production, the thickness of each layer was controlled b
a microcomputer using signals from quartz crystal oscillators
used as thickness monitors, purpose-made interfaces and software.

A deposition rate close to 0.1 nm was chosen for the growth o
the multilayered films.
The modulation wavelength was measured using an X-ray powde
diffractometer. In fact, the composition and strain modulation i
the film implies the existence of two periodicities in th
material, one corresponding to the basic lattice and the other t
the repeat distance between the layers. When an X-ray beam i
diffracted by the multilayer the double periodicity produce
appearance of satellite reflections on both angle sides of th
main (average) Bragg peak, MEYER et al./8/. Figure 1 is an exampl
of one such diffraction experiment using Cu K-alpha radiation.

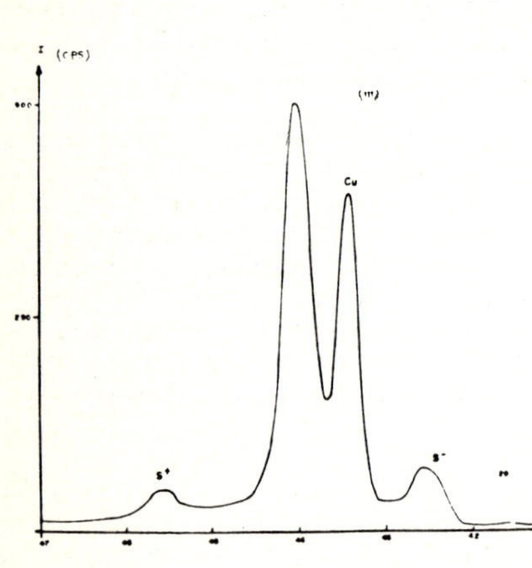

Figure 1. X-ray pattern

The angular position of th
satellite reflection with respec
to the main peak is related to th
modulation wavelength throug
Bragg's equation. There is a grea
deal of additional information i
the intensities and width of th
peaks in the diffractometer trace
but that analysis is not relevan
here.

The films were also examine
using Transmission Electro
Microscopy (TEM) at 100 and 300 kV
For standard TEM work through th
foils, the samples were prepared b
floating them off the mica and the
thinning down on both sides using
argon ion guns. The samples for
edge-on microscopy were prepared b
electroplating a strip of multilayered material, using a coppe
sulfate solution, then tranverse cutting discs with the films a
the center. Standard metallographic polishing and electrolytica
dishing were performed before the last step, which consisted i
ion gun thinning, till the right thickness was achieved.

3. Results

Modulation wavelengths, for the two films referred to in this i
these experiments, were calculated to be 2.8 and 6.0 nm fr
diffractometer traces like that in Fig 1. The overall thickness o
the films was about 700 nm as they contained 214 and 100 composit
layers, respectively. The multilayers had about the sam
composition, close to 50%, calculated from the angular position o
the main Bragg peak reflection.

Electron microscopy of the foils showed that the multilayer
growth was epitaxial with a (111) set of planes parallel to th
substrate. This is seen in the diffraction pattern inset in Fig.2
The micrograph in this figure also reveals a high density o
defects in the films and a subgrain structure of about 200 nm i
size. The micrograph and diffraction pattern which were taken fr
a multilayer having a small number of layers also show a sma

residual polycrystallinity and double positioned domains in the films. As the films grew thicker the polycrystallinity and the domains disappeared. This is concluded from single crystal patterns seen from such thick films.

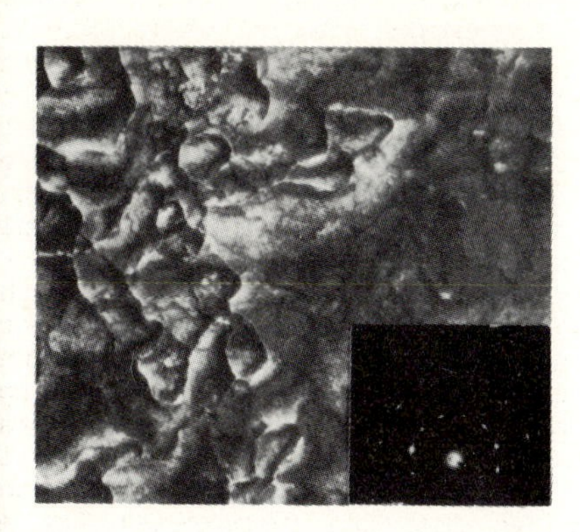

Figure 2. TEM of a multilayer

Figure 3 is a micrograph taken from the 6.0 nm modulation wavelength multilayer edge on and the corresponding diffraction pattern.The specimen was oriented in the microscope with the electron beam parallel to one of the 110 directions. Focussing was adjusted to show the contrast between the layers. The diffraction pattern shows that the material is twinned, the twinning plane being parallel to the multilayer planes.

Figure 4 is a micrograph showing twin domains. Experiments made using an electron probe of about 30nm diameter, in the nanoprobe mode of the microscope, allowed us to see that the diffraction pattern belonged to either, the matrix one or the twin one, when the beam was scanned across the specimen layers from one domain into the other. It is also noticed that the twin thickness is greater than the thickness of a monolayer.

4. Conclusions

All TEM micrographs show a high density of defects, even for the shortest modulation wavelength material; this makes very difficult Burgers vector analysis in the TEM. There is little doubt that the dislocations were produced in the growth process of the multilayers. The dislocations might have two origins, firstly they could be normal misfit dislocations, but, as the layer thickness is so small one does not expect to see such a high density of this kind of defects. Secondly, the dislocations could be generated according to the following simplified model: Let us assume that the growth mode is very near layer-by-layer type, as is suggested by the micrograph in Fig 3. Nevertheless, in an evaporation process one can never be sure of having deposited exactly the amount of material necessary to make an integer number of atomic layers. Therefore, most of the time one has an incomplete last monolayer at the end of the deposition of a layer. When the next layer starts to be deposited the impinging atoms will find areas of

Figure 3. Multilayer edge on

valleys and high planes. The new layer begins to grow at different places, but atoms find the lowest free energy places in the valleys near the edges. If a nucleation site starts with atoms in the right stacking position one can expect a coherent interface but when the nucleation site is not at the right lattice place a stacking fault develops, it will be surrounded by dislocations at the edges. Those dislocations having Burgers vector in the growth plane are imaged in the micrographs taken when several diffracting beams are active, as it is the case in Fig 2. However, the dislocations generated by this mechanism do not help to relieve the strain across the interface. If misfit dislocations are to appear they would do it positioned in the flat areas. Subsequent growth on a stacking fault could be the origin for the growth of twin domains, as Fig. 4 seems to indicate. The edges of these domains constitute walls parallel to the growth direction. Due to the difference in stacking sequence between the twin and the matrix, there should be also microtwinning and defects across these walls. Evidence of such microtwinning was found in diffraction patterns of very thin multilayers, because reflection spots corresponding to thirds of reciprocal lattice distances were found. The defect contrast can also be seen pronounced in some areas of the edge-on micrographs. Finally, twinning was also observed in (111) planes inclined with respect to the layer plane, in high resolution electron micrographs.

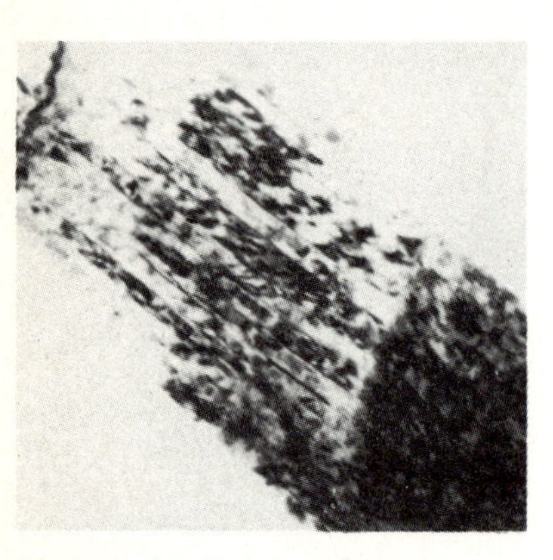

Figure 4. Twins edge on

5. Acknowledgements

The author is indebted to Dr. David Dingley and his group a Bristol, U.K., for constant advice and encouragement during thi work. The financial support from Universidad Nacional de Colombi and the IAEA is also gratefully acknowledged.

6. References

1. J.E. Hilliard: Artificial layer structures and their properties. In Modulated Structures, ed by J.M. Cowley, J.B. Cohen, M.B. Salomon and B.J. Wuench, AIP Proc. 153, New York 1976
2. J.W. Lynn et al.: J. App. Crystallogr. 9, 454 (1976)
3. J.H. Underwood et al.: App. Opt. 20, 3027 (1981)
4. J.W. Mattews: In Epitaxial Growth, Academic Press, New York (1979)
5. G.E. Henein et al.: J. App. Phys. 55, 2895 (1984)
6. P.J. Orozco: PhD. thesis, unpublished.
7. G.E. Henein et al.: J. App. Phys. 54, 728 (1983)
8. K.E. Meyer et al.: J. App. Phys. 52, 6608(1981)

Optical Properties and Enhanced Raman Scattering of Adsorbates on Rough Surfaces

T. Lopez-Rios

Laboratoire d'Optique des Solides, Université Pierre et Marie Curie,
4, place Jussieu, F-75252 Paris Cédex 05, France

1. Introduction

Since the pioneering work of FANO /1/ explaining the Wood anomalies of gratings, it has been realized that surface plasmons of rough surfaces can be excited by light. The wave vectors of the excited surface plasmons are given by the Fourier transform of the roughness. Much experimental and theoretical work on this topic has been done in recent years, most of which has been devoted to the study of plasmons of nearly flat surfaces. This restriction is probably due to the difficulty of performing realistic calculations of pronounced roughness, i.e. with large slopes. It is clear that, in general, a rough surface can support electromagnetic modes quite different from those of a flat surface. Lord RAYLEIGH /2/ studied the reflection of acoustic waves at a flat surface with holes and in a posthumous paper it is reported that, for resonant frequencies, he found an enhanced absorption of the acoustic waves in the holes. He suggested that similar resonances may exist with light for a metal surface with deep pits and that this could explain the abnormal colors of alkali metals thoroughly studied by WOOD /3/ at the beginning of this century. More recently, HESSEL and OLINER /4/ in an important paper described the diffraction behaviour of gratings of arbitrary shapes in a rather general form and found that such structures could support resonant guided modes of a type different from that of a flat surface. Nevertheless, the implications of the existence of electromagnetic resonances of rough surfaces remained quite limited until the discovery of Surface Enhanced Raman Scattering (SERS). This spurred a host of investigations on the surface plasmons of small particles intensively studied before in connection with their optical properties /5/. Both systems, nearly flat surfaces and isolated particles of various shapes, are well known at present. On the other hand, our knowledge of interacting particles or surfaces with large protuberances is in some respects very poor. From the experimental point of view the difficulty is to recognize in the optical absorption the contribution of the surface modes, which is weaker the more the mode is localized at the surface (large wave vectors). It is known today that surface mode excitation gives rise to large electromagnetic fields at the surface and leads to a tremendous increase of the Raman scattering of adsorbed molecules and to a lesser extent luminescence, resonant Raman and photochemical reactions. The experimental and theoretical aspects of these effects have already been reviewed /6-10/. All these problems deal with the interaction of a molecular dipole with a surface and with local field amplification by the excitation of surface plasmons.

In this paper I will review some unusual features of the optical properties of rough surfaces, in particular, I will discuss optical absorption and Raman scattering and illustrate, with a few examples obtained in our laboratory, how metal submonolayers strongly modify the optical absorption and Raman scattering of adsorbed molecules. A discussion on the basis of interacting spheres or gratings of narrow channels will be presented to illustrate the possible role of localized modes in SERS and other enhanced electromagnetic phenomena.

2. Optical absorption and Raman scattering of molecules on rough surfaces

HUNDERI and MYERS /11/ were probably the first to report that Ag films quenched on a cold substrate (120 K) present an optical absorption at 2.5 eV which does not exist for well-crystallized Ag and completely disappears with annealing at room temperature. The optical properties of quenched Ag films were later studied by several others workers /12-16/. Similar abnormal optical absorption is produced in alkali metals /5/ and is, as for Ag, due to the bad crystallographic structure, very probably in the form of pores or grain boundaries.

Figure 1, taken from /16/, shows the changes of reflectivity at normal incidence $\Delta R/R = 2(R - R_0)/(R_0 + R)$ induced by increasing deposits of Ag on a flat surface of a thin Ag film made at room temperature and cooled at 120 K. R_0 is the reflectivity of the flat surface and R, that of the surface with the quenched Ag deposits. Negative values of $\Delta R/R$ indicate a larger absorption of the surface with the quenched Ag deposits than of the flat surface. The two sharp peaks with opposite signs at about 3.8 eV are essentially due to the small values of Ag reflectivity at these frequencies and are of no interest in the problems discussed here. For the thinner deposits an absorption at about 3.6 eV is unambiguously assigned to the surface plasmons of a perturbed flat surface. The intensity and width of the resonance are roughly speaking indicative of the amplitude and correlation function of the roughness /17/. For increasing thickness of the Ag deposits an absorption at about 2.5 eV develops. The relative intensity of both absorptions depends very much on the thickness of the deposits and also on the experimental conditions. In Fig. 1 the absorption at 3.6 eV is easily understood to be due to the excitation of a flat surface mode by means of the momentum given by the roughness. The origin of the 2.5 eV absorption is, however, not well understood. HUNDERI /18/ explains this absorption with a Maxwell-Garnett model, assuming that the samples are inhomogeneous with regions within the grain boundaries having a dielectric constant quite different from that of the bulk Ag of each microcrystal.

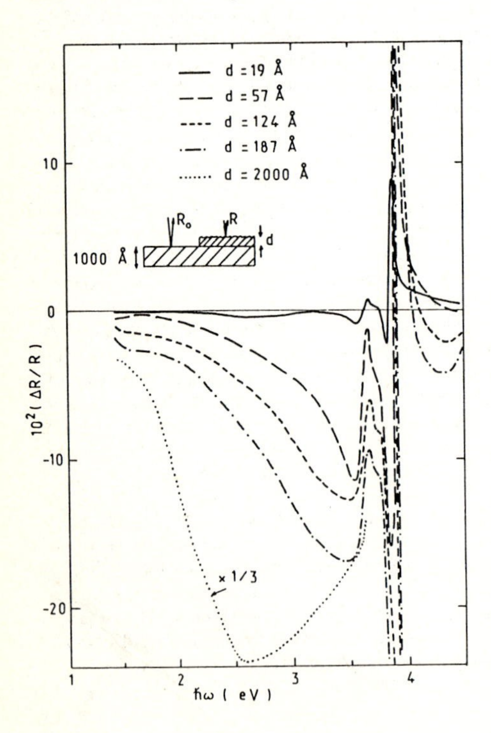

Fig. 1 Differential reflectivity at normal incidence $\Delta R/R = 2(R - R_0)/(R + R_0)$ for T = 120 K. R_0 is the reflectivity of the a flat Ag film 1000 Å thick made at room temperature, and R is the reflectivity with quenched deposits of Ag of thickness d.

This model leads to dielectric resonances with high average fields as was shown by ASPNES /19/ in a somewhat equivalent approach. The point to stress here is that the optical absorption at 2.5 eV always exists when the quenched Ag films are SERS active, and that both the absorption and SERS disappear with annealing. Unexpected optical absorptions and photoemission yields are observed on alkali quenched films /5,20/ which also present SERS /10/. The relationship between both effects, still to be clarified, is only tentatively discussed in this paper.

The structure of cold films is not well known. Studies by electron /21/ and tunnelling scanning microscopy /22/ at room temperature of Ag cold films, covered with very thin metal layers of Al or Cu or exposed to pyridine, give some information about the original surface, the adsorbates preventing a complete recrystallization. Nevertheless, this is not very useful in connection with SERS, since these surfaces no longer produce SERS. Some indirect information obtained by Auger, work function and photoemission studies of pyridine and Xe on cold Ag films indicate that the latter are made of a set of microcrystals with small channels or pores /23,24/. Several investigations of Ag cold films with different techniques /25,26,27/ evidence, as first pointed out by ALBANO et al. /28/, that SERS is mainly produced by molecules in the pores. Hunderi's idea that the abnormal optical absorption is induced by the grain boundaries is consistent with the experiments by Albano et al., i.e. pores are capable of generating SERS and producing an optical absorption.

For small particles, in particular colloids and island films, the relationship of SERS to optical absorption has been widely discussed in the literature /6, 29-32/. In connection with SERS the main point to elucidate is the intensity and distribution of the electromagnetic field at the surface. This is a difficult task because the standard optical methods measure the averaged field to the skin depth of light. One way to get information about the intensity of electromagnetic fields at the surface is to investigate the changes of reflectivity induced by very thin surface layers, their absorption being indeed dependent on their optical constants but also on the fields at the surface.

Figure 2 taken from /33/ shows the changes in reflectivity at normal incidence $\Delta R/R$ produced by different deposits of Cu on a Ag cold film. The shoulder at 3 eV is due to the combined effect of the optical absorption of Cu and the excitation of surface plasmons, as in the case of high-energy absorption in Fig.1. The absorption at 1.9 eV is particularly interesting. It does not exist at all for Cu on a flat surface

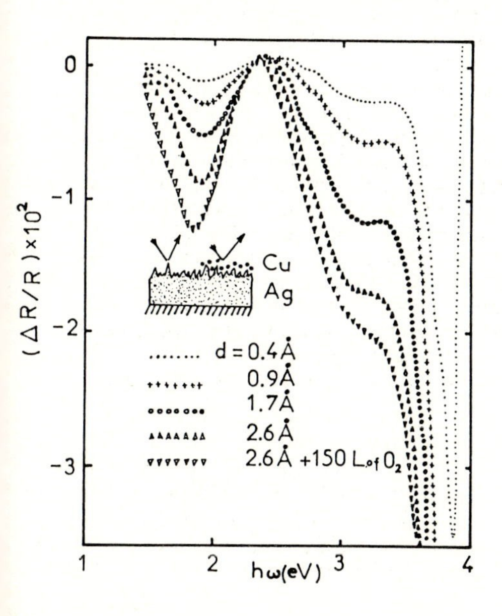

Fig. 2 Differential reflectivity at normal incidence $\Delta R/R$ for Cu overlayers of different thickness d on a surface prepared by condensing 1660 Å of Ag on a substrate at 150 K. $\Delta R/R$ is also indicated for a sample (with a Cu overlayer 2.6 Å thick) exposed to 150 L of oxygen.

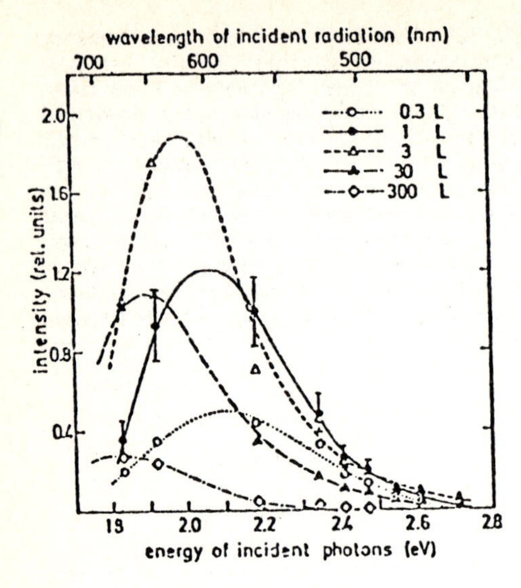

Fig. 3 Excitation spectra of the ν_1 mode (1006 cm^{-1}) for different coverages of pyridine adsorbed on a quenched Ag film. After Pockrand /34/.

of Ag /33/ and it is very likely due to the excitation of localized surface plasmons producing very high electromagnetic fields concentrated at the surface. This absorption was observed for other adsorbates such as Al, Pd and, as will be shown later, for pyridine as well (Fig. 6), clearly indicating that it is a characteristic of the surface rather than of the adsorbate. It seems reasonable to conclude that the excitation of surface modes leads to high fields at the surface and therefore to large absorption by the adsorbates.

Figure 3 shows the SERS excitation spectra for the breathing mode ν_1 (1006 cm^{-1}) of pyridine obtained by POCKRAND /34/ for different coverages of pyridine on cold Ag films. Pockrand finds that the excitation spectra are independent of the probe molecule (pyridine, C_2H_4) and of the chosen vibration when the excitation profile is represented as a function of the Stokes frequency /13/. He finds a resonance at about 2 eV with a width of 0.5 eV quite similar to those of Fig. 2, suggesting a relationship between them.

An important point to underline is that the absorption of Cu adsorbates in Fig. 2 does not correspond to bulk absorption of the quenched Ag film. On the contrary, the minimum on the surface absorption (2.5 eV) in Fig. 2 corresponds to a maximum of absorption for the cold film. The reason for this peculiar behaviour is still unclear. GAROFF et al. /31/ for rhodamine on Ag island films and GAO /35/ for Pd on Ag island films also found that the maximum of the adsorbate absorption does not correspond to a maximum of the film absorption. The absorption of the film is given by the fields inside the metal that are not necessarily maximum at the same frequency as the external fields. WEITZ et al. /32/ explain their results by the fact that at the resonance the field is nearly normal to the surface and the external and internal fields are related by the frequency-dependent dielectric constant. For cold films the situation is more complicated and it seems that a rather elaborate model will be necessary to understand the relationship between Fig. 1 and Fig. 2.

3. Quenching of SERS by metal adsorbates

One characteristic of SERS is the tremendous sensibility to surface modifications namely to the adsorption of metal or semiconductor or even gas adsorbates. Large modifications of SERS of silver electrodes by underpotentially deposited Tl, Pb and Cu have been investigated by several groups /36,37,38/. MURRAY /39/

investigated the quenching of the Raman scattering of CN on Ag island films by Au deposits, and explained his results with an electrostatic calculation basically describing the damping of the dipolar resonance of a single spheroid by the surface deposits. In ultrahigh vacuum, PETTENKOFER et al. /40/ reported a quenching of SERS of pyridine by oxygen adsorption. In our laboratory we have investigated the modifications produced by submonolayer coverages of Al /41/, Pd /42/, Si /43/ and Cu, Au /35/ on cold films and island films of silver. In all cases, the main effect of foreign atoms is to reduce SERS, this reduction being much more important at the beginning of the deposition, i.e. at lower coverages. This was always found to be accompanied by a decrease of the luminescence background characteristic of SERS /44/. For some metals (Cu, Al) new lines of pyridine were observed, produced by a small frequency shift due to a different chemical bonding of pyridine with adsorbates.

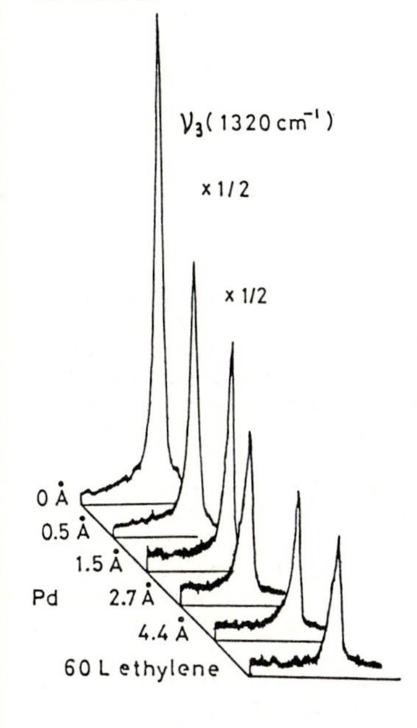

Fig. 4 The ν_3 mode (1320 cm^{-1}) of ethylene for a Ag cold film exposed to 60 L ethylene followed by deposits of Pd (from 0.5 Å to 4.4 Å) and a subsequent reexposure to 60 L of ethylene.

Figure 4 shows the scissors mode of ethylene ν_3 (1320 cm^{-1}) for a cold Ag film exposed to 60 L of ethylene for consecutive Pd deposits of increasing thicknesses and a reexposure of 60L. The Pd deposits produce a decrease of the Raman scattering of ethylene that cannot be attributed to a desorption of ethylene from the impinging Pd atoms, as can be inferred from the fact that new exposure of ethylene does not restore SERS. The surface of the Ag cold films is modified by the Pd deposits on an atomic scale, as was established by Auger spectroscopy /35,42/. Actually, it was determined that in such experiments Pd grows monolayer by monolayer in the same manner as on a flat surface of Ag, indicating that the external surface of a cold film is made of an ensemble of mostly flat regions, as previously shown by ALBANO et al. /28/.

MARINYUK et al. /45/ have reported the interesting and surprising observation that very thin Ag deposits on Pt electrodes induce a several-fold increase of the Raman scattering of pyridine. In our laboratory we have observed the same type of effect on SERS-active surfaces previously quenched by Pd or Cu deposits /41,35/. This effect was observed for cold films as well as for island films but the amount of Ag necessary to restore noticeable SERS (a fraction of a monolayer to ten monolayers) depends on the thickness of the Pd or Cu layer. Summarizing, it can be said that for different types of rough surfaces (electrodes, island films and cold films) SERS is extremely dependent on the nature of the top monolayer. This observation should lead to the conclusion that chemical effects are very important in SERS. Nevertheless, a more detailed analysis of such results show that, in fact, the quenching of SERS is dependent on the optical constants of the adsorbed metals rather than on their chemical nature and that the quenching is independent of the type of molecule giving the SERS. Even if an explanation on the basis of chemical theories is not excluded /7/, all these facts incite one to search for an explanation of quenching of SERS by metal deposits within the framework of electromagnetic theory.

4. Coverage dependence

One important step towards the understanding of SERS was made by MURRAY et al. /46/, who, with spacers of low Raman cross-section, were able to measure Raman enhancements as a function of the distance of the test molecule from the surface. Raman intensity as a function of molecular coverage could give the same kind of information. Figure 5 gives the intensity of the ν_1 mode of pyridine as a function of pyridine exposures for a cold Ag film, curve (a), and a cold Ag film previously covered by a Pd deposit 2.5 Å thick. Curve (a) is similar to the result of POCKRAND and OTTO. /47/. Auger measurements /35/ show that the sticking coefficient of pyridine on Ag and Pd are the same, the completion of the first monolayer being for about 6L. Figure 5 indicates that Pd strongly destroys the short range of SERS, i.e. the enhanced Raman scattering of the first monolayer. Similar results were obtained with island Ag films as active SERS surfaces.

Turning now to the optical properties of the systems already studied in Fig. 5, Fig. 6 shows for an angle of incidence of 50° and p - polarized light the changes of reflectivity induced by different amounts of pyridine on cold films. Figure 7 is as Fig. 6 but with the cold film previously covered by 2.5 Å of Pd. Both experiments were performed in the same experimental run, by translating the optical beam, on two halves of a sample (one part without Pd with reflectivity R_0 and the other covered by 2.5 Å of Pd with reflectivity R). For low coverages of pyridine there is an optical absorption at 1.9 eV which shifts to low frequencies with increasing pyridine coverage. A very similar red shift with increasing coverage is found in the excitation spectra (see Fig. 3), with a leveling off at about 10^2 L /34/. This absorption has the same spectral position and width as for Cu (see Fig. 2), clearly indicating that it is related to the electromagnetic properties of the rough surface and not to the electronic properties of the adsorbate. Moreover, we have not found experimentally any optical absorption at 2 eV for pyridine

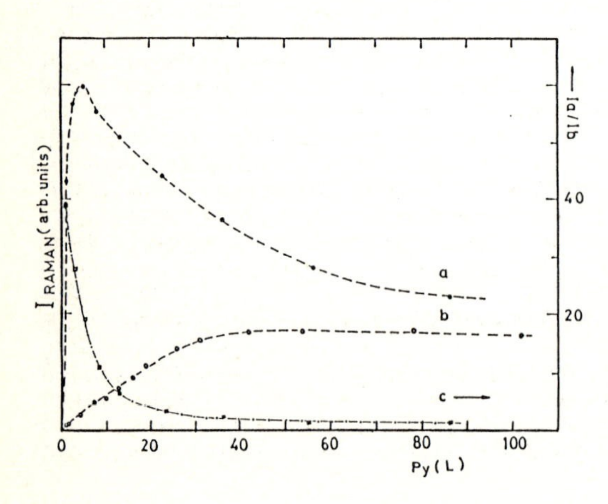

Fig. 5 Intensity of the ν_1 ($1006\ cm^{-1}$) mode of pyridine vs pyridine exposures for a Ag cold film (a); and for a Ag cold film covered by 2.5 Å of Pd (b). Curve (c) gives the ratios of curve (a) to curve (b).

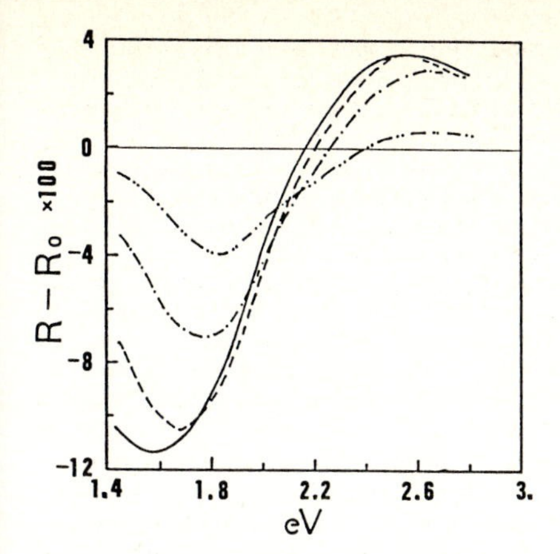

Fig. 6 Reflectivity R for p-polarized light incident at 50° upon a Ag cold film exposed to pyridine minus the reflectivity R_0 of the clean Ag film (T = 120 K). Two-dotted line: 11L; one-dotted line: 33L; dashed line: 81 L; and continuous line: 130 L.

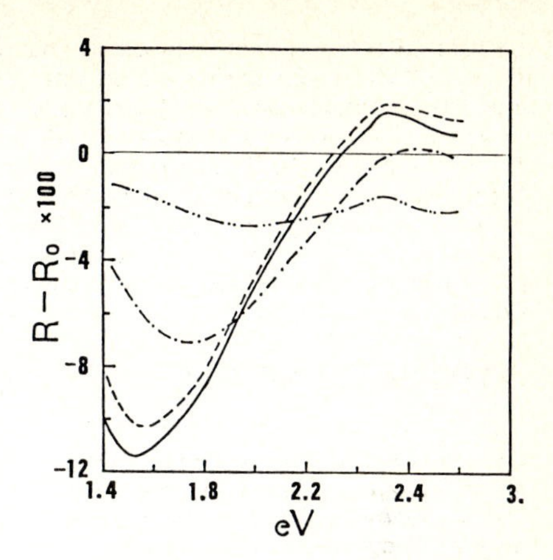

Fig. 7 The same as Fig. 6 but the Ag film previously covered by a Pd deposit 2.5 Å thick.

condensed at 120 K on a flat surface; this is in agreement with results for Au electrodes /48/ and indicates that the absorption at 1.9 eV is due to electromagnetic resonances. An important point to stress is that pyridine, which is not absorbing at 2 eV, produces a more important absorption than Cu (compare the absolute values of Fig. 2 and Fig. 6). When a molecule is near a metal surface the field due to its image could be very important and consequently the effective polarizability of the molecule becomes very large. This effect previously advanced to explain SERS /6/ was discarded by most workers because of unrealistic distances at which the molecule must be placed to have a noticeable effect at the laser frequency. A molecule driven by an electric field can excite surface plasmons by non-radiative energy transfer to a nearby metal surface with a consequent field amplification at the molecular location. Actually the surface plasmons to be considered are not those of a flat surface but the localized plasmons of small wave vectors on rough surfaces, namely in pores or between interacting particles. A discussion of SERS on the basis of spherical cavities has already been given in /27,49/.

Inspecting Figs. 6 and 7 in detail, we see, on one hand, that the changes of reflectivity $R-R_0$ are highly non-linear, $R-R_0$ being relatively more important for small rather than large coverages and, on the other, that the Pd layer greatly modifies the absorption of the first monolayer of pyridine and much less the following monolayers, so as to give rise to the same spectra for 130 L in both cases. It should be noticed that Pd produces the same qualitative results for the reflectivity changes as for SERS.

It is not possible to give a satisfactory explanation of the totality of the results that I have already presented and it seems to me that many other results showing short range SERS are not well understood. To clarify the idea that short range effects could have an electromagnetic origin, let me consider a molecular dipole of polarizability α placed in vacuum at a distance d from a flat surface of Ag and driven by an extended field E_0 perpendicular to the surface. The dipole moment is given by

$$\mu = \alpha (E_o + E_{im}),$$

where E_{im} is the image field given by /50/

$$E_{im} = \mu \cdot \int_0^{\infty} R_{//} (u^3 / l_1) \exp (- 2 l_1 d \, \omega / c) \, du ;$$

$l_1 = (u^2 - 1)^{1/2}$; $l_2 = (u^2 - \epsilon)^{1/2}$ and $R_{//} = (\epsilon l_1 - l_2)/(\epsilon l_1 + l_2)$ is the reflection coefficient of the Ag surface. In the electrostatic approximation ($u \rightarrow \infty$) equation (2) reduces to $E_{im} = \mu \cdot (\epsilon - 1)/4d^3(\epsilon +1)$. When a thin film is on the surface the expression (2) is still valid provided $R_{//}$ is replaced by the appropriate reflection coefficient. Even if the molecule is not absorbing, its image, which is a damped oscillating dipole, will lead to a dissipative effect that is evidenced by the imaginary part of the effective polarizability. Figure 8 gives $|E_{im}/E_o|^2$ [from formulas (1) and (2)] for a dipole of polarizability 10 Å^3, typical of pyridine placed 2 Å above the surface. One should note the effect on E_{im} of a Pd layer 1 Å thick which is also represented in the figure. The local field that the molecule really sees, responsible for the optical absorption and SERS, is in the present example greatly modified by the Pd layers. Local field effects have been thoroughly investigated /51,52/ to explain the optical properties of adsorbates on flat surfaces, but to my knowledge only a few results are available for rough surfaces /53/. The example of Fig. 8 is rather academic but illustrates that for other types of surfaces the quenching of SERS could be explained by the same mechanism. From this point of view, in equation (2) the exponential term of the integral is dominant for wave vectors $< (2d)^{-1}$ /54/. The Fresnel coefficient $R_{//}$ has a pole for $u = [\epsilon /(\epsilon +1)]^{1/2}$ which corresponds to surface plasmon dispersion, then with increasing separation of the molecule from the surface the surface plasmons of smaller wave vector will contribute to the local field. For surfaces made of crystalline planes we have some idea of at least two features having well-defined surface plasmons. One is the edged formed by two planes with surface plasmons propagating in the direction of the edge /55/. The other is two interacting parallel planes supporting an antisymmetric plasmon mode with very large field for close planes /56/. The surface plasmons of grain boundaries or pores could be of this type.

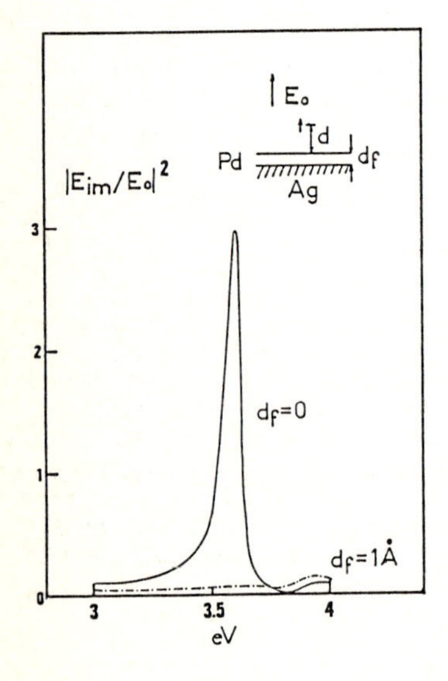

Fig. 8 $|E_{im} / E_o|^2$ for a dipole of polarizability $\alpha = 10$ Å^3 located at d = 2 Å from a flat Ag surface ($d_f = 0$); and at d = 2 Å from a Ag surface covered by a Pd layer 1 Å thick (dotted line). The external field E_o is oriented perpendicularly to the surface.

5. Surface localized modes

Plasmons of a flat surface have the electromagnetic field concentrated at the surface according as how large the wave vector is. A spherical particle has an infinite number of modes; the electric fields are increasingly concentrated at the surface as the order is increased. For small spheres (radius << wavelength) only the dipolar mode can efficiently be excited by light. In other words, the dipolar mode is the important one in the interaction of light with small spheres. In particular, it is well known that when the frequency of light coincides with the frequency of the dipolar mode a resonant absorption and large fields are produced. Most of the experimental work in SERS has been explained by this mechanism /6-10/.

When two spheres are brought together a splitting of the degenerate modes is produced with the appearance of one mode at nearly the same frequency as for the isolated sphere and another, red shifted, which is strongly dependent on the distance between the spheres. The first one can be visualized as the dipole of each sphere oscillating in antiphase and the latter as both dipoles in phase. The optical properties of interacting spheres or of one sphere with its image in a plane have been studied by several workers /57-61/. ARAVIND et al. /57/ showed with an electrostatic calculation that for interacting particles at the resonance, the field is highly concentrated in the region between the particles. Figure 9, taken from work by Aravind et al. /57/, shows the intensity of the enhancement $\tilde{I} = |E_{loc}/E_o|^2$, where E_{loc} is the local field at the observation point and E_o the incident field. The observation point and the incident field are in the x-z plane. The geometry is shown in the inset of Fig. 9. As the spheres approach each other the splitting becomes more and more important. The high-frequency peak remains at $\sim$3.48 eV whereas the low-frequency peak shifts downwards when the spheres approach. Figure 10, taken from the same work, is related to Fig. 9 and gives $\tilde{I}$ at differents points of the surface for an isolated sphere (λ =0) and interacting spheres. It should be noticed that in the region between the spheres (θ =180° or -180°) $\tilde{I}$ is about

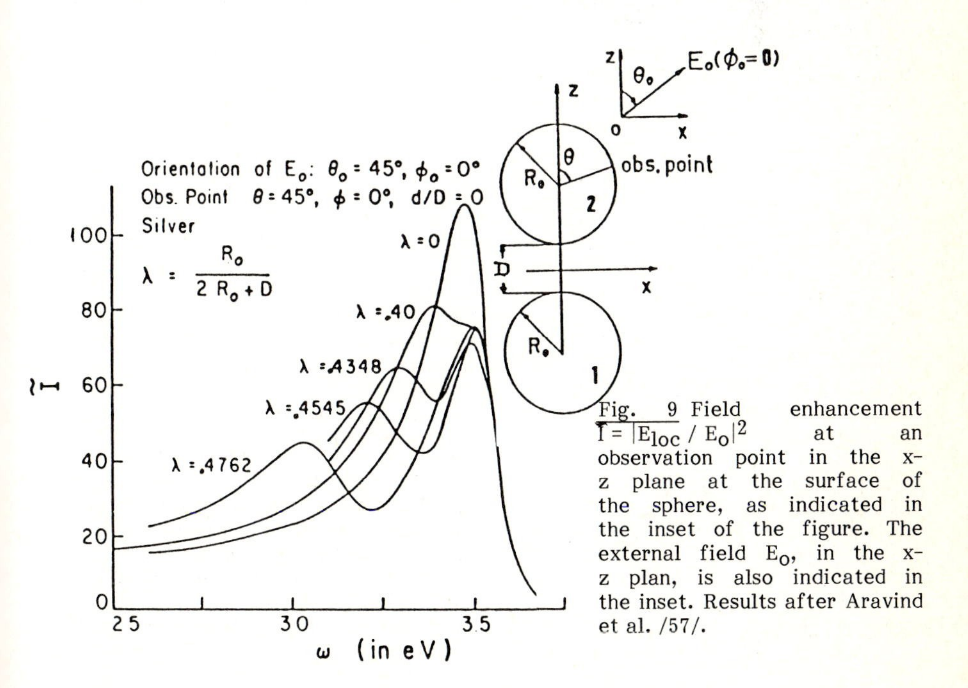

Fig. 9 Field enhancement $\tilde{I} = |E_{loc} / E_o|^2$ at an observation point in the x-z plane at the surface of the sphere, as indicated in the inset of the figure. The external field E_o, in the x-z plan, is also indicated in the inset. Results after Aravind et al. /57/.

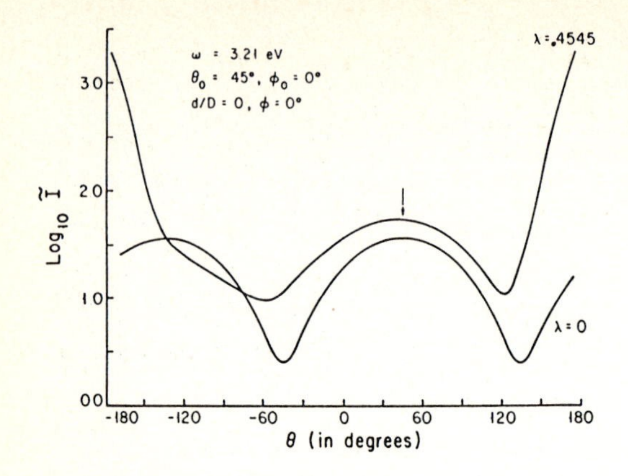

Fig. 10 Values of $\tilde{I}$, for different points at the surface of the sphere, corresponding to the calculation of Fig. 9.

50 times larger than for an isolated sphere. INOUE and OHTAKA /62/ have shown that for an infinite chain of interacting spheres the near field is essentially the same as for just two spheres.

Optical and Raman measurements as a function of pyridine coverage, presented in Figs. 5 and 6, could be tentatively explained on the basis of calculations by Aravind et al.. In some respects channels or pores on cold films may be considered as interacting surfaces and in a crude approximation may be modelled as two close particles. That the optical absorption in Figs. 2 and 6 takes place at 1.9 eV could indicate that the distance between the surface bumps or in the wall making the pores is less than the values of D employed in the calculations. Aravind et al. pointed out numerical difficulties in computing closer spheres. In the framework of this model, active sites and short range enhancements can be explained by pyridine being in the pores or between the bumps of island films. One may assume that very thin Pd deposits, even when placed outside the interacting region between the spheres, lead to a reduction of SERS and of the optical absorption. Because the fields between the spheres is highly multipolar, the optical response is expected to be very sensitive to surface modifications and to give rise to an apparent short range behaviour. Again, in terms of this model, the spatial long range behaviour of SERS is given by the field enhancement of the non-interacting part of the particles.

A rather different approach to the description of rough surfaces was proposed by WIRGIN and LOPEZ-RIOS /63/ and WIRGIN and MARADUDIN /64/. Basically, it consists in considering a lamellar grating as displayed in the inset of Fig. 11. An exact solution of such a problem for a medium of finite dielectric constant was given by PING SHENG et al. /65/, reducing the problem to the determination of the eigenvalues of a complex secular equation. In the WIRGIN model the vertical walls of channels are considered as perfectly conducting. With such an approach a modal representation of the fields in the channels always obeying the boundary conditions is straightforward. A Rayleigh representation of electromagnetic fields outside the channels is employed, combined with an impedance boundary condition at the horizontal plateau of the grating. One advantage of this model is that it leads to simple analytical expressions for the electromagnetic field and reflectivity, allowing one to distinguish easily the contributions of two different surface modes. For light polarized with the electric vector in the plane of incidence, one mode is the usual propagating plasmon excited by the momentum transferred by the grating. The other modes correspond to localized resonances in the channels with very large fields inside the channels. Figure 11 shows the reflectivity of one of these gratings and the enhancement of the field intensity $\tilde{I}$ at a point A in the middle of the channel and at point B in the middle of the external plateau. We see, on one hand, that very large fields could exist inside the channels and, on the other hand, that by changing the frequency we can turn the electromagnetic fields to be concentrated

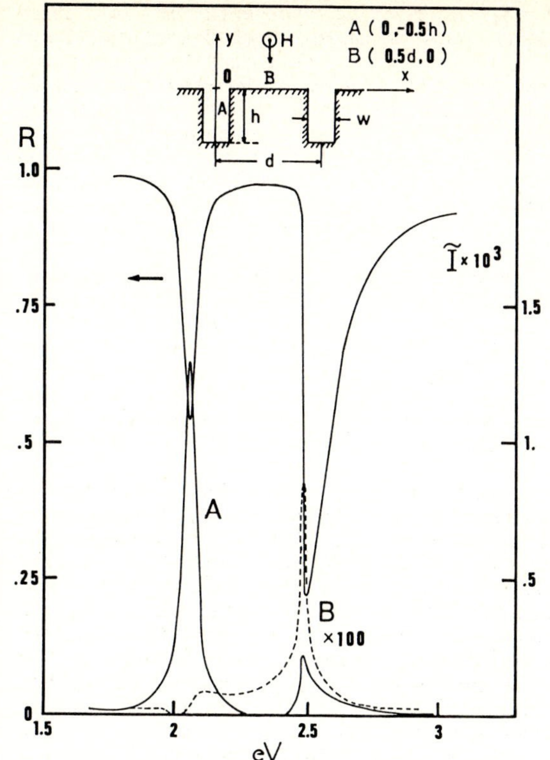

Fig. 11 Reflectivity (left side) and field enhancement $\tilde{I}$ for a point A (0, -0.5h) in the middle of the channel (continuous line) and for a point B (0.5d, 0) in the middle of the plateau for a Ag grating d = 0.5 μm; h = 0.1 μm and W = 0.02 μm. The light is impinging normal to the surface with a magnetic vector parallel to the channels of the grating.

into the channels or outside the channels. Note that the largest electric fields do not correspond to the largest absorption (notice that curve B is multiplied by 100). The absorption at about 2 eV corresponds to a localized resonance and that at 2.5 eV to a propagating resonance. In fact, an interaction between both types of modes is produced and both resonances have a hybrid character. This model is an alternative approach to those of Aravind et al. and could also explain why pyridine inside the pores gives rise to strong SERS. It could also explain why an important absorption of the surface atoms is produced for frequencies where the bulk absorption is small.

It must be realized that the spectral position, width and intensity of the fields depends on the shape of the surface and on the nature of the material. Very probably large field amplifications also exist at the rough surfaces of ionic crystals in the "reststrahl" region. In fact, in the high-energy side of the reststrahl band of NaCl, abnormal absorptions are observed. One of them is probably due to a multiple phonon effect but another one has been interpreted by Berremann with a model of grounded bumps as due to roughness /66/. Very probably this absorption of ionic crystals is the counterpart of abnormal optical absorption of quenched metals discussed at the beginning of this paper, and enhanced fields could exist at the surface of ionic crystals in the infrared.

6. Conclusion

Several effects : abnormal optical absorption and photoemission yields of badly crystallized films of free-electron metals, abnormal reflectivity of rough ionic crystals and SERS share the feature of not being well understood. Several reasons lead one to think that these effects could be different aspects of the same phenomena, i.e. localized electromagnetic resonances of rough surfaces. I have summarized herein some known pieces of the puzzle; it remains to join these pieces together.

Two models are reviewed : interacting spheres and lamellar gratings with deep grooves. For such systems two resonances of different natures exist. The first one, similar to dipolar resonance of an isolated sphere or to a surface plasmon of a flat surface. The second type of resonance corresponds to highly localized electromagnetic fields between the spheres or in the grooves of the gratings and might explain an optical absorption of cold Ag films (at 2.5 eV) and an observed surface absorption at 1.9 eV having the same spectral position and shape as the SERS excitation spectra.

The dependence of SERS on the coverage of pyridine on cold Ag films and coated by very thin Pd deposits constitutes evidence of long and short range contributions to SERS. The Pd deposits destroy only the short range contribution. This finding, apparently in favour of chemical theories, does not exclude an explanation of short range Raman enhancements by an electromagnetic effect. Some arguments of this type are proposed, namely a possible amplification of image fields by excitation of localized plasmons at the surface protuberance by the oscillating dipole.

Acknowledgement

I have greatly benefited from discussions with A. Wirgin, Y. Gao, Y. Borensztein and F. Abelès.

References

1. U. Fano, J. Opt. Soc. Am. 31, 213 (1941).
2. L. Rayleigh, Phil. Mag. 39, 225 (1920).
3. R.W. Wood, Phil. Mag. 38, 98 (1919).
4. A. Hessel and A.A. Oliner, Appl. Opt. 4, 1275 (1965).
5. P. Rouard and A. Meessen, in Progress in Optics XV, ed. by E. Wolf (North-Holland, Amsterdam 1977).
6. Surface Enhanced Raman Scattering, ed. by R.K. Chang and T.E. Furtak (Plenum, New York 1982).
7. A. Otto, in Light Scattering in Solids IV, ed. by M. Cardona and G. Güntherodt, Topics in Applied Physics, Vol. 54 (Springer, Berlin 1984).
8. R.K. Chang and B.L. Laude, in CRC Critical Reviews in Solid State and Materials Sciences, Vol 12, 1 (CRC Press Inc. 1984).
9. I. Pockrand, Surface Enhanced Raman Vibrational Studies at Solid/Gas Interfaces, Tracts Mod. Phys., Vol. 104 (Springer, Berlin 1984).
10. M. Moskovits, Rev. Mod. Phy. 57, 783 (1985).
11. O. Hunderi and H.P. Myers, J. Phys. F 3, 683 (1973).
12. M. Moskovits and P.H. McBreen, J. Chem. Phys. 68, 4992 (1978).
13. I. Pockrand, Chem. Phys. Lett. 92, 514 (1982).
14. P.H. McBreen and M. Moskovits, J. Chem. Phys. 54, 329 (1983).
15. T. Lopez-Rios, Y. Borensztein and G. Vuye, J. Phys. (Paris), Lett 44, L99 (1983).
16. T. Lopez-Rios, Y. Borensztein and G. Vuye, Phys. Rev. B 30, 659 (1984).
17. E. Kretschmann and E. Kröger, J. Opt. Soc. Am. 65, 150 (1975).
18. O. Hunderi, Phys. Rev. B 7, 3419 (1973).
19. D.E. Aspnes, Phys. Rev. Lett. 48, 1629 (1982).
20. J. Monin, Thesis, Paris (1972); J. Monin and G.A. Boutry, Phys. Rev. B 29, 1309 (1974).
21. T. Lopez-Rios, G. Vuye and Y. Borensztein, Surf. Sci. 131, L367 (1983).
22. J.K. Gimzewski, A. Humbert, J.G. Bednorz, B. Reihl, Phys. Rev. Lett. 55, 951 (1985).
23. E.V. Albano, S. Daiser, R. Miranda, K. Wandelt, Surf. Sci. 150, 368 (1985); Surf. Sci. 150, 386 (1985).

24. J. Eickmans, A. Otto, A. Goldmann, Surf. Sci. 171, 415 (1986).
25. H. Seki, T.J. Chuang, Chem. Phys. Lett. 100, 393 (1983).
26. H. Seki, T.J. Chuang, M.R. Philpott, E.V. Albano, K. Wandelt, Phys. Rev. B 31, 5533 (1985).
27. H. Seki, T.J. Chuang, J.F. Escobar, H. Morawitz, E.V. Albano, Surf. Sci. 158, 254 (1985).
28. E.V. Albano, S. Daiser, G. Ertl, R. Miranda, K. Wandelt and N. Garcia, Phys. Rev. Lett. 51, 2314 (1983).
29. C.Y. Chen and E. Burstein, Phys. Rev. Lett. 45, 1287 (1980).
30. J.G. Bergman, D.S. Chemla, P.F. Liao, A.M. Glass, A. Pinczuk, R.M. Hart and D.H. Olson, Opt. Lett. 6, 33 (1981).
31. S. Garoff, D.A. Weitz, T.J. Gramila and C.D. Hanson, Opt. Lett. 6, 245 (1981).
32. D.A. Weitz, S. Garoff and T.J. Gramila, Opt. Lett. 7, 168 (1982).
33. T. Lopez-Rios, Y. Borensztein and G. Vuye, J. Phys. (Paris) 44, C10-353 (1983).
34. I. Pockrand, Chem. Phys. Lett. 92, 509 (1982).
35. Y. Gao, Thesis, Paris (1986).
36. J. Watanabe, N. Yanagihara, K. Honda, B. Pettinger and L. Moerl, Chem. Phys. Lett. 96, 649 (1983).
37. T.E. Furtak and D. Roy, Phys. Rev. Lett. 104, 59 (1983).
38. B. Pettinger and L. Moerl, J. Phys. (Paris) C10, 333 (1983).
39. C.A. Murray, J. Electron Spectrosc. Relat. Phenom. 29, 371 (1983).
40. C. Pettenkofer, J. Eickmans, U. Ertük and A. Otto, Surf. Sci. 151, 9 (1985).
41. T. Lopez-Rios, Y. Gao and G. Vuye, Chem. Phys. Lett. 111, 249 (1984); Surf. Sci. 164, L819 (1985).
42. Y. Gao and T. Lopez-Rios, Phys. Rev. Lett. 53, 2583 (1984); J. Vac. Sci. Technol. B 3, 1539 (1984).
43. Y. Gao and T. Lopez-Rios to be published in Solid State Commun.
44. Y. Gao and T. Lopez-Rios, Surf. Sci. 162, 976 (1985).
45. V.V. Marinyuk, R.M. Lazorenko-Manevich and Y. M. Kolotyrkin, Solid State Commun. 43, 721 (1982).
46. C.A. Murray, D.A. Allara and M. Rhinewine, Phys. Rev. Lett. 46, 57 (1981).
47. I. Pockrand, A. Otto, Solid State Commun. 35, 861 (1980).
48. F. Kusy and K. Takamura, Surf. Sci. 158, 633 (1985).
49. H. Chew and M. Kerker, J. Opt. Soc. Am. B 2, 1025 (1985).
50. R.R. Chance, A. Prock and R. Silbey, in Adv. Chem. Phys. eds. I. Prigogine and S.A. Rice (Wiley-Interscience, New York, 1974) 37, 543.
51. A. Bagchi, R.G. Barrera, R. Fuchs, Phys. Rev. B 25, 7086 (1982).
52. W.L. Mochan, R.G. Barrera, Phys. Rev. Lett. 55, 1192 (1985); Phys. Rev. Lett. 56, 2221 (1986).
53. W. H. Weber and G.W. Ford, Phys. Rev. Lett. 44, 1774 (1980).
54. G.W. Ford and W.H. Weber, Phys. Reports 113, 195 (1984).
55. L. Dobrzynski and A.A. Maradudin, Phys. Rev. B 6, 3810 (1972).
56. E.N. Economou and K.L. Ngai, in Adv. Chem. Phys. eds I. Prigogine and S.A. Rice (John Wiley, 1974) 27, 265.
57. P.K. Aravind, A. Nitzan, H. Metiu, Surf. Sci. 110, 189 (1981).
58. R. Ruppin, Phys. Rev. B 26, 3440 (1982); Surf. Sci. 127, 108 (1983).
59. P.K. Aravind and H. Metiu, Surf. Sci. 124, 506 (1983).
60. J.I. Gersten and A. Nitzan, Surf. Sci. 138, 165 (1985).
61. F. Claro, Phys. Rev. B 30, 4989 (1985).
62. M. Inoue and K. Ohtaka, J. Phys. Soc. Jpn. 52, 3853 (1983).
63. A. Wirgin and T. Lopez-Rios, Opt. Commun. 48, 416 (1984) and 49, 455 (1984); T. Lopez-Rios and A. Wirgin, Solid State Commun. 52, 197 (1984).
64. A. Wirgin and A.A. Maradudin, Phys. Rev. B 31, 5573 (1985).
65. Ping Sheng, R.S. Stepleman and P.N. Sanda, Phys. Rev. B 26, 2907 (1982).
66. D.W. Berreman, Phys. Rev. B 1, 381 (1970).

Structure and Electrical Properties of rf Sputtered $Cd_{1-x}Fe_xTe$ Thin Films

O.A. Fregoso[1], *J.G. Mendoza-Alvarez*[2*], *and F. Sánchez-Sinencio*[2*]

[1]Instituto de Investigación en Materiales, Universidad Nacional Autónoma de México, México, D.F., México
[2]Centro de Investigación y de Estudios Avanzados del IPN, Departamento de Física, A.P. 14-740, 07000 México, D.F., México

Structural, electrical and magnetic properties of II-VI compounds doped with a controlled quantity of magnetic ions (semimagnetic semiconductors), have been widely investigated, mainly on single crystal compounds /1,2/, and less extensive studies have been made on semiconductor thin films /3,4/.

In this paper we report the preparation of Cd(1-x)Fe(x)Te thin films grown by rf sputtering, and the characterization of their structural and electrical properties as a function of substrate temperature.

The films were prepared in a diode rf sputtering system with a 1cm-diameter target mounted on an air-cooled cathode. The target was made by vacuum pressing of 99.999% pure CdTe powder from Cerac Inc., with a 7x2 mm^2 Fe^{57} foil added. The films were deposited onto Corning 7059 glass substrates. The target-to-substrate distance was 3cm. Prior to film deposition the chamber was evacuated at a pressure of $3x10^{-6}$ Torr; 99.999% pure Argon was used in the sputtering process. We prepared a set of samples under the following conditions: background pressure=$3x10^{-6}$Torr; Ar pressure=3mTorr; rf power=100 watts; deposition time=60 min.; and substrates temperatures Ts=25,100,150,200,250, and 300ºC. Just before deposition, the target was usually presputtered for 15 minutes under the same deposition conditions.

Film thickness was measured on test samples using a Dektak Sloan profile analyzer, and in this way the deposition rate was determined. Resistivity measurements were performed at room temperature by using the four electrodes technique /5/. Square samples of 1 cm-side with thermally evaporated aluminum coplanar electrodes were used in the resistivity measurements. The samples were routinely examined for structural information using X-ray diffraction (XRD). The compositional analysis was performed in a Scanning Auger Microscope (SAM)-ESCA system. The SAM measurements showed that the distribution of Fe^{57} in the films is homogeneous, and we could not detect the presence of Fe clusters. The Fe^{57} doping was around 5at% and it was independent of the deposition time. The film deposition rate, for Ts=200ºC was around 6.4 Å/sec.

The deposition rate, r, changes depending on the substrate temperature. In Figure 1 the logarithmic dependence of r on 1/Ts for a fixed rf power and Ar pressure is shown. From this curve we got an activation energy of 0.28eV for the growth process; this energy corresponds to the nucleation-condensation mechanism obtained in the Walton theory /6/ and in the Walton experimental results /7/. The growth rate decreases gradually with increasing substrate temperature due to the reduction of the sticking coef-

---
* Also at the Escuela Superior de Física y Matemáticas del IPN. Mexico

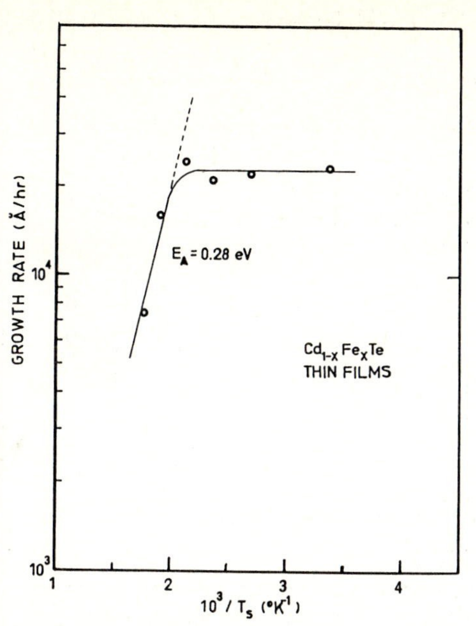

Fig.1 Deposition rate for the sputtered films as a function of substrate temperature. At Ts=200ºC, there is a maximum in the growth rate.

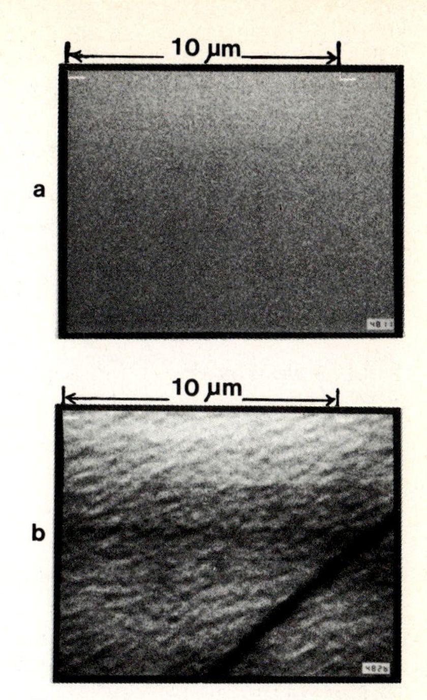

Fig.2 Scanning electron micrographs showing the surface texture for films grown at: a) Ts= 25ºC, b) Ts=200ºC

ficients of Cd, Fe and Te, and also because of the assisted resputtering of the film by Ar and electron species /6-9/.

The morphology of the sputter-deposited films is fine-grained polycrystalline with a grain size that depends on Ts. In Figures 2a and 2b, scanning electron micrographs for two different films are shown: the first one was grown at Ts=25ºC and the second one at Ts=200ºC. As it can be seen in these micrographs, the film surface features change, showing larger grain sizes for the film grown at a higher Ts.

The crystal structure of the films was studied through the XRD patterns (shown in Figure 3) of the films grown at different Ts. The film grown at room temperature shows an amorphous-like pattern. For Ts=100ºC, the XRD pattern begins to show peaks related to crystalline structure, that is, we have polycrystalline material. For Ts=150ºC, the pattern shows evidence of the presence of the cubic and hexagonal CdTe phases with a preferential orientation along the (110) direction. For a Ts=200ºC, the XRD pattern indicates that the film is growing in the cubic (zincblende) phase with a preferential orientation growth along the (111) direction. For higher temperatures of growth, the patterns show less intense and broader diffraction peaks; this is partially due to the fact that for higher Ts the films are thinner, as shown in Fig. 1; and also probably because at high Ts there is an increase in Fe diffusion causing disorder and defects in the films. From X-ray data, we were also able to estimate the grain size in the films /10/, obtaining values of ~50Å (Ts=100ºC), ~500Å(Ts=200ºC), and ~250Å(Ts=250ºC).

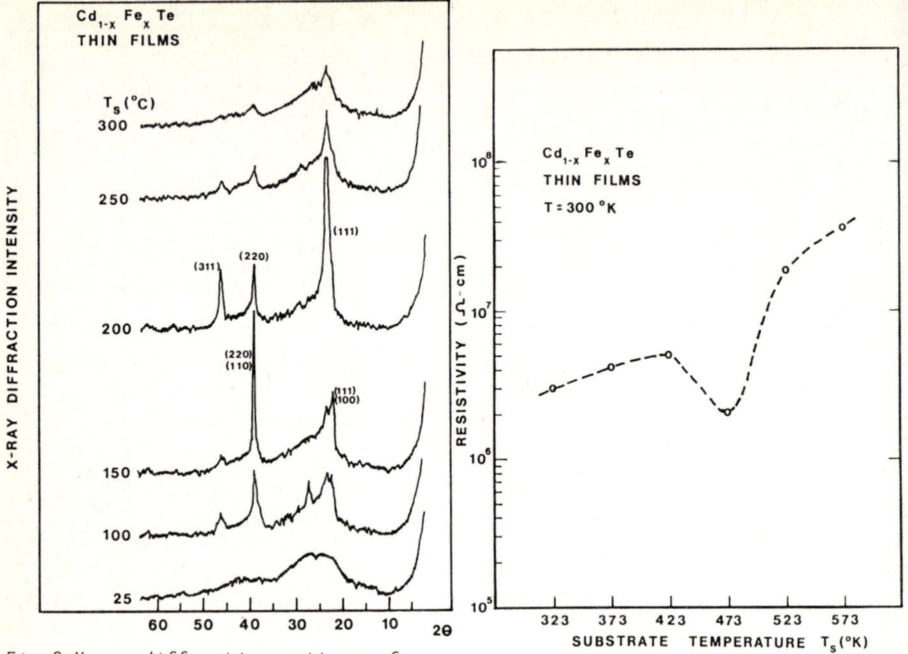

Fig.3 X-ray diffraction patterns for films grown at different substrate temperatures. At Ts=200ºC, the film grow in the zincblende structure

Fig.4 Room temperature electrical resistivity dependence on the film substrate temperature

From the XRD patterns, we see that the best crystalline structure is obtained for a substrate temperature of 200ºC.

From hot probe measurements we found that the films show n-type conductivity, which could be due to Fe(3+) entering substitutionally in Cd sites. Resistivity measurements were made for the different films deposited at several Ts. The dependence of the resistivity on the substrate temperature is shown in Figure 4. In this Figure it can be seen that there is an increase in resistivity as we increase Ts, except for a minimum at Ts=200ºC. The increase in resistivity could be due to a decrease in the film carrier density that occurs if Fe(2+) is placed in Cd sites instead of Fe(3+) (which can be interstitial), at high Ts. There is evidence, from Mossbauer measurements on these films that this is the case/11/. However, the resistivity also depends on the mobility and this will increase as the grain size increases; then, as we see from Fig. 3, the film deposited at Ts=200ºC will have a higher mobility and, as a consequence, lower resistivity. This could explain the minimum in resistivity at Ts=200ºC observed in Fig. 4.

In summary, we have studied the structure and electrical resistivity of sputter-deposited Cd(1-x)Fe(x)Te films, and have found that these properties are strongly influenced by the substrate temperature. The film composition was independent of Ts, with an Fe content of 5 at% without clusters. The structural characteristics and grain size are optimal at Ts=200ºC. Also, the resistivity presents a minimum for this particular Ts=200ºC as a result of larger grain sizes.

ACKNOWLEDGEMENTS

We want to thank Dra. Leticia Baños for providing the use of the X-ray diffraction equipment, M.A. Vidal for his assistance with the XRD amorphous pattern, and S.Romero and J.L. Peña for the SAM analysis. This work has been partially supported by CONACyT, the OAS and the Fundación Ricardo J. Zevada.

REFERENCES

1. J.A. Gaj: J. Phys. Soc. Japan 49, Suppl. A, 797 (1980)
2. G.A. Slack, S.Roberts and J.T. Vallin: Phys.Rev. 187, 511(1969)
3. K. Lischka, G. Brunthaler and W. Jantsch: J.Cryst. Growth 72,355(1982)
4. L.F. Schneemeyer, B.A. Wilson, W.P. Lowe, F.J. DiSalvo, S.E. Spengler, J.V. Waszczak, and J.F. Dillon,Jr.: Materials Research Society, 1986 Spring Meeting, Abstract Book, p. 43
5. W.R. Runyam: In Semiconductor Measurements and Instrumentation, McGraw-Hill,New York 1975)
6. D. Walton: J. Chem. Phys. 37,2182(1962)
7. D.Walton, T.Rhodin, and R.W.Rollins: J. Chem. Phys. 38,2698(1963)
8. J.A. Thornton and D.G. Cornog: J. Vac. Sci.Technol. A 18, 199(1981)
9. Nobuo Matsumura, Takefumi Ohshima, Junji Saraie and Yutaka Yodogawa: J.Cryst.Growth 71, 361(1985)
10. D. Cullity: in Elements of X ray Diffraction (Addison-Wesley, New York 1956)
11. R. Scorzelli, private communication

Characterization of Sputtered CdTe Thin Films by XPS, Auger and ELS Spectroscopies

I. Hernández-Calderón, J.L. Peña, and S. Romero

Centro de Investigación y de Estudios Avanzados del IPN, Departamento de Física, A.P. 14–740, 07000 México, D.F., México

The study of CdTe thin films and CdTe related heterostructures is of considerable scientific and technological interest due to their potential application in the manufacture of optoelectronic devices /2/. A variety of processes have been employed to prepare CdTe films. In particular, preparation of CdTe polycrystalline films by rf sputtering offers the possibility of wide range control of n and p dopant concentrations. However, a systematic study of the influence of deposition parameters and growth mode is necessary for the production of films with the required optical and electrical properties for practical applications /3/. In this work, we present the results of X-ray diffraction and X-ray photoemission (XPS), Auger and electron energy-loss (EELS) spectroscopies obtained from CdTe sputtered thin films.

A standard rf diode sputtering system with a five inch target of sintered high purity CdTe was employed to deposit CdTe thin films. The sputtering chamber had a base pressure of less than $5x10^{-6}$ Torr. Ar pressures varied between 1 and 6 mTorr and rf powers were in the 100 to 400 watts range. The temperature of the Corning glass (7059) substrates was varied between 50 and 200Cº. Substrate temperature (Ts) was measured directly on the exposed substrate surface using a chromel-alumel thermocouple. Growth rates varied between 0.5 and 2.5 Å/s. Films of 3000 to 7000 Å thickness were produced. Before deposition 15 minutes of pre-sputtering were allowed.

Figure 1 shows a diffraction pattern with typical characteristics of sputtered films. The scattering geometry allows to detect only those planes parallel to the substrate. Examination of the pattern indicates the presence of the cubic (zinc blende) and hexagonal (wurzite) phases with preferential growing in the <111> and <002> directions, respectively. The metastable hexagonal phase of CdTe seems to exist only in films /2,4,5/. The coexistence of both structures is not surprising if one considers a peculiarity of atoms in the zinc blende or wurzite structures corresponding to the (111) or (002) planes, respectively: they are singly or triply bonded to adjacent planes /6/, Fig. 1. This means that arriving atoms will be singly or triply bonded to the surface. It can thus happen that an atom that has been singly attached to the surface is rotated 60 degrees around the [111] direction in relation to the zinc blende position, Fig.1. In these conditions growth of the hexagonal phase is initiated. Variation in substrate temperature and heat treatment induced changes in the relative intensities of the peaks, but growth habits remained the same. Due to overlaping of the main diffraction peaks of the cubic and hexagonal phases, it is not possible to obtain a quantitative relation of phase composition from this kind of patterns.

Following experiments were performed in a PHI560/ESCA-SAM system. Figure 2 shows typical XPS spectra of Cd and Te 3d core levels of CdTe films before and after surface cleaning. Figures 2a,b correspond to Cd spectra of the contaminated surface with photoemitted electron directions normal and near

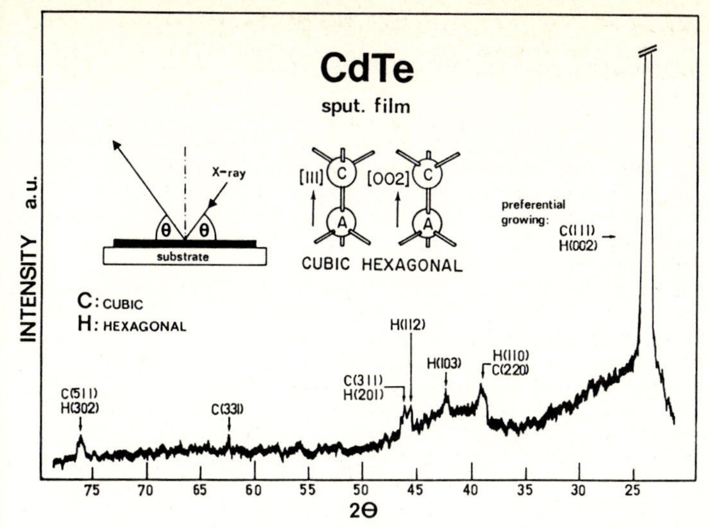

Fig. 1. X-ray diffraction pattern of a CdTe sputtered film. The insets show the scattering geometry and bonding in the growth direction.

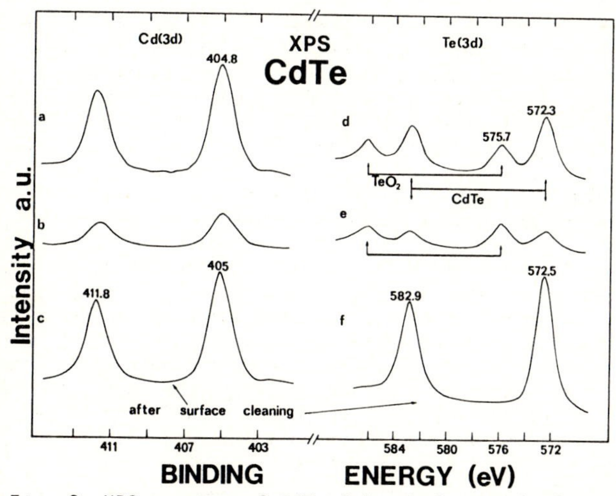

Fig. 2. XPS spectra of CdTe films before and after surface cleaning.

the surface, respectively. In the last case we have increased surface sensitivity. One observes only a relative change in intensity. After short Ar^+ bombardment of the surface the spectrum of Fig. 2c is observed. Peaks are shifted 0.2 eV to higher binding energies and coincide with those of a bulk CdTe single crystal. Te core levels of the contaminated surface are shown in Figs. 2d,e. One doublet is related to Te in CdTe and the other to TeO_2 The ratio of the oxide peaks to the CdTe peaks is 1:3 for normal emission (bulk sensitivity), Fig. 2d, and 4:3 for near surface emission (surface sensitivity), Fig. 2e. After surface cleaning the spectrum of Fig. 2f is observed. CdTe-related peaks are also shifted 0.2 eV to higher binding energies and coincide with those of CdTe single crystal. Above results indicate that a superficial tellurium oxide layer exists in films exposed to air and that Cd remains unaltered. This conclusion is also confirmed by investigations on CdTe(110) surfaces /7/.

Auger surveys of films exposed to air indicated C, O, S and Cl as main surface contaminants. Short Ar^+ bombardment was enough to obtain a spectrum of the clean surface. The main peak for Cd was located at 377 eV kinetic energy and for Te at 484 eV in the dN/dE spectrum. Auger profiles and scanning showed that the films were homogeneous and free of contaminants. Having as a reference a CdTe single crystal, measurements of the peak to peak amplitude in the dN/dE distribution gave 50.5±1% atomic concentration of Te for films grown with Ts in the 50 - 200ºC range. Excess tellurium can be attributed to cadmium losses in the growing film /8/. Examination of the Cd region showed minor changes before and after surface cleaning. However, the Te region presented considerable modifications. These are attributed to the existence of a TeO_2 layer, in agreement with previous XPS results. Changes were very similar to those of elemental Te. Figs. 3a,b,c show the Auger spectra of clean, slightly contaminated and very contaminated Te, respectively. The peak at 478 eV in Fig. 3c is assigned to tellurium oxide. The peak in 510 eV is due to oxygen, however it also contributes to the peak at 490 eV. Note the changes in the ratio of the amplitudes of the peaks at 482 and 490 eV from the contaminated to the clean surface.

EELS experiments were performed using electrons of 1200 eV inciding at 60 degrees to the normal surface. Non-specular scattered electrons were collected by the cylindrical mirror analyzer with pass energy of 25 eV (FWHM of

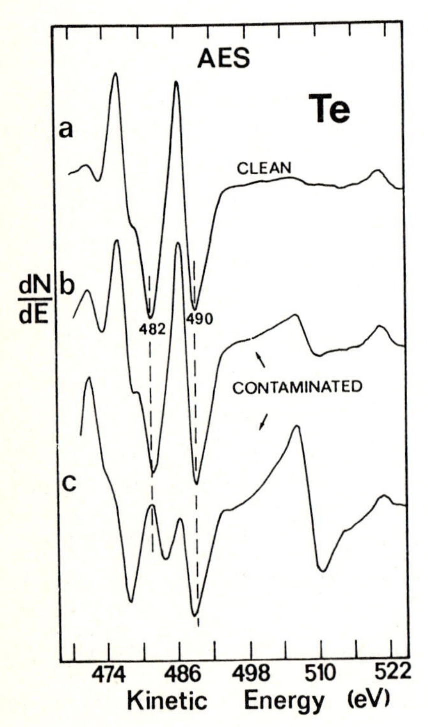

Fig. 3. Auger spectra of clean and contaminated elemental tellurium

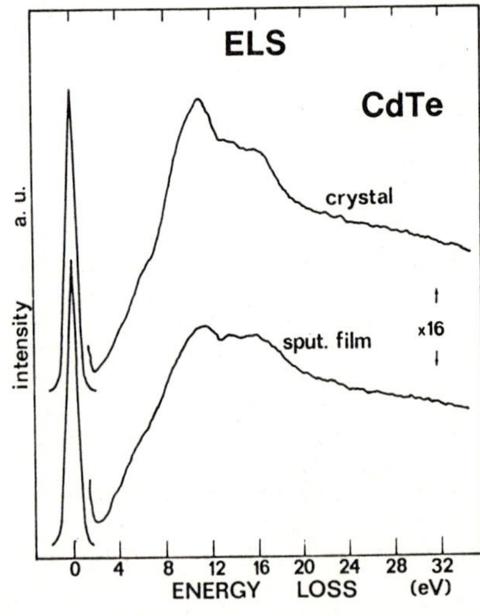

Fig. 4. Electron energy-loss spectra of a polycrystalline bulk sample and of a CdTe sputtered film

the reflected elastic peak equal to 1 eV). Figure shows the EELS spectra of a sputtered CdTe film grown with Ts=100ºC and of a CdTe bulk polycrystal. The film spectrum shows main features around 5.6, 11.2, 13.8 and 16.7 eV and the polycrystal around 5.5, 11.1, 14 and 16.5 eV. Characteristic energy losses are due to bulk and surface plasmons and interband transitions. Due to the equivalence of the transverse and longitudinal dielectric constants in this energy range, optical data are useful in the interpretation of interband transitions /9/. From the reflectivity measurements of CARDONA and GREENAWAY (CG) /10/, the peak around 5.5 eV in the EELS spectrum can be attributed to X_1-X_5 interband transitions; CG observed also a feature around 10.1 eV that PHILLIPS /11/ assigned to L_3'-L_2' transitions. The respective loss can be related to the shoulder at the low-energy side of the 11.1 eV peak. The peak around 16.6 eV is assigned to the bulk plasmon. In transmissions EELS experiments with 30 KeV electrons, GAUTHE /12/ found a value of 16.9 eV for the bulk plasmon in thin CdTe films. The feature around 11.2 eV is assigned to the surface plasmon; this gives a value of 1.49 for the ratio of the energies of the bulk to the surface plasmon. Since the 4d core levels of Cd in CdTe have binding energies around 11 eV below the top of the valence band /13/, the peak around 13.8 eV is assigned to transitions from the Cd(4) core levels to the conduction band, in agreement with the observations of CG /10/. Above assignments of the EELS spectra are also consistent with investigations on CdS and CdSe by BRILLSON /14/.

Summarizing: The investigated CdTe sputtered films are polycrystalline, containing a mixture of cubic and hexagonal phases which are nearly stoichiometric. A tellurium oxide overlayer is always formed under exposition to air. Cd does not oxidize. The electronic properties investigated by XPS, Auger and EELS are very similar to those of CdTe bulk material.

AKNOWLEDGMENTS

We thank S. Jiménez-Sandoval for his invaluable assistance in preparing the samples and to Dr. Leticia Baños for providing the X-ray diffraction patterns. I.H.C. wishes to thank the R. J. Zevada Foundation of México for partial financial support.

REFERENCES

1. Also at Escuela Superior de Física y Matemáticas, IPN, México.
2. T. H. Myers, Y. Lo, R. N. Bicknell, J. F. Schetzina, Appl. Phys. Lett., 42, 247(1983).
3. M. B. Das, S. V. Krishnaswamy, R. Petkie, P. Swab, K. Vedam, Solid-St. Electron., 27, 329(1984).
4. P. N. J. B. Webb, D.E. Brodie, Can. J. Phys. 54, 446(1976).
5. J. Buch, D. Valentovic, Thin Sol. Films, 51, 349(1978).
6. I. Hernández-Calderón, H. Höchst, Surf. Sci. 152, 1035(1985).
7. J. G. Werthen, J. P. Haring, R. H. Bube, J. Appl. Phys., 54, 1159(1983).
8. R. F. C. Farrow, G. R. Jones, G. M. Williams, I. M. Young., Appl. Phys. Lett., 39, 954(1981).
9. D. L. Greenaway, G. Harbeke: Optical Properties and Band Structure of Semiconductors (Pergamon Press, Oxford, 1968).
10. M. Cardona and D L. Greenaway, Phys. Rev., 131, 98(1963).
11. J. C. Philipps, Phys. Rev., 133, 452A(1964).
12. B. Gauthe, Phys. Rev., 114, 1265(1959).
13. M. Pessa, O. Jylhä, P. Huttunen, M. A. Herman, J. Vac. Sci. Tech. A2, 418(1984).
14. L. J. Brillson, J. Vac. Sci. Tech., 20, 652(1982).

Growth of Polycrystalline CdTe Films by Hot Walls Close-Spaced Vapor Transport Technique

O. Zelaya, F. Sánchez-Sinencio, J.G. Mendoza-Alvarez, and J.L. Peña

Centro de Investigación y de Estudios Avanzados del IPN,
Departamento de Física, A.P. 14-740, 07000 México, D.F., México

1. INTRODUCTION

The close-spaced vapor transport method (CSVT) is currently used to grow II-VI /1,2/ and III-V /3,4/ compound semiconductors. Film epitaxial growth has been achieved on single crystalline materials and polycrystalline films with large grain size have been successfully grown on amorphous substrates. Unlike other deposition systems, the CSVT is a simple and inexpensive technique. The transport process involved in the CSVT has been studied by Anthony et al./5/, and they have proposed a diffusion-limited transport (DLT) model. In this model, the molecular flux (Cd and Te_2), J, can be expressed as proportional to $\exp(-E_a/kT_{so})$, where E_a = 1.99 eV is an activation energy, T_{so} is the source temperature; also, J depends inversely on both the control gas pressure, P, and the separation, h, between source and substrate.

In this work, polycrystalline CdTe films were grown on glass substrates, by a modified CSVT technique (hereafter HW-CSVT technique), in which we use thermal conductive graphite walls in the growing chamber, instead of the conventional thermal insulator quartz walls. Our experimental results for J as a function of P, are in good agreement with the DLT model. However, the influence of T_{so} on J can not be explained by the DLT model. The slight deviation is probably due to molecular transport by a convection process.

2. EXPERIMENTAL

The experimental arrangement used for the HW-CSVT technique, has been described elsewhere /6/. The Ar gas used during the film growth, provided by Mathesson (TM*), was 99.999% pure. At the source of the growing chamber, was placed CdTe powder 99.999% pure, provided by Cerac (TM*). Films with an area of 4 cm^2 were grown on 7059 Corning glass substrates. In order to measure the amount of evaporated CdTe, the source was weighed before and after each growth, using a Sauter-414 microbalance with a sensitivity of one tenth of a miligram. The distance h was fixed at 3.5 mm. The samples were grown under the following conditions: a. P in the range 0.5 - 1000 Torr, T_{so} = 700°C and substrate temperature T_{sub} = 600°C. b. T_{so} in the range 600-750°C, T_{sub} = 550°C and 500°C, and P = 20,50,200 and 560 Torr.

3. EXPERIMENTAL RESULTS AND DISCUSSION

Figure 1 shows the relationship between log J and log 1/P, for T_{so} = 700°C and T_{sub} = 600°C. The solid line corresponds to the DLT model. It can be observed that the experimental points follow approximately a straight line for P > 10 Torr. For P < 10 Torr, the experimental points depart from the linear behavior and tend to a saturation region. This last region is due to the fact that the evaporation rate is insufficient to keep up the flux rate required by the DLT model at low gas pressures/5/.

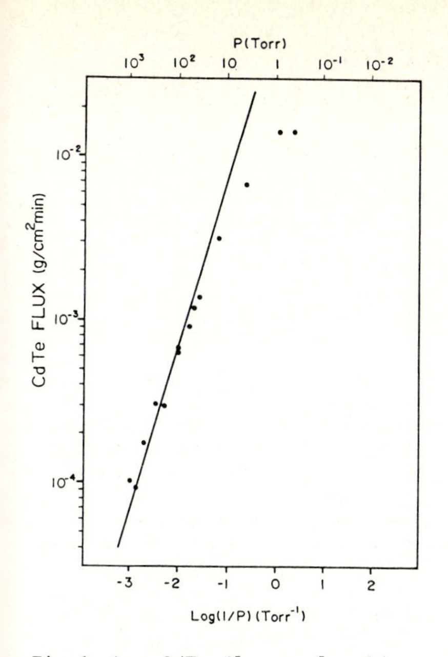

Fig.1 Log CdTe flux vs log 1/p for T_{so} = 700°C and T_{sub}=600°C

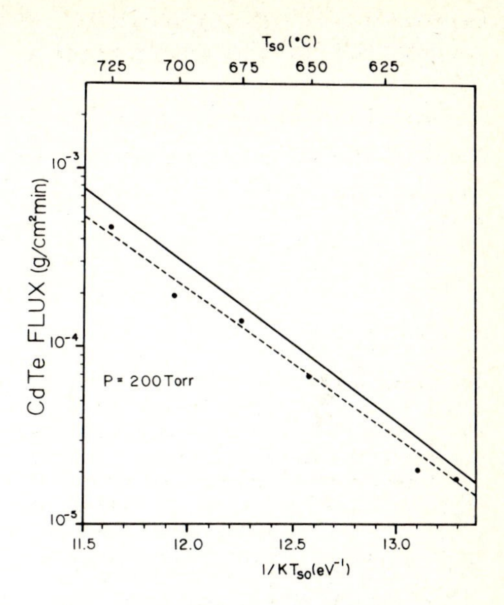

Fig.2 Log CdTe flux vs. $1/KT_{so}$ for P = 200 Torr and T_{sub} = 550°C

Figure 2 shows the relationship between log J and $1/T_{so}$ for P = 200 Torr and T_{sub} = 550°C. The solid line corresponds to the DLT model. The dashed line is the least squares fit to the experimental points. An activation energy $E_a = 1.98 \pm 0.04$ eV has been determined from the slope of the dashed line. The dependence of log J on $1/kT_{so}$ was also studied for the following two cases: a. P = 50 Torr and T_{sub} = 500°C, b. P = 20 Torr and T_{sub} = 550°C; the experimental results are shown in Figures 3 and 4 respectively. The activation energies obtained from the plots in Figures 2, 3 (E_a = 1.86 $\pm$ 0.08 eV) and 4 (E_a = 1.96 $\pm$ 0.06 eV) are in good agreement with the theoretical value of 1.99 eV /5/.

When the Ar pressure was increased, as it is shown in Figure 5 for P = 560 Torr, there is a large scattering in the experimental points. The linear behavior of log J vs $1/kT_{so}$ predicted by the DLT model, is not followed by the experimental points. The main deviation happens at low T_{so}, where molecular flux values up to 3 times higher than those predicted by the DLT model were obtained. Under these circumstances, one is led to discuss two possibilities: 1. The hot walls are influencing the diffusion process. 2. Another process, like convection, is playing an important role. The first possibility can be ruled out, since the influence of the hot walls should be the same on the slope of the straight line in the cases shown in Figures 2 to 5. Although in chambers with thermal insulator walls, natural convection only occurs when the Rayleigh number is over a critical value of 1700/7/, Olson and Rosenberger /8/ have observed that natural convection can occur at Rayleigh number as low as 230, when hot walls cylinders are used in convection chambers and Xe is the fluid. In our hot walls system, we have calculated a Rayleigh number of about 400 for P=560

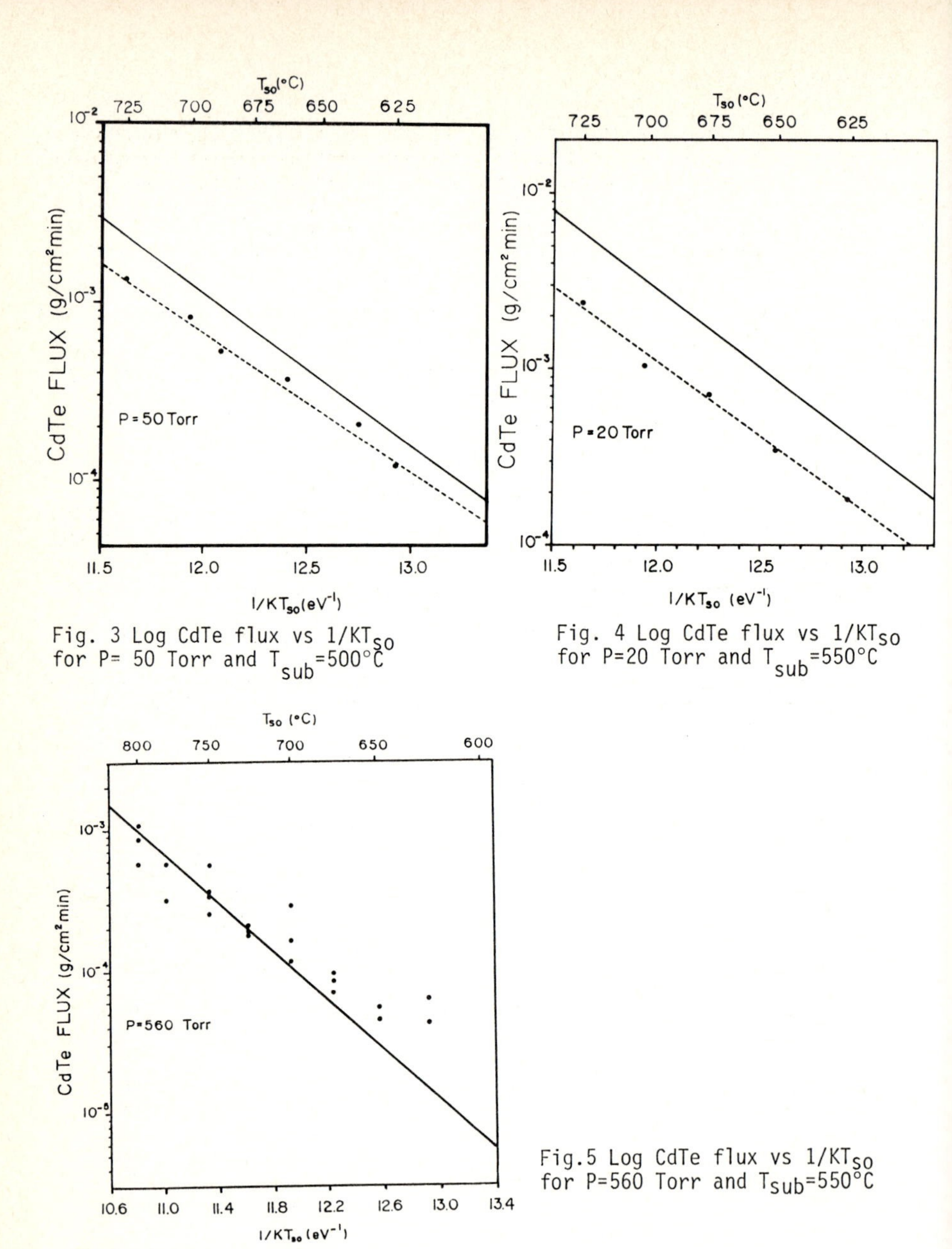

Fig. 3 Log CdTe flux vs $1/KT_{so}$ for P= 50 Torr and T_{sub}=500°C

Fig. 4 Log CdTe flux vs $1/KT_{so}$ for P=20 Torr and T_{sub}=550°C

Fig.5 Log CdTe flux vs $1/KT_{so}$ for P=560 Torr and T_{sub}=550°C

Torr and h=3.5 mm. Furthermore, the Rayleigh number depends linearly on the gradient temperature, ΔT, between source and substrate and goes as the square of the gas density, ρ^2, in the growing chamber. For constant volume and pressure, the density decreases inversely as the temperature increases. Thus, for higher source temperatures, the decreasing effect of ρ^2 dominates on the increasing effect of T, and we obtain a smaller Rayleigh number. This result would explain why in Figure 5, the effect of natural convection can be more important at low temperatures than at high temperatures.

4. CONCLUSION

We have characterized the dependence of the growth of polycrystalline CdTe films by hot-walls close-spaced vapor transport technique on gas pressure and source temperature. The dependence of the growth on the gas pressure can be explained by the diffusion-limited transport model. The dependence of the growth on the source temperature at high pressures ($P > 500$ Torr), can not be explained by the DLT model, and there is some evidence that natural convection process can not be disregarded.

ACKNOWLEDGEMENTS

This work was supported by a grant from CONACyT/Mexico, the Organization of American States (OAS) and the Fundacion R.J. Zevada/Mexico.

REFERENCES

1. F.H. Nicoll: J. Electrochem. Soc. 110, 1165 (1963)
2. F. Buch, A.L. Fahrenbruch and R.H. Bube: J. Appl. Phys. 48, 1596 (1977)
3. R.G. Shulze: J. Appl. Phys. 37, 4295 (1966)
4. O. Igarashi: J. Appl. Phys. 41, 3190 (1970)
5. T.C. Anthony, A.L. Fahrenbruch and R.H. Bube: J. Vac. Sci. Technol. A 2, 1296 (1984)
6. C. Menezes, C. Fortmann and S. Casey: J. Electrochem. Soc. 132, 709 (1985)
7. C. Normand, Y. Pomeau and M.G. Velarde: Rev. Mod. Phys. 49, 581 (1977)
8. J.M. Olson and F. Rosenberger: J. Fluid Mech. 92, 609 (1979)

Effect of Laser Fluence, at and near Threshold, on CdTe Compound Formation

N.V. Joshi[1], *H. Galindo*[1], *L. Mogollón*[1], *A. Pigeolet*[2], *L.D. Laude*[2], *and A.B. Vincent*[1]

[1]Facultad de Ciencias, Universidad de los Andes, Mérida, Venezuela
[2]Lab. de Physique de l'Etat Solide, Université de l'Etat, B-Mons, Belgium

An effect of laser fluence on CdTe synthesis formed by laser-beam solid interaction has been examined systematically. Combined study of Raman and optical spectra shows that segregation of constituent elements decreases and crystal size increases as the laser power increases up to its threshold value.

1. INTRODUCTION

Much interest has been generated recently in investigating the effects of the interactions of high intensity laser beams with matter [1,2,3]. Laser induced phenomena such as melting [4], recrystallization [5], incorporation of dopants [6] and formation of binary or ternary compounds by intermixing constituent elements, have been reported [1]. The last one, namely, semiconducting compound formation (binary or ternary) opens new possibilities of device fabrication and with this view, a thin layer of $Al_xGa_{1-x}As$ on GaAs has been successfully obtained [7].

As expected, there exists a threshold power for which such interaction or intermixing of the layers takes place. In spite of recent progress, the information about the changes, which are induced near the threshold power has not been systematically examined so far. The purpose of the present investigation is to explore this fundamental aspect of the laser-solid interaction in a compound formation process.

CdTe, a binary semiconducting and technologically important compound, is a suitable candidate for this purpose. BAUFAY et al [3] have demonstrated its formation by the laser interaction technique. Moreover, precise information about optical absorption and lattice vibrational modes is available both for a single crystal and for polycrystalline thin films.

Naturally, these structure-dependent properties will be a great asset to understand the crucial role of incident power on the process of compound formation.

2. EXPERIMENTAL

A thin layer of telurium and on top of it, a layer of cadmium, was grown on sputter clean glass surface by using electron gun technique. The obtained two layer system was irradiated with a laser beam of wavelength 488 nm obtained from an argon ion laser (Spectra Physics model 177). The sample was irradiated by the laser beam and scanned with velocity of 6.55 mm/sec.
In this way, CdTe thin films were obtained. The details are given in an earlier publication [3]. Optical absorption spectra were recorded by using a Spex Datamate system and the Raman spectra were recorded with a Spex Ramalog 5. Experimental details are given in earlier publications [8].

3. RESULTS AND DISCUSSION

Figure 1 shows Raman spectra of CdTe thin films obtained by laser irradiation for various values of incident power (26 watts/cm^2 to 44 watts/cm^2). It can be seen from this figure that even for low incident power (26 watts/cm^2), lattice vibrational modes corresponding to CdTe are clearly observed. Transverse optical mode (ω_{TO}) and longitudinal optical mode (ω_{LO}) are situated at 145 cm^{-1} and 167 cm^{-1} respectively. These values are in close agreement with earlier reported and also with theoretically calculated values (139 cm^{-1} & 167 cm^{-1}) [9]. In addition to these, a Raman line is observed at 206 cm^{-1} which can be attributed to the combination mode $2\omega_{LA}$ as the longitudinal acoustical mode observed at 101 cm^{-1}. Then, even at 26 watts/cm^2 of incident power, less than the threshold of 39 watts/cm^2, intermixing between the layers of cadmium and tellurium takes place in such a way that every cadmium (or tellurium) atom has tetrahedral sites with T_d symmetry and the force constants matrix has the same value for the CdTe single crystal.

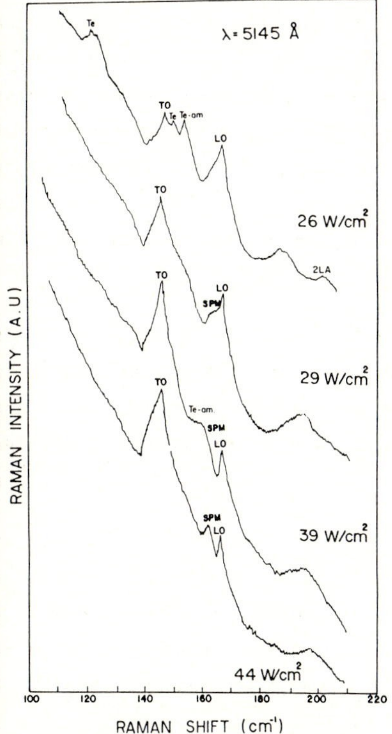

Fig. 1. Raman scattering spectra of CdTe synthetised by laser with 75 μ of separation between two scans and several laser fluences.

A close look at these spectra shows that weak but noticeable peaks appear at 123 cm^{-1}, 147 cm^{-1} and 155 cm^{-1} for the sample formed at low intensity. Earlier experimental study reveals that the vibrational modes of crystalline and amorphous tellurium are located at 123 cm^{-1}, 143 cm^{-1} and 157 cm^{-1} respectively [10]. Close agreement between the present experimental values and the reported values suggests the presence of a mixture of amorphous and crystalline tellurium. As the laser power for compound formation increases, these modes disappear, indicating that segregation of tellurium diminishes gradually.

A broad peak located at 196 cm^{-1} could be tentatively attributed to the oxide of tellurium which is expected to be formed. Both forms of oxides, TeO and TeO_2 may exist in principle. However, the present limited study does not permit to be confident about the oxidation state and hence, about the origin of this peak.

A detail examination of the spectra reveals that the broadening of Raman modes decreases as the fluence of the laser increases. This means that the size of the crystallites increases with the fluence of the laser [11]. An additional support to this view can be obtained from the extra mode located at 161 cm^{-1}. This mode lies in the band gap and could be caused by surface phonon mode (SPM).
In order to obtain direct confirmation, the frequency of surface phonon mode was calculated using a model proposed by RUPPIN & ENGLMAN [12] for a thin slab and the frequency of SPM was found to be 161cm^{-1}. The other surface mode lies very close to the ω_{LO} mode and hence it was not possible to detect.

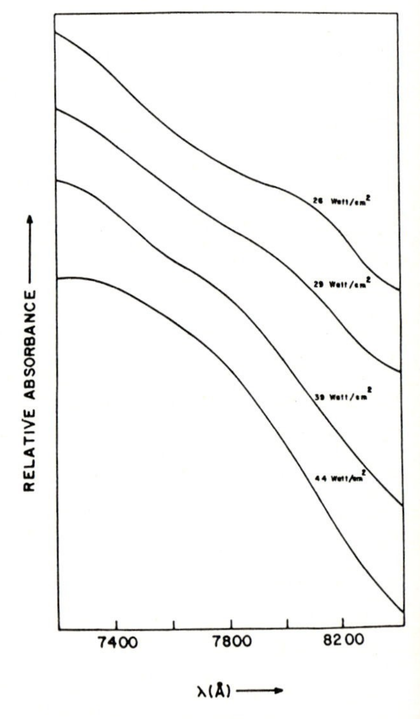

Fig. 2. Absorption spectra for the same samples of Fig. 1.

Band structural properties can be evaluated with the help of the absorption spectra shown in Figure 2. The optical absorption edge is located in the expected range; however, there is a strong hump in the absorption edge. Inte sity of the hump reduces for the samples which are grown by higher fluences and disappears at the threshold power value greater than 39 watts/cm^2.

Such hump is generally observed in a thin film, which is not a single phase material, but rather a mixture of two different materials. The present experimental data suggests that at low laser fluences microcrystallites of CdTe are separated by clusters of cadmium or tellurium. At threshold power, the hump disappears and the Raman lines become sharper, indicating absence of two phases and the presence of large crystallites.

When the sample is irradiated with the laser beam, the radiated part (or part of it) is rapidly heated by electron-photon interaction. At the threshold power, the temperature of the sample rises sufficiently to melt the film surface. This activates the diffusion process and intermixing of the layers takes place leading to a compound formation. When the laser power is below threshold, the energy is not enough to melt and activate the diffusion in the entire irradiated region; but is sufficient to melt only part of the irradiated surface. Thus the isolation of small crystallites in the matrix of cadmium (or tellurium) is observed. The same conclusion, namely the segregation effect, depends on laser fluence and it disappears at threshold power.

This also suggests that segregation can be avoided by selecting proper fluence, scanning velocity and separation between the two scans. To confirm this, the separation between two scans was increased from 75 μ to 150 μ, all the other experimental conditions were kept constant. The Raman spectra are shown in Fig. 3. These lines are sharp and surface modes are absent, indicating the presence of large crystallites (>1000 A°).

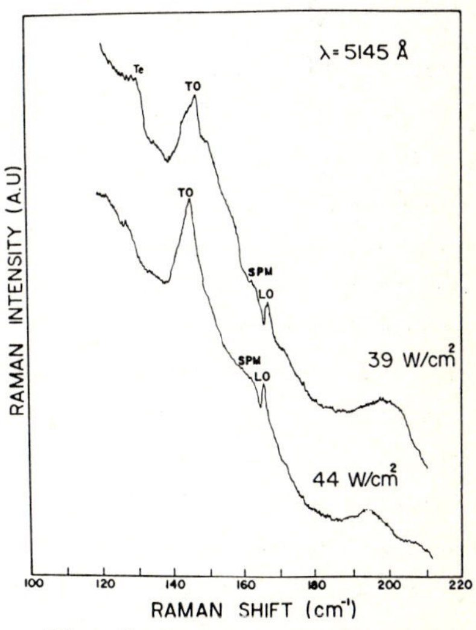

Fig. 3. Raman scattering spectra of CdTe synthetised by laser with 150 μ of separation between two scans and laser fluences of 39 watts/cm^2 and 44 watts/cm^2.

4. SUMMARY

The present experimental investigation shows that

1) Even slightly below the threshold power, the laser-induced interaction and intermixing of the constituent elements takes place maintaining the right site symmetry for each atom, but the size of the crystallites is small and they are isolated in the matrix of cadmium (or tellurium).

2) At the threshold power, the segregation is eliminated and good quality large crystallites are formed (>1000 A°).

3) It is possible to control the size of the crystallites by adjusting scanning velocity, separation between two scans and power of the laser beam.

5. REFERENCES

1. L.D. Laude: Prog. in Cryst. Growth and Characteriz., 10, 141 (1984).
2. W. Pamler, E.E. Marinero and M. Chen: Phys. Rev. B, 33, 5736 (1986).
3. L. Baufay, D. Dispa, A. Pigeolet and L.D. Laude: Jour. Cryst. Growth, 59, 143 (1982).
4. R.F. Wood and G.E. Giles: Phys. Rev. B, 23, 2923 (1981).
5. H.J. Leamy, W.L. Brown, G.K. Celler, G. Foti, G.M. Gilmer and J.C. Fan: In Laser and Electron Beam Solid Interaction and Materials, ed. by J.F. Gibbons, L.D. Hess and T.W. Sigmon, (North-Holland, 1981.) p. 89.
6. M.O. Thompson, J.W. Mayer, A.G. Cullis, H.C. Webber, N.G. Chew, J.M. Poate and P.C. Jacobson: Phys. Rev. Lett., 50, 896 (1983).
7. N.V. Joshi and J. Lehman: Proc. of Beam-Solid Interact., 1986. In Press.
8. N.V. Joshi & L. Mogollón: Prog. in Cryst. Grow. and Charac., 10, 65 (1984).
9. K. Knuc, M. Balkanski & M.A. Nusimovici: Phys.Stat.Sol.(b), 72, 229 (1975)
10. M.H. Brodsky: In Light Scattering in Solids, ed. by M. Cardona (Springer-Verlag Berlin, Heidelberg, New York, 1975.) p. 236.
11. S. Hayashi and H. Kanamori: Physica, 117B, 520 (1983).
12. R. Ruppin and R. Englman: Rep. Prog. Phys., 33, 149 (1970).

Optical Studies of Cu_2S Thin Films Topotaxially Grown on Evaporated $Zn_xCd_{1-x}S$ Thin Films

G. Gordillo

Departamento de Física, Universidad Nacional, Bogotá, Colombia

Thin Cu_2S layers were topotaxially grown on $Zn_xCd_{1-x}S$ thin films by means of a wet conversion (dipping) method. The influence of the Zn concentration of the thermal evaporated $Zn_xCd_{1-x}S$ layer on the optical properties of the Cu_2S layer were studied. For this purpose the $Zn_xCd_{1-x}S$ films were evaporated on a conductive transparent indium-tin-oxide (ITO) layer, prepared by sputtering. A decrease of the ion exchange reaction velocity on increasing the Zn content of the $Zn_xCd_{1-x}S$ layer was observed. The measurements permit one to obtain the optimal dimensions of the Cu_2S layer which is the active component of the $Cu_2S/Zn_xCd_{1-x}S$ solar cell. Also, these measurements help to interpret transport and surface properties of $Cu_2S/(ZnCd)S$ solar cells.

1. Introduction

The utilization of $Zn_xCd_{1-x}S$ instead of CdS in the fabrication of $Cu_2S/Zn_xCd_{1-x}S$ thin film solar cells leads to an increase of the open circuit voltage [1] due to improvement of the lattice mismatch and electron affinity difference between the Cu_2S and $Zn_xCd_{1-x}S$ layers. Conversion efficiencies greater than 10% (by $x = 0.1$) [2] and open circuit voltages of 0.74 V (by $x = 0.4$) [3] have been obtained with $Cu_2S/Zn_xCd_{1-x}S$ solar cells.

The $Cu_2S/(ZnCdS)$ solar cells were fabricated by topotaxial formation of the Cu_2S layer on a (ZnCd)S layer evaporated from coaxial sources [4] using a wet exchange reaction (dipping process) between a CuCl solution and the underlying $Zn_xCd_{1-x}S$ layer [5]. Because the Zn concentration of the $Zn_xCd_{1-x}S$ layer influences the exchange reaction velocity and the performance of the post-treated solar cells, optical studies of the Cu_2S layer and spectral response measurements were performed in order to explain the behavior of the cells.

2. Experimental Results

The fabrication process of the $Cu_2S/Zn_xCd_{1-x}S$ solar cells is described in detail elsewhere [4,6,7]. After fabrication of the $Cu_2S/Zn_xCd_{1-x}S$ heterojunction it

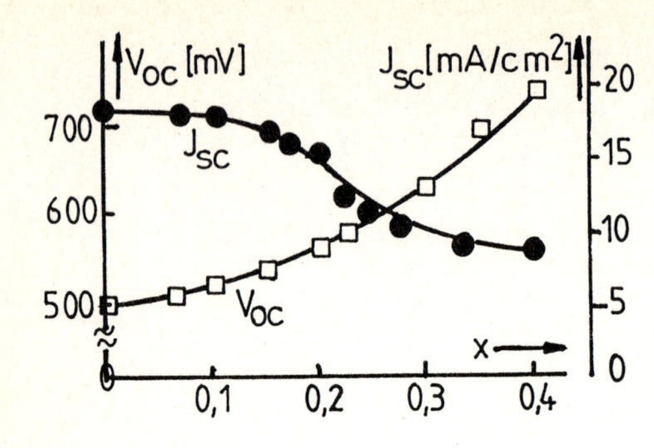

Figure 1:
Short circuit current and open circuit voltage of $Cu_2S/Zn_xCd_{1-x}S$ solar cells as a function of the Zn concentration of the $Zn_xCd_{1-x}S$ layer

is specially treated (as described in [8]) in order to improve the efficiency and stability of the cells. However for Zn concentrations greater than 10% the mentioned postfabrication treatment is not effective [3]. The most important experimental results associated with $Cu_2S/Zn_xCd_{1-x}S$ solar cells are summarized in Fig.1. The Zn concentrations of the films have been determined by polarograp (electrochemical method), energy dispersive X-rays and photoluminescence measurements. The increase of the V_{oc} with increasing Zn concentration is in agreement with the theoretical predictions [9], however the degradation of the shor circuit current was not expected. There are three possible causes for the observed degradation of the J_{sc}:

a) degradation of the optical properties of the Cu_2S layer

b) decrease of the diffusion length of the minority carrier of Cu_2S

c) increase of the surface recombination losses at the Cu_2S surface as a co sequence of the Zn diffusion from the $Zn_xCd_{1-x}S$ layer to the Cu_2S surface.

2.1 Optical Measurements on the Cu_2S Layer

In order to study the optical properties of the Cu_2S layer an indium-tin-oxide (ITO) layer with a transmission coefficient of 85% and a resistivity of 10 ohm cm, prepared by rf sputtering, was used as a back-contact of the sollar cell.

2.1.1 Transmission Measurements

Representative transmission spectra of the Cu_2S layers grown topotaxially on $Zn_xCd_{1-x}S$ (followed by postfabrication treatment) are shown in Fig.2a. The thi ness of the Cu_2S layer corresponds to the following formation parameters: 7 g/ CuCl concentration, 7 s dipping time, pH = 3, 90°C solution temperature. An increase of transmission coefficient is caused by a decrease of the thickness of the Cu_2S layer which occurs when the Zn content of the $Zn_xCd_{1-x}S$ layer increas This observation is related to the different exothermic free energy of formati

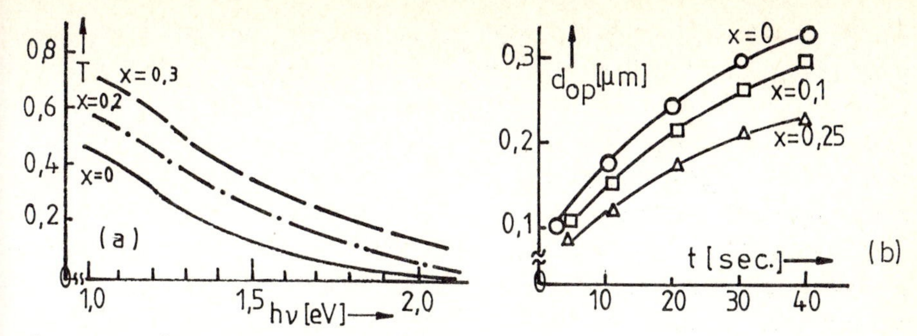

Figure 2: a) Transmission coefficient of Cu_2S layer grown on $Zn_xCd_{1-x}S$ layer with Zn concentration as parameter
b) Optical thickness of Cu_2S layers as a function of the dipping time for various Zn concentrations

ZnS and CdS which are 0.55 eV/molec. and 1.4 eV/molec. respectively [10]. Due to this difference, the ion exchange reaction $Zn_xCd_{1-x}S + 2CuCl \rightarrow Cu_2S + xZnCl + (1-x) CdCl_2$ proceeds slower at high X values. This dependence can be derived from Fig.2b.

2.1.2 Absorption Coefficient

The absorption coefficient of the Cu_2S layer was measured by transmission and reflexion measurements. Figure 3 shows the absorption coefficient of Cu_2S layers grown on $Zn_xCd_{1-x}S$ layers with different Zn concentrations.

From the results shown in Fig.3 it can be concluded that the Zn concentrations of the $Zn_xCd_{1-x}S$ layers do not influence the absorption coefficient of the Cu_2S layer grown topotaxially on $Zn_xCd_{1-x}S$.

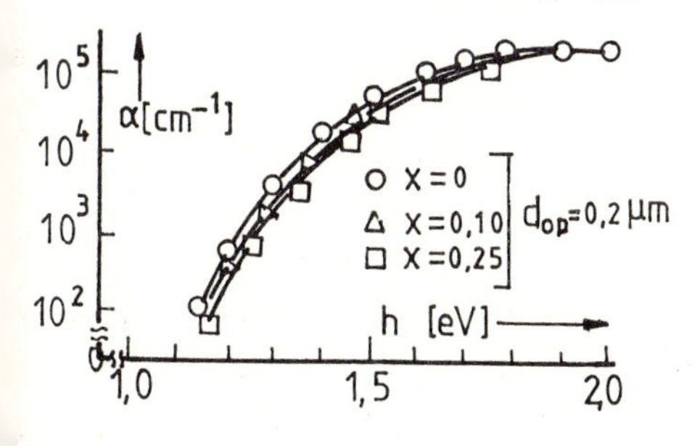

Figure 3: Absorption coefficient of Cu_2S layers topotaxially grown on $Zn_xCd_{1-x}S$ layers with different Zn concentrations

2.2 Surface Recombination

The previously presented experimental results indicate that the degradation of the J_{sc} shown in Fig.1 is not caused by a decrease of the absorption coefficient when the Zn content in the $Zn_xCd_{1-x}S$ increases. Also, measurements and theoretical calculations of the spectral responses of the $Cu_2S/Zn_xCd_{1-x}S$ solar cells

[4] indicate that the diffusion length of the minority carriers in Cu_2S does not depend on the Zn content of the $Zn_xCd_{1-x}S$ layer either and is approximatel 0.3 μm.

It is known that the formation of a Cu_2O window layer from metallic Cu deposited on the Cu_2S surface followed by heat treatment in air reduces the surfa recombination losses [8]. Before the heat treatment the surface recombination is strong, and so there is a very small J_{SC}. After optimal heat treatment, forma tion of a Cu_2O window reduces surface recombination losses and J_{sc} reaches the maximum value. Prolonged heating causes formation of Cu_xO (with $1 < x < 2$) and thus the surface recombination increases, leading to a decrease of J_{sc}. The influence of the heat treatment on the surface recombination becomes more important with increasing Zn concentration, as shown in Fig.4.

This behavior of the $Cu_2S/Zn_xCd_{1-x}S$ solar cells is caused by Zn diffusion from the $Zn_xCd_{1-x}S$ layer to the Cu_2S surface during the topotaxial formatior of Cu_sS and during the heat treatment [9]. The presence of Zn on the Cu_2S surface leads to formation of Zn_xO [10] which decreases the surface recombinatior and then the J_{sc} decreases, as shown in Fig.1.

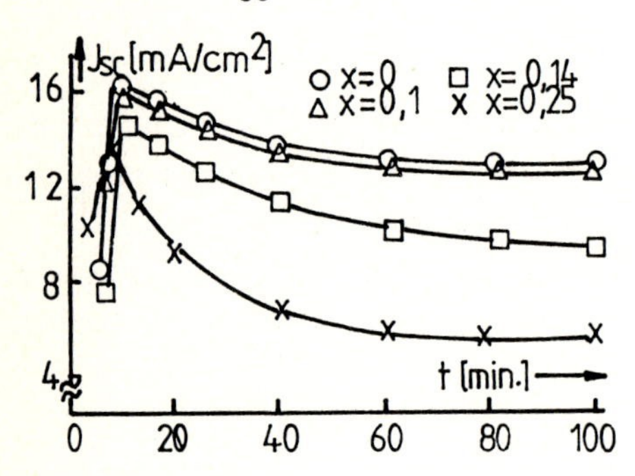

Figure 4:
Dependence of the J_{sc} of the $Cu_2S/Zn_xCd_{1-x}S$ solar cells of the annealing time at 180 C in air: Parameter Zn concentration

3. Conclusions

The absorption coefficient and diffusion length of Cu_2S thin films formed on $Zn_xCd_{1-x}S$ thin films evaporated from coaxial sources, using a wet exchange re action between a CuCl solution and the underlying $Zn_xCd_{1-x}S$ layer, are not ba sically influenced by the Zn diffusion onto the Cu_2S layer during the ion exchange reaction and during the heat treatment. However, diffusion of Zn onto the Cu_2S surface increases the surface recombination losses leading to a degradation of the short circuit current of the $Cu_2S/Zn_xCd_{1-x}S$ solar cells wher the Zn concentration is greater than 10%.

References

1. G. Gordillo: 5th International Solar Forum, Berlin 1984, p.760
2. R.B. Hall, R.W. Birkmire, J.E. Phillips, J.D. Meaking: 15th IEEE Photov. Spec. Conf., Orlando 1981, p.777
3. G. Gordillo: Solar Cells **14**, 219 (1985)
4. G. Gordillo: PhD Dissertation (Univ. Stuttgart 1984)
5. F.A. Shirland: International Workshop on CdS Solar Cells (Univ. Delaware 1975) p.465
6. G. Gordillo, H.W. Shock: Seminare sur les Cellules Solaires Cu_2S/CdS, Montpellier 1983
7. W.H. Bloss, F. Pfisterer: 5th EC Photov. Solar Energy Conf., Athens 1983, p.728
8. F. Pfisterer, H.W. Schock, J. Woerner: 3rd EC Photov. Solar Energy Conf., Cannes 1980, p.762
9. L.C. Burton: Solar Cells **1**, 159 (1979/1980)
10. L.C. Burton, P.N. Uppal, D.W. Dwight: J. Appl. Phys. **53**,3, 1538 (1982)

ZnSe Films Doped with TbF_3

C. Falcony and A. Ortiz [+]

Centro de Investigación y de Estudios Avanzados del IPN,
Departamento de Física, A.P. 14–740, 07000 México, D.F., México

The photoluminescence characteristics of ZnSe films doped with TbF_3 are reported. This films were deposited by a modified close-spaced vapor transport method on either polycristalline conductive oxide or amorphous SiO_2 layers. The appropriate value of deposition parameters such as substrate and source temperatures depends strongly on the type of substrate. The photoluminescence spectra present also different characteristics for the two types of substrates used. Films deposited on conductive oxide show a strong peak at $\sim$ 520 nm while films deposited on SiO_2 have a photoluminescent peak at $\sim$ 610 nm.

1. Introduction

Wide band gap materials such as ZnS and ZnSe have been the focus of extensive work because of their potential use in the fabrication of blue light emitting devices/1-5/. The possibility of preparing good quality single crystal or polycristalline thin films of these materials is of major interest for the development of flat panel, large area displays. Several deposition techniques have been experimented to achieve this goal. Among them molecular beam expitaxy (MBE) /4/ and Metal-Organic Chemical vapor deposition (MD-CVD) are the most common /5/. In the present work we have used a modified close-spaced vapor-transport (CSVT) technique/6/ which allows us to obtain low-cost polycrystalline films. The use of rare-earth fluorides as luminescent centers in ZnS has been reported previously by other authors/7,8/. In this paper we report the luminescent characteristics of ZnSe films doped with TbF_3. We have also studied the effect of the substrate on the deposition parameters and on the luminescent characteristics of the films.

2. Experimental Details and Results.

The modified CSVT technique used for the preparation of the samples has been described earlier/6/. In this technique the source material (powder) is placed in a graphite or boron nitride chamber and the substrate is located on top in such way that evaporant species reaching the substrate are always surrounded by a high-temperature envelope. The temperature of the substrate and the source can be controlled independently of each other. During deposition this arrangement is kept in Ar gas ambient at $\sim$500 mTorr. The source material was a mixture of ZnSe and TbF_3 powder (typically 0.1gr. of ZnSe to 0.02 grs. of TbF_3). The ZnSe powder was previously heated at 900°C in flowing hydrogen atmosphere to reduce to excess of Se. Some metallic Zn was added to the mixture in order to compensate for losses of this element during evaporation. After deposition, the films were found to have $\sim$ 3.4 at % of Tb in them. Two types of substrates were used; tin oxide coated glass slides or thermally grown SiO_2 layer on crystalline silicon. The tin oxide layers were polycrystalline. The optimum substrate (Ts) and source (Tf) temperature were different for each case. For tin oxide substrates Ts = 475°C and Tf = 710°C. In the case of SiO_2 substrates T_S=730°C and Tf=850°C. A film 0.5 µm thick was obtained under these conditions and a deposition time $T_D = 3$ and 9 min for SnO_2 and SiO_2 respectively.

The photoluminescence from this film was excited with the 325 nm line of He-Cd laser, the light emitted was collected with optical fibers and the spectral resolution was accomplished with a an optical multichannel analyzer. The sample was placed in a commercial cryostat with quartz windows to achieve temperatures in the range of 10 to 300 °K.

Fig. 1 shows the photoluminescence spectrum for a ZnSe:TbF_3 film deposited on SnO_2 coated glass slide. It presents a strong broad peak at ~520 nm. The dashed line is the photoluminescence spectra of the TbF_3 powder. This last spectra has a defined structure associated with well-identified transitions/7/. If an envolvent is traced over this spectrum the average peak intensity is at larger wavelengths (~ 550 nm) than the ZnSe:TbF_3 peak. The luminescence spectrum from the ZnSe:TbF_3 films was independent of temperature in the 10-300°K range. This was not the case for films deposited on amorphous SiO_2 as shown in Fig. 2 where the luminescence spectra are plotted for two different sample temperatures (15° and 292°K). At 292°K a single broad peak appears at ~ 610 nm. At lower temperatures (15°K) a second peak is present at ~ 540 nm. For comparison purposes the luminescence spectra for undoped ZnSe films on conductive oxide is shown in Fig. 3 for 15°K and 292°K (room temperature).

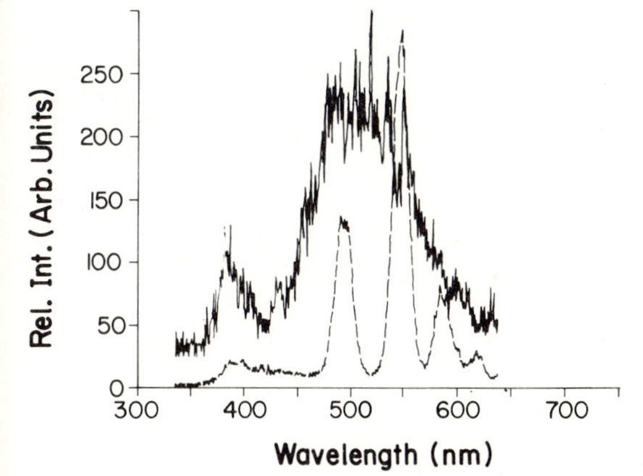

Fig. 1. Photoluminescence spectra for ZnSe:TbF_3 film on SnO_2. The dashed line spectrum corresponds to TbF_3 powder.

3. Discussion and Conclusions.

The luminescence spectra show large differences between both types of substrates used for this work. From the spectra for those films deposited on SnO_2 we observed that the light emission peak is placed at shorter wavelengths than the photoluminescence peaks for TbF_3 powder and it also has different characteristics. This fact rules out the possibility of two separated phases of ZnSe and TbF_3, since in such a case a composition of a ZnSe plus TbF_3 spectra would be more likely to be observed. Therefore, we believe that TbF_3 is incorporated as a dopant in the ZnSe polycrystallites. It is possible that as in the case of ZnS/7/ the TbF_3 enters as a molecular impurity into the ZnSe. In this case the spectra is independent of tempera-

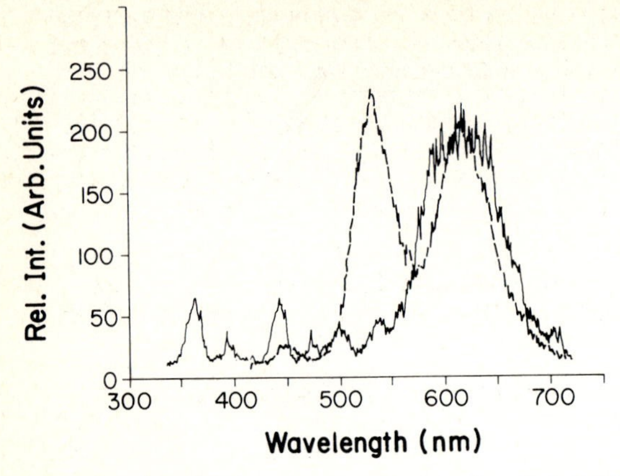

Fig. 2. Photoluminescence spectra for ZnSe:TbF_3 films on SiO_2 at room temperature (-) and at 15°K (---).

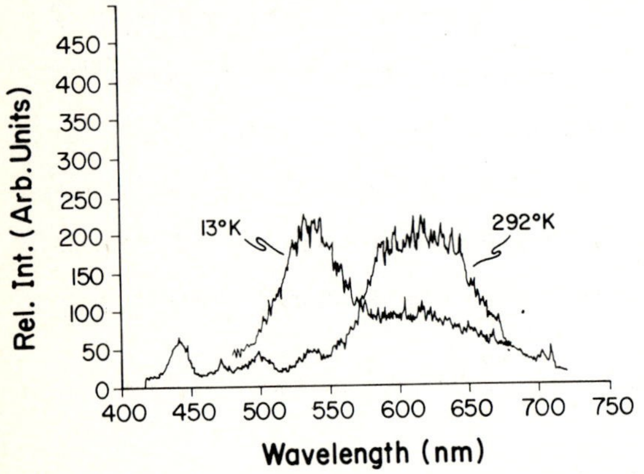

Fig. 3. Photoluminescence spectra for undoped ZnSe at two different temperatures.

ture. The luminescence spectra from samples deposited on SiO_2 show a rather different picture. The emission peak is located at longer wavelengths (∿ 610 nm) than those samples deposited on SnO_2. Also, at lower temperatures a second luminescent peak appears at ∿ 540 nm. This second peak might be associated with the Cu-green peak/9/ observed in undoped zinc selenide films (Fig. 3). The first peak (610 nm) is located in wavelengths close to the self-activated peak (620 nm) also observed in undoped ZnSe/9/. However it is far more intense at room temperature and has less temperature dependence than the so-called self-activated or copper-red emission. The differences between the luminescence characteristics for both types of substrates might be due to the high temperatures required for film deposition on the amorphous SiO_2 substrates. One possibility would be that the TbF_3 molecules are broken during deposition changing the nature of the luminescent center.

In summary we have deposited polycristalline ZnSe films doped with TbF_3 by a modified close spaced vapor transport technique. We have observed strong dependence on the deposition parameters and luminescent characteristics of the films with the type of substrate used (polycrystalline SnO_2 or amorphous SiO_2). In the case of SnO_2 substrates a green photoluminescence emission was observed. We believe that this emission is due to molecular impurification of ZnSe with TbF_3. The films deposited on SiO_2 show yellow-orange luminescence emission. This difference in spectra might be associated with the high temperatures required to deposit on the amorphous SiO_2 substrates.

The authors thank Dr. Feliciano Sánchez-Sinencio for helpful discussions. Also the technical assistance of J. García Coronel, M. Pérez de Izarrarás and M.A. Vega Colín is acknowledged. This work was partially sponsored by CONACyT and CONACyT-NSF.

+Permanent address: Instituto de Investigaciones en Materiales, Universidad Nacional Autónoma de México.

4. References

1. P.J. Dean, A.D. Pitt, P.J. Wright, M.L. Young and B. Cockayne, Physics B116, 508 (1983).
2. H. Ohnishi, H. Yoshino, K. Ieyasu, N. Sakuma and Y. Hamakawa, Proc. of the Soc. for Inf. Disp. 25, 193 (1984).
3. T. Mishima, W. Quan-Kun and K. Takahashi, J. Appl. Phys. 52, 5797 (1981).
4. N. Mino, M. Kobayashi, M. Konagai and K. Takahashi, J. Appl. Phys. 58, 793 (1985).
5. N. Mino, M. Kobayashi, M. Konagai and K. Takahashi, J. Appl. Phys. 59, 2216 (1986).
6. C. Falcony, F. Sánchez-Sinencio, J.S. Helman, O. Zelaya and C. Menezes, J. Appl. Phys. 56, 1752 (1984).
7. E.W. Chase, R.T. Hopplewhite, D.C. Krupka and D. Kahng, J. Appl. Phys. 40, 2512 (1969).
8. J.M. Stewart and M.J. Dresser, J. Appl. Phys. 50, 5950 (1979).
9. K.M. Lee, L.S. Dong and G.D. Watkins, Solid State Commun. 35, 527 (1980).

Photoconductivity in Posthydrogenated Amorphous Silicon Thin Films

E. Bacca, L.F. Castro, M. Gómez, and P. Prieto

Departamento de Física, Universidad del Valle, A.A. 25360 Cali, Colombia

Amorphous silicon thin films prepared by evaporation in high vacuum and posthydrogenated in a hydrogen dc glow discharge, show photoconductor properties. The films were analyzed using TEM in order to show that they were really amorphous and with EPR to obtain an order of magnitude of the density of dangling bonds that produced states in the gap region. The photoconductivity was measured using a standard phase sensitive detection system. The behavior with the temperature and intensity of incident radiation was determined.

1. INTRODUCTION

Amorphous silicon a-Si thin films prepared by evaporation in high vacuum and posthydrogenated by a hydrogen dc glow discharge show photoconductor properties. Films which are prepared in such way have a high density of dangling bonds ($\sim 10^{19}$ cm^{-3}) a photoconductivity comparable to the dark conductivity and high conductor and photoconductor stability. Photoconductivity studies allow the investigation of the optic and electronic properties, as well as the density and distribution of states in the gap region of these films /1, 2, 3, 4 /.

Measurements of photoconductivity as a function of light intensity and temperature were obtained. These showed the existence and distribution of defects in the pseudogap, which are responsible for the sublinear behavior found ($\gamma < 0.5$) /2/. The obtained activation energies range between 0.11 and 0.37 eV.

2. EXPERIMENTAL PROCEDURE

Amorphous silicon thin films were prepared by evaporation in high vacuum over glass substrates, maintained at a deposition temperature of 523 K and at pressure of about 10^{-6} mbar. They were posthydrogenated by a hydrogen dc glow discharge at the same deposition temperature.

For photoconductivity measurements, coplanar gold contacts were evaporated over the samples, separated 0.5 mm in order to guarantee a good ohmic contact. The voltages applied through the electrodes were 20 and 40 volts.

The photocurrents were generated using a high-pressure Hg polychromatic light source with intensities froom 0.6 to 15 mw/cm^2 and measured in a temperature range from 300 to 380 K. The measurements were made using the standard phase sensitive detection system.

The evaporated films showed stable behavior upon heating, changing the light intensity and atmospheric contamination.

3. EXPERIMENTAL RESULTS AND DISCUSSION.

The properties of a-Si films before hydrogenation are typically as follows: (i) the EPR spin density is about $5\times10^{19}cm^{-3}$, (ii) electrical conductivity is about 10^{-8} - 10^{-7} $\Omega^{-1}cm^{-1}$, without significant photoconductivity, and (iii) the obtained activation energies for them are 0.50 and 0.60 eV. Previous investigations showed that the conduction mechanism of these films at room temperature is by means of extended states. The large EPR signal indicates that there are many unpaired electrons. The numerous gap states act as traps /3/.

After hydrogenation, the films show a typical dark conductivity σ_D of 10^{-7} and 10^{-6} $\Omega^{-1}cm^{-1}$. Under steady illumination by white light, photoconductivity σ_p, can reach 2×10^{-8} and 1×10^{-6} $\Omega^{-1}cm^{-1}$, i.e., a change in conductivity of nearly same order of magnitude. Such films show a residual EPR spin density of $3\times10^{19}cm^{-3}$. It is assumed that hydrogenation is incomplete at deeper layers.

The photoconductivity depends upon light intensity F according to $\sigma_p \sim F^{\gamma}$ where γ is a constant which depends on the excitation level, the energy of the absorbed photon, the temperature and the conditions of preparation of the samples. These results for the photoconductivity of our films are correlated with a continuous distribution of gap states |1,2,4|.

The values obtained for γ in the samples S3 and S4 posthydrogenated at temperatures T_H are indicated in fig. 1.

They show a sublineal behavior with F ($\gamma < 1$) due to the high excitation level /1/. The value of γ reflects an exponential or quasiexponential distribution of recombination center and reveals a saturation of recombination centers. Such transitions can lead to values of $\gamma < 0.5$ /6/. Since a drastic

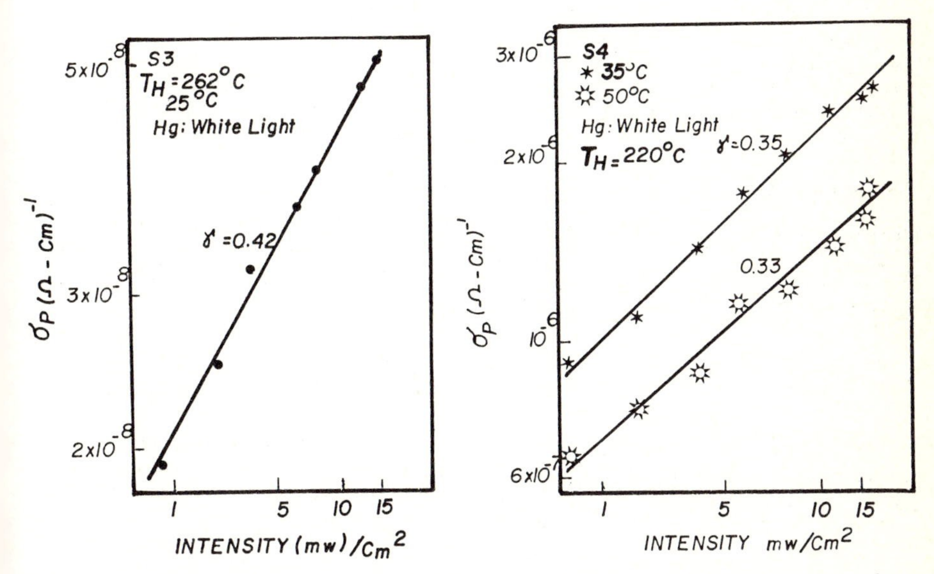

Figure 1. Photoconductivity σ_p vs intensity of white light for samples S3 and S4 at different tmeperatures.

decrease in electron lifetime can occur even when a small number of center with large electron capture cross-sections become involved in the recombination traffic /2/.

For temperature above room temperature, the conduction is by extended states, as ahown by the dependence of σ_p with temperature illustrated in Fig. 2 /5/. The obtained activation energies indicate that the electrons are the major carriers. The activated behavior depends on the preparation conditions and reveals the existence of additional defects for the S4 sample.

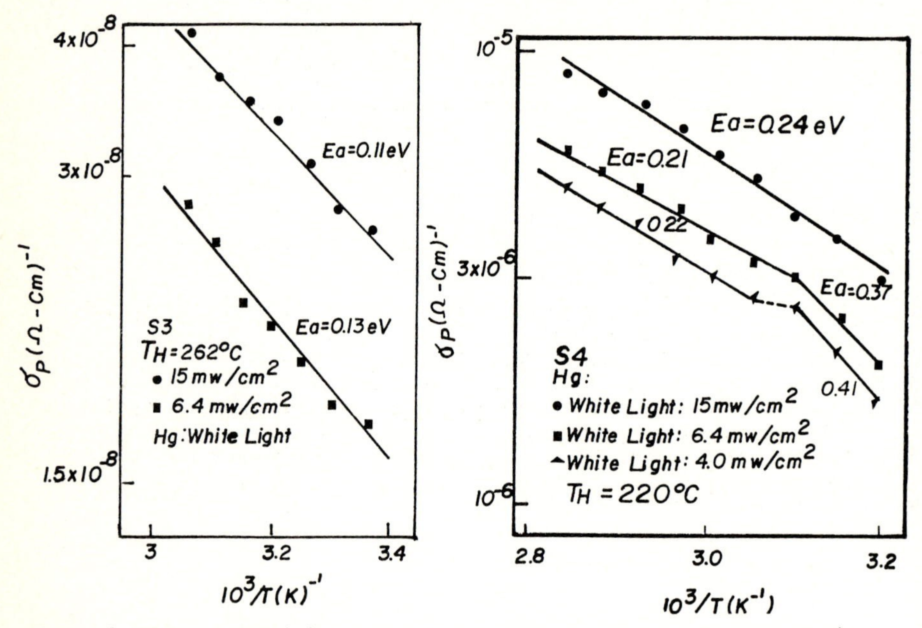

Figure 2. Photoconductivity as a function of temperature for samples S3 and S4 at different intensities of white light.

The different dependence of σ_p on the intensity illumination has been related to the recombination through the deep centers. The previously discussed values of γ less than 0.5 have been correlated with the continuous distribution of gap states above midgap and the displacements of the quasi-Fermi level E_{Fn}. Such behavior can be attributed to the strong dependence of γ with the energy of the incident photons.

The results obtained indicate that the densities of states in these films are $\sim 10^{18}$ cm^{-3} in the region of $\sim$0.28 eV from the conduction band E_c. The behavior of σ_p with temperature indicates that the conduction is due to the extended states, due to the high density of states in the energy gap.

The experimental results are reported in the table 1.

Table 1.

Sample	E_g [eV]	E_a(dark) [eV]	σ_D(300K) 10^{-8} $[\Omega cm]^{-1}$	Light Source	E_a [eV]	σ_p(300k) 10^{-8} $[\Omega cm]^{-1}$
S3	1.95	0.44	20	Hg	0.13	2.74
					0.11	1.79
S4		0.35	310	Hg	0.41-0.22	91.3
					0.37-0.21	117.0
					0.24	221.0

This work was supported by Colciencias number: 903-7094.

4. REFERENCES.

1. A. Rose: Concepts in Photoconductivity and Allied Problems, (Interscience, John Wiley, New York, 1960).
2. C.R. Wronski and D.E. Daniel: Phys. Rev. B 16, 794 (1981).
3. B.Y. Tong, P.K. John and S.K. Wong: Appl. Phys. Lett. 38, 10 (1981).
4. D.E. Carlson and C.R. Wronski: In Amorphous Semiconductors, ed. by M.H. Brodsky (Springer-Verlag, 1979).
5. P. Nagels: In Amorphous Semiconductors ed. by M.H. Brodsky (Springer-Verlag, 1979).
6. A. Rose: Phys. Rev. 97, 322 (1955).

Deposition and Properties of Silicon-Rich Silicon Dioxide Films Using CO_2 or N_2O as Oxidant Compound

A. Torres J.[1], *W. Calleja A.*[1], *M. Aceves M.*[1], *and C. Falcony*[2]

[1]INAOE Tonantzintla, A.P. 51 and 216, C.P. 72000, Puebla, México
[2]Centro de Investigación y de Estudios Avanzados del IPN, Departamento de Física, A.P. 14–740, 07000 México, D.F., México

Silicon-rich oxide films have been obtained in a hot wall atmospheric CVD system from SiH_4 and CO_2 as oxidant. For comparison purposes some films were prepared using N_2O as oxidant. Both type of layers exhibit a non-ohmic behavior and high dielectric constant. When these layers are used in a dual SiO_2/Si-rich oxide (SiO_2:Si) structure an improvement of the yield of the devices was obtained due to the field screening to non-uniformities of the thermal oxide. No difference was observed between the layers deposited using N_2O as oxidant agent and the ones prepared with CO_2.

1. Introduction

The switched capacitor approach has gained strong popularity due to the use of smaller area capacitors than resistors in filtering circuits. However the capacitor area can be improved using high dielectric constant materials instead of SiO_2. The use of dielectrics other than SiO_2 in MOS capacitors present some problems,such as no compatibility of the process of fabrication of integrated circuits with the deposition technique of the most common dielectric materials for this application. DONG et al./1/ have reported a preparation method of silicon-rich silicon dioxide (SiO_2:Si) which is compatible with the MOS process and has a high dielectric constant. This material used in combination with thermal oxide is suitable for the fabrication of EEPROM memories as reported by DIMARIA et al. /2/. In this work we present the experimental results of SiO_2:Si layers CVD deposited from SiH_4 using two different oxidant agents like CO_2 and N_2O. The electrical characteristics were determined using MIS capacitors with an SiO_2:Si layer as dielectric film. Also the C-V characteristics of a double dielectric structure consisting of thermal oxide and SiO_2:Si stacked layers are reported.

2. Experimental Details.

The SiO_2:Si films were obtained in a hot wall atmospheric pressure CVD system. The material was deposited on (100) silicon wafers of 0.3-3 Ω-cm p-type. The deposition temperature was 700 °C and the reaction gases were SiH_4 and N_2O or CO_2 using N_2 as a carrier gas. The total flux was maintained around 40 liters/min and the ratios $Ro = (N_2O)/(SiH_4)$ or $Co=(CO_2)/(SiH_4)$ were varied in order to change the silicon content in the material. The stream velocity obtained with these conditions was about 15 cm/sec. Prior to deposition of the film, the wafers were cleaned with the standard procedure reported by KERN et al. /3/. Thermal SiO_2 was grown on the silicon wafers at 1000 °C using trichloroethane (TCA) as an additive for the case of double dielectric MIS capacitor (SiO_2-SiO_2:Si). The thickness of the SiO_2 layer was ∿ 750 Å. The layers were annealed in N_2 at 1 000°C for 30 minutes after deposition and then aluminum dots of 0.0204 cm^2 of area were vacuum deposited with a e-gun to define an MIS structure. Three different types of capacitors were fabricated: using thermal oxide SiO_2:Si and a SiO_2-SiO_2:Si double layer. All of them were syntered in forming gas at 425°C during 30 minutes after metallization.

3. Results

The SiO_2:Si films reported in this paper have a Ro and Co equal to 10, the flux of reactant gases used was less than 0.5 liters/min, all the samples had a mirror-like appearence after deposition. No difference was observed when N_2O or CO_2 was used as an oxidant agent. The deposition rate was found to be bewteen 220 and 280 Å/min. The thickness uniformity is in the 10% range which is typical of atmospheric pressure CVD systems. Figure 1 shows the high-frequency C-V curves for SiO_2:Si films deposited with the two types of oxidant gases. Similar characteristics are observed for the same Ro and Co values regardless of the oxidant used in the sample preparation process. Fig. 2 shows the C-V curves for a double dielectric capacitor. The accumulation capacitance of the dual structure is comparable to the thermal oxide but the total thickness is larger. Also, the SiO_2:Si layers exhibit a high dielectric constant. From the capacitance measurement we estimate a dielectric constant value of 13. This value of the dielectric constant is higher than the one reported by DONG /1/. Although larger values were reported later by LAI et al. /5/. Fig. 3 shows histograms of the accumulation capacitance for a) SiO_2:Si layers; b) thermal SiO_2 layer and c) SiO_2/SiO_2:Si dual capacitor structure. The uniformity of the capacitance values of the double dielectric structure indicates that the non-uniformities due to the irregularities of the thermal oxide are screened by the SiO_2:Si layer deposited on top of the oxide. This improves the yield of the devices. The same results were observed for either oxidant gas in the deposition of the films.

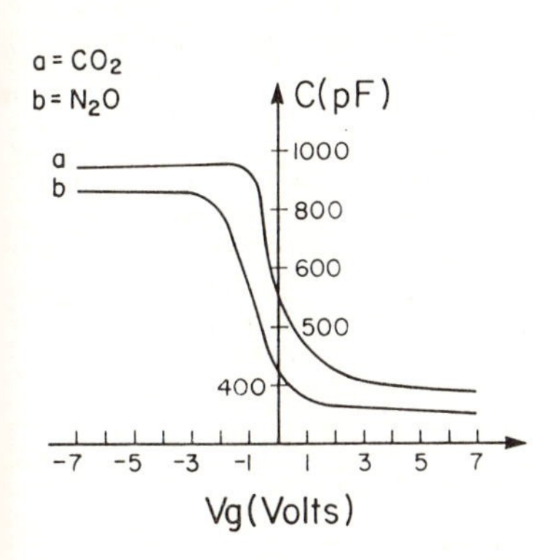

Fig. 1. C-V curves for Ro = 10 and Co = 10

The I-V characteristic of SiO_2:Si films exhibit a non-ohmic behavior and this is illustrated in the Fig. 4. The insert shows schematically the measurement circuit. In this case Ro = 10 and the current injected is controlled at low and moderate electric fields without destructive break-down of the films. The current injected is function of the electric field and the silicon content in the oxide /6,7/. The conduction increases as the silicon content is increased. The I-V characteristics are similar for samples prepared with CO_2.

4. Conclusions

The SiO_2:Si films obtained using CO_2 or N_2O as an oxidant agent show similar characteristics, and both processes of deposition are compatible with all

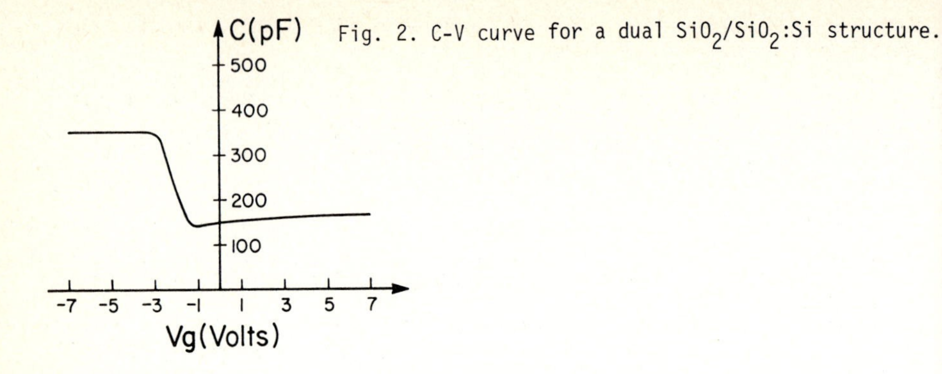

Fig. 2. C-V curve for a dual SiO_2/SiO_2:Si structure.

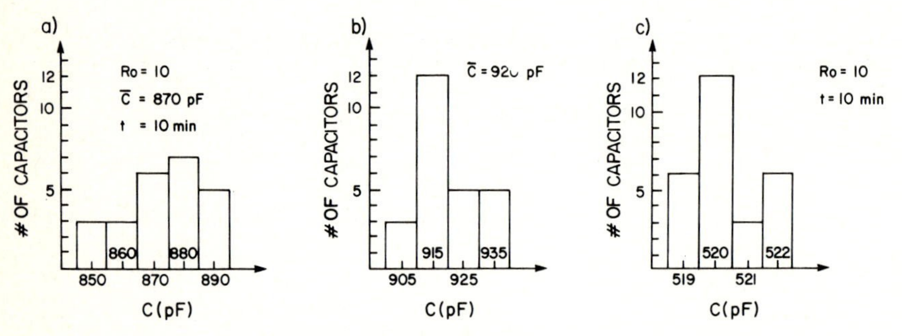

Fig. 3. Histograms of capacitance values for a) SiO_2:Si (Ro = 10); b) SiO_2 thermal and c) dual structure SiO_2/SiO_2:Si.

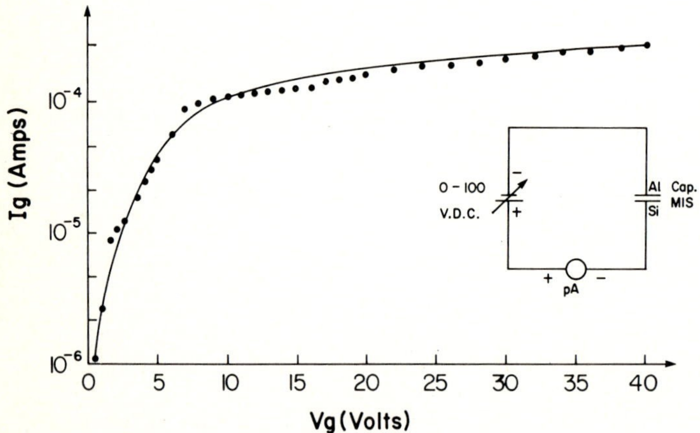

Fig. 4. Characteristic I-V curve for a SiO_2:Si layer (Ro=10). Inset shows the experimental arrangement.

the actual MOS processes. The layers deposited with either oxidant exhibit similar non-ohmic behavior in the dark I-V measurement. The dual layers show more uniform capacitance values than simple SiO_2 layers. This indicates a screening effect of the interface irregularities. The material obtained with both oxidants for the same value of Ro and Co has a high dielec tric constant value (∿ 13). The thickness variation is 10% typical of the atmospheric pressure CVD systems.

The authors would like to acknowledge Dr. G. Castellanos for useful discussions and the technical assistance of J. García-Coronel, M. Pérez de Izarrarás and M.A. Vega-Colin. This work was partially supported by CONACyT and CONACyT-NSF.

5. References

1. D. Dong, E.A. Irene and D.R. Young, J. Electrochem. Soc. 125, 819 (1978).
2. D.J. DiMaria, K.M. DeMeyer, C.M. Serrano, and D.W. Dong, J. Appl. Phys. 52, 4825, (1981).
3. W. Kern, D.A. Puotinen, RCA Review, p. 187, (1970).
4. E.J. Jansson, G.J. DeClerck, J. Electrochem, Soc., 125, 1696 (1978).
5. Stefan K.-C. Lai, D.J. DiMaria, and F.F. Fang, IEEE Trans. on Elect. Dev. 30, 894, (1983).
6. DiMaria, D.W. Dong, C. Falcony T.N. Theis, J.R. Kirtley, J.C. Tsang, D.R. Young, F.L. Pesavento and S.D. Brorson, J. Appl. Phys. 54, 5801 (1983).
7. C. Falcony and J.S. Helman, J. Appl. Phys. 54, 442 (1983).

Production of Semiconductor Oxide Films

G. Arenas and C. Jiménez

Departamento de Física, Universidad Nacional, Bogotá, Colombia

A direct current cathodic sputtering system, supplied with a heat pipe to cool the target, has been built. Thin films of tin, indium and tin-indium oxides were produced through reactive deposition. One aim of the experiments was to study these films in order to find technological applications. We looked for a simple method to produce conductive, transparent foils. Prospective work is mentioned in the communication.

1. Introduction

A broad research program on production, properties measurement and application of semiconductors is currently under development at the Universidad Nacional de Colombia, in Bogota. We have tried to produce conductive and transparent films of metal oxides. Typical technological uses of these films are transparent contacts and infrared mirrors [1,2].

Some workers have obtained sputtered indium-tin oxides from the corresponding oxide targets [1,3]. We experimented with metallic cathodes with different compositions, because we rather wanted to study the growth of oxide films produced under different operating conditions. Reactive sputtering from metallic cathodes allowed the production of metallic, semiconducting or isolating films when different gas mixtures were used. Direct current sputtering was chosen as the operating method because of its simplicity; it could be more easily scaled up into a routine production method for transparent contacts on glass than, for example, rf sputtering. We tried to find out how to predict the film properties as a function of the growth conditions.

2. Experimental Details

We started with a standard vacuum evaporator station used earlier to coat samples for electron microscopy. In order to change it into a sputter coater, we added a glass bell jar to permit installation of a vertical feedthrough where

Table 1

Cathode	Composition
Indium	pure In 99.99%
Tin	pure Sn 99.99%
Sn/In 0.1	Sn 10% In 90%
Sn/In 0.19	Sn 19% In 81%

the sputter cathodes were fixed. Low melting temperature cathodes (tin, indium and alloys of these metals) were constructed. To avoid water-cooling the feed-through at a high voltage, it was designed as a thermosyphon using ammonia as the working fluid and cooling fins on the outside to release the heat. The metal cathodes, summarized in Table 1, were vacuum melted in aluminum cylinders (40 mm in diameter), ready to be fixed to the high voltage feedthrough. The use of aluminum as a cathode holder guarantees low contamination of the films produced, because of its very low sputter yield. The purity of the metals and the composition of the alloys were measured with a conventional SEM-EDAX.

The metal oxide films were obtained through reactive sputtering in an argon-oxygen mixture. Gases 99.99% pure were mixed in a preevacuated container; then the mixture was injected into the sputtering chamber. The partial pressures reported were measured in the mixing container.

The total pressure during operation was kept between 2 and 7 Pa. The current density changed between 0.09 and 0.29 $mA \cdot cm^{-2}$.

The films were investigated for transparency, using light with wavelength in the range 500 - 1000 nm, and for dc electrical conductivity. The production conditions as well as the gas mixtures were systematically changed until the best compromises between transparency and conductivity were obtained.

To determine the transmittance values, the coated substrates were placed at the output aperture of a monochromator (with a tungsten lamp) and the intensity measured. These values were compared with the values obtained with a clean substrate. The electrical conductivity of the films was measured using a four contact method, with gold electrodes previously deposited on the same glass substrates.

High energy electron diffraction (80 kV) allowed the identification of the resulting phases and species. For these experiments, the films were deposited on freshly cleaved NaCl surfaces and then floated off in water. The thickness of the foils was measured using multiple beam interference techniques.

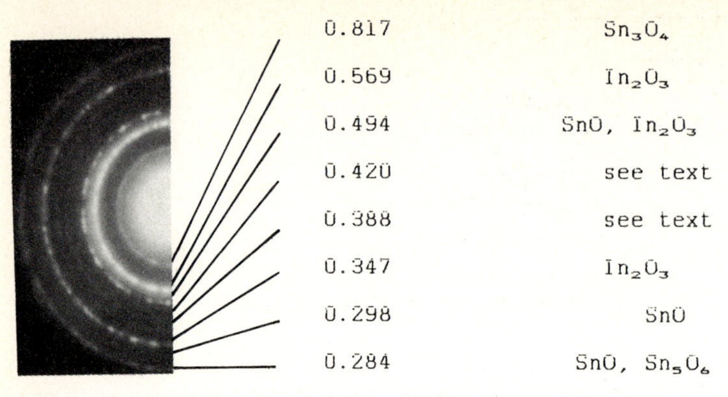

Figure 1.- Electron diffraction pattern of an indium-tin oxide foil (U = 80 kV, with 3 μm diameter sad)

Table 2

Cathode	$\log(PO_2/Ptot)$ during sputtering	Transparency (thickness 95 nm)	Conductivity $(ohm \cdot m)^{-1}$
In	-1.27	57%	1.9×10^4
Sn	-1.05	96%	500
Sn/In 0.1	-1.04	73%	6.3×10^3
Sn/In 0.19	-0.93	76%	1.9×10^3

3. Results

Representative results of the best compromise between transparency and conductivity are summarized in Table 2. The oxide species and phases found were different from others previously reported [1,3,4,5]. Electron diffraction pattern of films prepared from the Sn/In 0.1 cathode by our method show (Fig.1) the presence of In_2O_3, SnO, and Sn_3O_4. The diffraction patterns did not show any traces from SnO_2, or amorphous metallic phases. The crystallites had typical diameters of 100 nm without any preferred growth directions. Two weak lines were not fully identified; however fitting seems to indicate the presence of Al_4O_4C, an impurity likely to originate through reaction with the aluminum holder of the cathode; these two diffraction rings may also come from the presence of a complex indium-tin-oxygen compound. But the diffraction pattern alone did not allow a positive identification.

Table 2 does not give by any means all the information obtained. The argon to oxygen proportion was kept constant, but for different total pressures between 2 and 7 Pa, we observed abrupt changes in conductivity and transparency

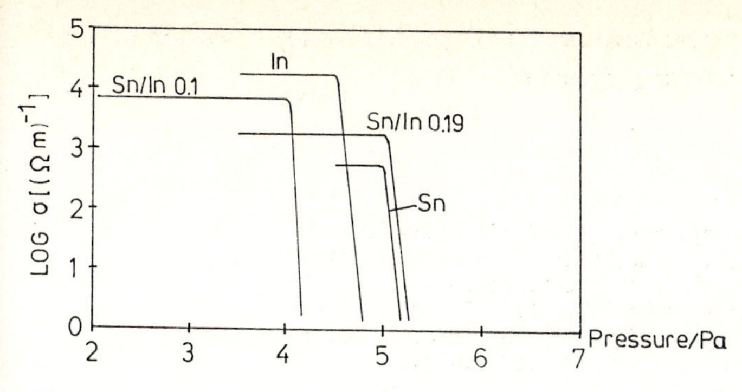

Figure 2.- Conductivity of transparent In/Sn oxide foils as a function of the pressure during sputtering (schematic)

of the films. In order to obtain tin oxide films with the desired properties, it was necessary to keep the pressure within very narrow limits (4.5 to 5 Pa). The indium oxide films and the oxides of the (Sn/In) 0.1 alloy were easier to obtain, the last one was the easiest (Fig.2).

A very important part of the procedure was to carefully oxidize the cathode, within the discharge, in order to obtain proper operating conditions. These conditions, as well as the characteristics of the necessary metal oxide film may depend on the particular sputter apparatus, so we do not give any more details.

We obtain opaque, metallic-like films, using very low oxygen partial pressures (with high sputtering yield). Those films went transparent when annealed in warm air. With low sputtering rates, when we used high oxygen partial pressures, the films were stable and behaved as n-type semiconductors. These results could be compared with a model proposed by Heller [6], who examined the building up and transport mechanisms of the oxides from the metallic cathode towards the anode. The properties and behavior of the discharge and the films obtained do agree, at least qualitatively, with that model.

4. Conclusions and Perspectives

It can be concluded from the results that we found an easy method to produce films with specified conductivity and transparency. The films are non-stoichiometric and their good conductivity could be explained from the relatively large crystallite size. Transparent, conducting films on glass substrates were obtained; they can be used as back electrodes for many applications.

The Sn/In oxides open up the field of photovoltaic devices: the films will soon be tested as contact electrodes for solar cells. They, as well as the oxides of Zn-Sn metal mixtures, could be doped with metal atoms and will be adapted as smoke and humidity detectors.

References

1. J.R. Bosnell, R. Waghorne: Thin Solid Films **13**, 149 (1973)
2. G. Frank, E. Kauer, H. Koestlin: Thin Solid Films **77**, 107 (1981)
3. J.M. Siqueiros et al. : This conference
4. E. Leja et al.: Thin Solid Films **67**, 45 (1981)
5. L.L. Kazmerski, D.M. Racine: Thin Solid Films **30**, L19 (1975)
6. J. Heller: Thin Solid Films **17**, 163 (1973)

Part II

Theory of Clean Surfaces and Chemisorption

The Use of Clusters to Calculate Physical and Chemical Properties of Transition Metal Surfaces

G. Blyholder

Department of Chemistry, University of Arkansas,
Fayetteville, AR 72701, USA

1. Introduction

The main focus here will be on the use of molecular orbital theory to calculate the equilibrium structure, geometric and electronic, and reactivity of adsorbed atoms and molecules. In order to put these calculations in perspective, the calculations for bare clusters will be considered first. While there is a large literature on clusters of alkali, alkaline earth, and main group metals, this review will deal only with transition metals. Experimental results will be treated only to the extent that they are used to check calculations. Experimental methods will not be discussed at all. Calculational results and not details of calculational methods will be dealt with, although methods will be connected to results so that a feel for the appropriateness of various methods can be gained. Organometallic chemistry, coordination chemistry, and the inorganic chemistry of cluster compounds (i.e. one to six or so metal atoms completely covered with ligands) are closely related to the present topic and are the source of many ideas and empirical correlations [1], but they will not be explicitly dealt with here.

The justification for the cluster approach rests on the assumption that the interaction of a molecule with a surface is largely localized to a small number of metal atoms which may be treated as a surface compound independantly of the rest of the extended surface. The great strength of some chemisorption bonds and the variation in behavior of different crystal faces [2] suggest localized bonding. Some theoretical justification for using a surface compound model has been given [3-5]. Perhaps more importantly, the experimental work of Sachtler [6] has shown that the energy and charge distribution for CO chemisorbed on Pd and Ni atoms in Pd-Ag and Ni-Cu alloys are not greatly affected by the presence of Ag or Cu atoms adjacent to the chemisorption site. Thus, in at least some cases, a localized model of bonding should be quite satisfactory. The approach to surface chemistry through band theory and calculations for semi-infinite slabs [7] is of such complexity that few calculations have appeared.

Experimental data has not led directly to a knowledge of the structure of adsorption sites and of the metal-adsorbate bond angles and distances until quite recently. There are now appearing papers using LEED intensities, EXAFS, and angle-resolved emissions to define site geometries. Infrared [8] and electron energy loss spectroscopies [9] have given considerable information about the structure of adsorbed molecules but the surface metal atoms are seldom included in the structural information obtained. A theoretical understanding of these data is needed.

In addition to this interest in clusters as a practical means to gain insight into metal surfaces [10], bare clusters are of interest in developing an understanding of how properties change in going from isolated atoms to small collections of atoms to the bulk state. Data is now becoming available for small transition metal clusters (1 to 100 atoms) in the gas phase and these indicate that some properties, such as reactions with H_2 molecules, are not simple monotonic functions of particle size [11].

The most sophisticated molecular orbital calculations are based on ab-initio Hartree-Fock (HF) procedure. While the single determinant wave function procedures work quite well for molecules composed of low atomic number atoms, experience indicates that the next level of sophistication which is the inclusion of correlation energy via configuration interaction (CI) is necessary to get good binding energies for diatomic transition metal molecules [12]. While in principle HF-CI procedures can give the most accurate results, their complexity usually limits them to transition metal diatomics. Even in ab initio HF calculations, good results are not guaranteed since results are very dependent on the choice of basis set.

In the $X\alpha$ methods [13] the evaluation of many multicenter integrals is eliminated by the Slater statistical approximation to exchange integrals in which these integrals are replaced by an empirical constant times a function of the electron density. In the most common "muffin-tin" procedure, the molecule is treated in terms of spherical potential regions. Although often referred to as a "first principles" method the results depend on the choices made for the sizes of the spherical regions and the extent to which they overlap each other. In general the $X\alpha$ methods have given reasonable spectroscopic properties in calculations for molecules in their experimentally determined geometries but has not been useful to calculate molecular geometry.

The least complex of the molecular orbital models is the extended Huckel (EH) [14]. This model has been successful in treating trends in properties for organic molecules and in discussing orbital usage in organic and inorganic systems but does not produce reliable molecular geometries. Derivatives of the EH method have added repulsion terms to examine geometric problems but wide experience with their reliability is lacking.

The CNDO [15] and MINDO [16] procedures replace many integrals in the HF method with empirical parameters to gain computational simplicity and attempt to reproduce experimental energies and geometries by appropriate choice of parameters. These methods have been quite successful and useful in organic chemistry.

2. Bare Clusters

2.1 Introduction

The properties of bare clusters have been the subject of several reviews and conferences [17-19]. Unfortunately, the validity of calculations for the geometry, bond lengths, ionization potentials, and electronic density of states have usually lacked experimental verification. This should change in the near future because of the growing literature on the production and property measurements of small clusters of 2 to 100 atoms supported on various surfaces, embedded in rare gas and hydrocarbon matrices, and most recently, in the gas phase [19]. Mass spectral data for metal clusters produced in the nozzle expansion of molecular beams have given unexpected reactivities of metal clusters with H_2 as a function of the number of metal atoms in a cluster [20]. The status of experimental knowledge and theoretical calculations for both homonuclear and heteronuclear diatomic and triatomic transition metal molecules has been recently reviewed [14]. Although firm data for diatomics is very useful as a testing ground for theoretical methods because they are the simplest case of metal-metal bonding, they will not be dealt with further in the review. A continuing theme as various properties are considered is the nature of the transition from atomic to bulk properties as the number of cluster atoms increases.

2.2 Geometry

A large number of calculations have dealt with whether or not M_3 species are linear. There has been some consideration of the geometry of larger clusters and a limited consideration of the stability of an icosahedral versus an fcc structure for M_{13}.

For Ag_3, EH [21] predicts a linear structure while HF calculations using an effective core potential (ECP) and configuration interaction (CI) give a bent structure [22]. The ECP-CI calculations gave Ag_3^- as linear and Ag^+ as triangular. For Cu_3, EH [23] and CNDO [18] predict a linear structure. The experimental data for Ag_3 and Cu_3 has not yielded an entirely consistent interpretation [14]. For Cu_4, the EH [23] and CNDO [18] methods agree on a linear structure but for Cu_5 while the EH [23] calculation predicts linear, the CNDO [18] calculations predict 3-dimensional structures. There is no experimental data for Cu_4 and larger Cu clusters.

For Ni_n, the ECP-CI calculations [24] give a linear structure for n=3, a square structure for n=4 and 3-dimensional structures for Ni_5 and Ni_6. CNDO calculations for Pd_n [25] and Ni_n [26] with n up to 13 favor 3-dimensional structures. In comparing icosahedral to fcc structures for M_{13}, the icosahedral structure was preferred in HF calculations for Cu_{13} [27] and in EH calculations for Pd_{13} but not for Ag_{13} [28].

2.3 Bond Lengths and Energies

The absolute values of these properties are most important for diatomic molecules on which HF-CI methods have enough accuracy to make them meaningful. For clusters larger than diatomic, most of the calculations are semiempirical so trends, rather than absolute values are most meaningful. Generally bond lengths and binding energies per atom increase as cluster size increases. In HF calculations for Cu clusters the bond length increased from 2.34 Å for Cu_2 [29] to 2.43 Å for Cu_8 [27] and the binding energy per atom increased linearly [27]. For Ni clusters up to 6 atoms, ECP-CI calculations gave bond lengths increasing with size but remaining below bulk values [24]. Similar results for clusters up to Ni_{12} were found in CNDO calculations [26]. In EH calculations for Cu_{13} the binding energy was calculated to be 18 kcal per atom compared to the bulk value of 82 kcal per atom [23]. For Ag clusters of up to 55 atoms the binding energy increased with the number of atoms but only reached about one-third the bulk value [21].

2.4 Ionization Potentials

An interesting feature of calculations of ionization potentials (IP), usually taken as the energy of the highest occupied molecular orbital, is the frequent appearance of an alternation in a higher and a lower IP as the number of atoms increases by one. Experimentally, this effect is found for Na clusters where IP's are measured by photoionization mass spectrometry [30]. The effect is evident for cluster of 2 to 8 atoms, but not above 9 atoms. Since each Na atom contributes one electron to the valence shell, it is found as expected that the clusters with an even number of atoms, and thus all doubly occupied orbitals, have the higher IP values. It is reasonable to expect this type of behavior in calculations for Cu, Ag and Au which have a $d^{10}s^1$ configuration. In EH [25] and CNDO [18] calculations for Cu and Ag this alternation in IP is found and it extends in these calculations to clusters much above M_{10}. For Ni clusters in which we note that Ni is not d^{10}, a sawtooth variation of IP values is not found in CNDO calculations [26]. Experimental data for gas phase Fe clusters of up to 25 atoms shows a decrease in IP over this range which is not monotonic and also does not show a sawtooth behavior [31].

2.5 Density of States

There are two main issues that arise from the calculated density of states (DOS) for transition metal clusters. The first is whether or not the DOS for a cluster of about 13 atoms should closely resemble the bulk DOS. The other issue, which is closely related, is how large must a cluster be in order for calculations and experimental properties to be bulk-like. These issues have caused considerable discussion because Xα calculations produce bulk-like properties for 13 atom clusters whereas semiempirical and ab initio HF based calculations produce results for small clusters that are somewhat different from bulk properties.

The Xα calculations done for Ag_n(n=7 to 19) [32], Ni_{13}, Pd_{13} and Pt_{13} [33] have a number of features in common. The total d band widths are from 2 to 4 eV wide which is in reasonable agreement with the bulk band widths of 3 to 5 eV. An interesting feature of these Xα calculations is that the d orbital energies for the central atom,which has its full complement of nearest neighbors,lay below the main d band for all the other atoms which are on the surface of the cluster and so have fewer nearest neighbors. The maximum density in the d band occurs near the Fermi level.

The HF calculation for Cu_{13} [27] has the d band about 10 eV below the Fermi level and also has the d orbital energies of the central atom below the d band of the other atoms. It is a general feature of the simple HF method that the density of states is zero just above the highest occupied molecular orbital (Fermi level). In an ECP-CI calculation for Ni_6 [24] the d band starts just below the Fermi level but has its maximum about 5 eV below the Fermi level. Most EH [34] and CNDO [26,35] cluster calculations for Ag, Cu, Ni and Pd have a DOS similar to this Ni_6 ECP-CI calculation.

The question of the nature of the DOS of clusters of various sizes really is not settled. Photoemission spectra of Pd clusters supported on amorphous carbon show a d level at 2 to 3 eV below the Fermi level for single atoms [36]. The emission for the d levels broadens and approaches the Fermi level as the cluster size increases. The full width at half maximum for XPS spectra of Ir, Rh, Pt and Pd on carbon all show an increase as the particle size of clusters is increased [37]. This initial data suggests that small clusters do have a DOS different from the bulk. In EH calculations for Pd, Ru, Au and Ag more than 100 atoms were required for the d band width to reach a value near that of the bulk [38]. The generalized valence band calculations for Ni clusters required over 87 atoms for the IP values and s band width to become constant [39].

3. Chemisorption on Clusters

3.1 Embedded Clusters

One way to handle the problem of comparing adsorption on a cluster to adsorption on an extended surface is to examine adsorption properties as a function of cluster size and enlarge the cluster until no further change occurs. This approach is computationally very expensive. Another approach is to examine the effect of embedding the cluster in metallic-like surroundings. Calculations at the HF level for H adsorbed on cubrium metal, a theoretical construct with one s orbital and one electron at each lattice point, as a function of cluster size up to 30 metal atoms have been made [40]. A localized bond for the H adsorption was produced even for a broad band metal (valence band 15 eV wide). The chemisorption properties of a small cluster, HM_6, had a binding energy and equilibrium bond length for the H interaction close to those of the largest cluster. However to obtain metallic properties for the cluster a large cluster was required and the charge distribution for the cluster itself varied with size. The embedding approach has been used in the CNDO formalism to examine CO adsorption on graphite [41]. Comparing an isolated C_{13} cluster to an embedded cluster indicated that the qualitative description of the CO adsorption bond and the adsorption bond length were the same with and without embedding but the binding energy was about 70% greater on the isolated cluster. Since the bonding capacity of carbon (classically tetravalent) is fixed, whereas transition metals exhibit a wide range of bonding capacities at similar energies, the edge effect for carbon would be expected to be greater than that for a metal. These calculations support the use of cluster calculations for chemisorption even in the absence of embedding.

3.2 Hydrogen Chemisorption

Hydrogen chemisorbed on a nickel cluster system has been the subject of a variety of calculations and serves well to illustrate the types of information coming from cluster calculations. Because transition metals are such good catalysts, the bonding provided by the 3d orbitals in first row transition metals has historically received much attention. Among the earliest cluster calculation, EH calculations for H interacting with clusters of 8 to 10 Ni atoms indicated the Ni 4s orbitals were most important with some contribution from the 3d orbitals [42]. CNDO calculations [43] for H adsorbed on 10 atom Ni clusters gave the Ni 4s and 4p orbitals as being primarily responsible for the Ni_{10}-H bonding with the d orbitals playing a relatively minor role. Subsequent less approximate calculations including HF [44,45] and Xα [46] calculations confirm that the 3d electrons play a minor role in directly establishing the Ni-H binding energy. Although some would totally banish 3d orbitals from the valence shell [44], and one [47] has claimed an important direct effect of 3d orbitals in bonding, important indirect effects of the 3d orbitals will be discussed later. In the second and third transition series the M-H bond has been found to be primarily formed from a d orbital as shown in calculations for Pd_3H [48] and Pd_4H [49] at the HF level and in Xα calculations for Pd_4H and Pt_4H [46].

Except for early EH [43] calculations, H has been calculated to be more stable over a multicenter site than directly over a Ni atom [43,44,47,50]. A comparison of the H atom binding energy (BE) and the equilibrium Ni-H bond distance for H adsorbed on Ni clusters in two HF [44,47] and MINDO/SR [50] calculations is given in Table 1. These three calculations are in qualitative agreement with each other about the relative properties of different sites and with the experimental finding [51,52] that 3 center and 4 center H chemisorption sites are most stable.

The MINDO/SR method, which is the least approximate of the commonly used semiempirical methods, has been used in calculations for a H atom at various locations on a 14 atom, 2 layer Ni cluster. The equilibrium bond distances are determined by varying the H atom position on a line perpendicular to

Table 1. Calculations for H atoms chemisorbed on Ni clusters.

Site	Binding Energy [kcal/mole]				Re[Ni-H]			
	Exp.	HF1	HF2	MINDO/SR	Exp.	HF1	HF2	MINDO/SR
on top		37	57(45)	41		1.50	1.53	1.60
2 center		65	54(70)	63		1.59	1.64	1.70
3 center	63	74		69		1.63		1.90
4 center	64	69	58(64)	66	1.85(±.06)	1.78	1.74	1.95

the surface and passing through a symmetry site. The difference between the BE at the 4 center and the 2 center sites may be taken as the barrier to diffusion. This gives a calculated value of 3.8 kcal/mole compared to estimates from experimental data of about 5 kcal/mole. The actual potential minimum in the MINDO/SR calculation has the H atom displaced from the 4 center position by 0.25 Å toward a 2 center position in agreement with LEED and ELS results [52], from which it was concluded that H is adsorbed on 4 center sites with some attraction toward 2 Ni atoms. The obtaining of geometry optimized energies at all points on the surface can readily be done with semiempirical methods but is too costly in computing time for ab initio methods, even using core potentials. The H atoms were calculated to have a charge of about -0.1 e in the 4 center site and the negative charge on H increased as the coordination numbers of the bonding Ni atoms decreased. This provides an explanation for the remarkable feature that the work function change upon hydrogen adsorption on the Ni(110) face of 0.530 eV is almost 3 times the changes of 0.170 and 0.195 eV on the Ni(100) and Ni(111) faces. Because the 110 face has surface atoms with lower coordination numbers than the 100 and 111 faces, the charge transfer and hence the work function change will be greater. The negative charge for H on Ni is in contrast to Fe where MINDO/SR calculations [53], indicate the chemisorbed H atom is positive, neutral or slightly negative for 4 center, 2 center or on top sites, respectively. The larger negative charge for H on Ni could account for the greater catalytic hydrogenation activity for Ni catalysts over Fe catalysts because the negative charge on the H facilitates nucleophilic attack on adsorbed hydrocarbon fragments, where our preliminary calculations show the carbon atoms to be positively charged.

The calculation labeled HF1 [44] in Table 1 examines the interaction of H atoms with a 20 atom Ni cluster in an ab initio calculation at the HF level using an effective core potential. The calculated binding energies are seen to be in reasonable accord with experimental values and the difference in energy between the 4 center and 2 center sites of 7 to 9 kcal is comparable to the estimated 5 kcal/mole for experimental diffusion measurements. A unique feature of the HF1 calculation is the calculation of M-H vibrational frequencies at 589 cm^{-1} for the Ni(100) 4 center site and 1210 cm^{-1} for the 3 center site on Ni(111). These compare well with the experimental values of 597 and 1121 cm^{-1} [52]. The calculated energy levels indicate a H-cluster bonding orbital 7 to 9 eV below the Fermi level and an attenuation of state density near the Fermi level in qualitative agreement with UPS spectral observation of a level about 6 eV below the Fermi level [54] and an attenuation in levels about 0.25 eV below the Fermi level [55]. The effect of varying the number of atoms in the cluster within the particular framework of these calculations was examined by doing calculations with from 4 to 10 Ni atoms and a H atom. The calculated geometries and force constants varied only slightly with cluster size but the H binding energies changed from 0 to 4.5 eV in an irregular manner. Calculations for

H on a 28 atom Ni cluster gave properties similar of these for the 20 atom cluster. The basis set used in all of these calculations differs somewhat from those of other workers. The Ar core and the 9 3d electrons are replaced by an averaged effective core potential to leave only the 4s orbital to be obtained variationally. The restriction to one variational orbital per atom is necessary to allow calculations for large clusters. The effect of omitting 4p orbitals has been found by other workers to be large [47,56]. To examine the effect of including the 3d electrons in the core one calculation for $Ni_{20}H$ with the H atom in a bridge position was made in which the 4 Ni atoms closest to the H atom had a 10 valence shell electrons treated variationally while the other 16 Ni atoms used the d^9 averaged potential. The results of this calculation gave properties within 5% of those where all Ni atoms used the d^9 averaged potential. The conclusion in this paper [44] that the 3d orbitals can adequately be included in the core potential is at variance with some others [47,57], at least if problems other than properties of a single H atom are considered.

In the HF2 calculations [47] in Table 1, clusters of 13 or 14 Ni atoms were used with a basis set including 4s and 4p orbitals and a modified effective core potential (MCP) that includes the 3d electrons for all Ni atoms except for the one Ni atom to which the H or H_2 is bonded. For multicenter sites all Ni atoms are treated with the MCP. For HF2 the binding energies in parenthesis in Table 1 are for a 13 atom cluster while the others are for a 14 atom cluster. All 13 atom cluster and most 14 atom cluster calculations giving the binding energy data for HF2 in Table 1 were treated with all atoms using the MCP which includes 3d orbitals in the core. The variation in properties with cluster size pointed out in this paper may be due more to the inconsistency in calculational treatment than size variation. Variation in properties with size for small clusters is most prominent in calculations with only one electron per atom as in these Ni calculations [44,47] with a MCP that fixes a d^9 core to leave only 1 variational electron and in calculations for Cu and Ag with d^{10} s^1 configurations. The dissociation of H_2 over one Ni atom, with all electrons included, in a 13 atom cluster where all other atoms use the MCP was found to require a CI treatment to reduce the dissociation barrier to 4.4 kcal/mole from the 66 kcal found without CI. If all 13 atoms used the MCP the barrier was also very high, which led to the conclusion that the 3d electrons were necessary when treating H_2 dissociation. Calculations made for dissociation on a bridge site indicated a much higher barrier to dissociation but these calculations were done with all atoms treated with the MCP so the results are not really comparable for the different sites.

The adsorption and dissociation of H_2 on a 14 atom Ni cluster has been looked at in considerable detail using the MINDO/SR procedure [58]. The interaction begins with the formation of a physisorbed state in which the slightly stretched H_2 molecule has its axis at right angles to the bond axis of 2 Ni atoms on the edge of the cluster and is stable with respect to desorption by 2.5 kcal/mole. Such a precursor state has often been proposed as a prelude to chemisorption. The H atoms proceed to separate with an activation energy of 2 kcal/mole while maintaining bridge adsorption and then drop into chemisorbed states. The low barrier to dissociation is less than the physiosorption energy so it is easily surmounted by energy transfer from the H_2-metal bond to the H-H bond. In the physisorbed state there is a small charge transfer to the metal leaving the physisorbed H_2 positive and there is no donation of charge into the antibonding σ_μ orbitals of the H_2. The d orbitals do not contribute to the bonding in the physisorbed state. In the chemisorbed state the H atoms are negatively charged.

Within the MINDO/SR theoretical framework the difference in density of states (DOS) between $Ni_{14}H_2$ (H_2 dissociated) and the Ni_{14} cluster and the experimental UPS difference spectrum [59] for chemisorbed H relative to clean Ni(100) has been examined [60]. The main features of the experimental difference spectrum is a decrease in the d band DOS just below the Fermi level followed immediately by an increase in the d band DOS just below this and a band due to adsorbed H about 7.2 eV below the Fermi level. These features are reproduced by the theoretical curve. The d band changes may be explained as a lowering in energy of the d band as a result of the transfer of electron charge from the Ni cluster to the adsorbed H atoms. Because electron-electron repulsions are a large term in the d band energies a reduction in the number of electrons stabilizes the band. The theoretical spectrum has a band 7.7 eV below the Fermi level due to Ni sp-H s bonding levels in rough agreement with the experimental band at 7.2 eV. The calculated d band shift is important because it indicates that a d band shift in photoelectron spectra does not necessarily mean that the d orbitals directly participate in the adsorption bond as has sometimes been assumed. The model here has explained the d band shift as being a response to the sp DOS shifting to form bonds. This result suggests that ab initio calculations which do not include the 3d electrons in the valence shell cannot be totally correct.

The effects of hydrogen surface coverage on Ni(100) have been examined with MINDO/SR calculations for 2 hydrogen atoms located at different sites with respect to each other on the (100) face of a 14 atom Ni cluster [61]. Addition of a second hydrogen decreased the spin multiplicity state of the cluster which is in agreement with the experimental result that the magnetic moment of small Ni particles is decreased as hydrogen is adsorbed on them. When the second H atom is placed close to the first so that it strongly bonds with a Ni atom already bonding to the first H atom, the result is a large decrease in stability which does not occur if the second H is further away. Analysis of the energy terms showed that this result was not due to direct H-H interaction but was an indirect interaction through the metal. Surface models indicate that a filling up to one-half a monolayer would not significantly decrease the adsorption energy but that above one-half a monolayer the adsorption energy would be significantly decreased. This model is in accord with thermal desorption spectra which show H_2 desorbing from Ni(100) from two distinct surface states [62]. The effect of changing coverage was further investigated with calculations for $Ni_{14}H_n$ (n=1 to 5) [64]. Using a classical model of the work function change with adsorption being directly proportional to the surface dipole produced by adsorption indicated that the change in work function should be linear with coverage up to about one-half a monolayer and then should start leveling off. This is in approximate agreement with experiment [62].

A HF-ECP calculation [63] for the interaction of H with clusters of iron atoms containing 30, 36, 48, or 66 Fe atoms has attacked the problem of H migration into the interior of a cluster, which is important to considerations of hydrogen solubility in metals and hydrogen embrittlement. The calculations used a one-electron ECP based on a $4s^1 3d^7$ state for the Fe atom. In a bcc iron lattice the interior tetrahedral sites were a minimum on the potential surface while the octahedral sites were at a maximum, 0.2 eV higher. The barrier to diffusion between tetrahedral sites occurred at trigonal sites with a barrier height of about 0.1 eV. The volume expansion for a H atom at a tetrahedral site was calculated to be 21%.

There have been many other excellent publications about hydrogen interacting with Ni clusters and clusters of other metals, but due to space limitations this review only discussed the above ones, which were felt to illustrate the main types of information that can be obtained from cluster calculations.

3.3 Carbon Monoxide and Hydrocarbon Adsorption

Experimentally there are probably more studies of CO adsorption than any other adsorbate. As a diatomic adsorbed molecule it is more complex than an adsorbed H atom and so offers more structural variability in the adsorbed state. From the great number of publications dealing with calculations of CO interacting with clusters only a few that illustrate properties different from those considered for H adsorption will be discussed. A striking feature of molecularly chemisorbed CO is that it retains great similarity to gas phase CO and is very similar to CO in inorganic transition metal carbonyl complex chemistry. This is illustrated by the similarity in valence and core photoionization spectra of free CO, $Co_4(CO)_{12}$ and CO adsorbed on Co(0001) [65]. The valence photoemission spectrum of CO chemisorbed on Ni was first satisfactorily explained by a CNDO calculation for CO with a 10 atom Ni cluster [66]. The experimental spectrum shows 2 main peaks due to CO about 12.5 and 15.7 eV below the vacuum level. The calculations correlated the first band with the 5σ and 1π levels of CO which were shifted close together by chemisorption and the lower band with the 4σ level of CO. More detailed subsequent calculations have added information but the general picture remains essentially unchanged. These calculations indicate that like H chemisorption the CO bonding is due primarily to Ni s and p orbitals with the d orbitals playing a minor role in the direct bonding interaction.

Early CNDO calculations for CO adsorbed on a 13 Ni atom cluster gave a correlation for adsorbed CO stretching frequencies with calculated bond order for CO adsorbed on different surface sites, i.e. adsorbed directly over a Ni atom or in a multicenter bond [67]. Recently $X\alpha$ [68] calculations that use an extensive basis set rather than the usual muffin tin approximation for Ni_2CO gave $\nu(CO) = 1850\ cm^{-1}$ and $\nu(Ni_2\text{-}C) = 495\ cm^{-1}$. The good agreement of these values with ELS spectra at 1870 and 1810 cm^{-1} for the CO stretch and at 400 and 380 cm^{-1} for the metal-carbon stretch of chemisorbed CO suggests that the chemisorption bond is quite localized so that a very small cluster calculation is good for some properties.

ELS spectra of CO adsorbed at low coverage on Cr(110) has been interpreted as indicating the CO is lying flat on the surface rather than standing upright as in most other systems. Modified EH calculations [69] for a $Cr_{33}CO$ cluster give the most stable structure with the CO lying down in a 4 fold site. The stability of this configuration is attributed to strong π and 5σ donation to the metal and back donation into the CO π^* orbitals. The back donation weakens the CO bond to give a barrier to its dissociation of about 0.4 eV which is close to the estimated experimental value of 0.42 eV.

The surface effect of catalytic poisons and promoters has been investigated using $X\alpha$ calculations for $(CO)Ni_9S_4$ and $(CO)Ni_9Li_4$ [70]. In these calculations the CO was perpendicular to the surface over one Ni atom. Sulfur reduces the interaction of CO with the cluster by destabilizing the 5σ orbital (M-C-O bonding) with respect to other orbitals, reducing the participation of the central Ni atom in chemisorption, and changing the 5σ orbital to be less bonding for the Ni-C bond. Thus sulfur as a catalyst poison reduces the CO interaction with the surface. The effect of Li is to lower the $2\pi^*$ orbital (antibonding in the C-O bond) to near the Fermi level so that it becomes partially occupied to weaken the C-O bond and facilitate its decomposition.

The cluster calculation for the interaction of hydrocarbons with metals will be illustrated by two calculations for acetylene (H-C≡C-H), one a HF calculation for a small cluster and the other a EH calculation for a large cluster. The HF calculation [71] was carried out using Gaussian-80 on a planar 4 atom Ni cluster representing a (111) plane with the C_2H_4 in a bridge position between 2 touching Ni atoms and having the C-C axis perpendicular to the Ni-Ni axis and parallel to the Ni atom plane; a position suggested by LEED studies. The absorption energy was calculated to be 89 kcal/mole in comparison with the experimental value of 67 kcal/mole. Geometry optimization produced a bending up of the H atoms away from the plane of the surface by about 55° and a corresponding carbon $sp^{2.5}$ hybridization. These results generally agree with suggestions from UPS and ELS data that the carbon hybridization is $sp^{2.5-2.8}$. While another ELS study suggested the H atoms were not bent in a plane normal to the surface, the calculations gave higher energies to these structures. In modified EH calculations [72] for Pt_{19} clusters interacting with various C_2H_n (n=1-4) fragments the most stable position for C_2H_2 was calculated to be at a 3 fold site with each C bonded to two Pt atoms with the C-C-H angle 155° and with the C-C-H plane 30° from the normal in agreement with ELS and LEED data. The C-H bonds were activated by the surface interaction in that the barrier to dissociation of the C-H bond by the H atom transferring to the closest Pt atom was 35 kcal/mole, which is much less than the C-H bond energy. The Pt d orbitals made a strong contribution to the bond energies.

3.4 Conclusions

Cluster calculations do a surprisingly good job of describing the adsorption interaction considering that even in clusters of 20 atoms almost all atoms are on the surface. Chemisorption energies are affected by the presence of edges, i.e., by the number of nearest neighbors of the atoms comprising the adsorption site. Calculations for multicenter sites, which seem to occur most often, will require clusters large enough that all the surface atoms at the site at least have all their nearest neighbors. Calculations with large clusters (20 or more atoms) are needed to depict catalytic reaction processes on surfaces. These will most likely be done with semiempirical methods that have established their credibility by matching calculations on small clusters by ab initio methods.

References

1. E.L. Muetterties, R.R. Burch, A.M. Stolzenberg: Ann. Rev. Phys. Chem. 33, 89 (1982)
2. G.A. Somorjai: Cat. Rev. 7, 87 (1972)
3. R.P. Messmer: In Semiempirical Methods of Electronic Structure Calculation, Part B, ed. by G.A. Segal (Plenum, New York 1977)
4. T.B. Grimley: In Molecular Processes on Solid Surfaces, ed. by E. Grauglis, R.D. Gretz, R.I. Jaffee (McGraw-Hill, New York 1969) p.299
5. T.L. Einstein: Phys. Rev. B11, 577 (1975)
6. Y. Somo-Noto, W.M.H. Sachtler: J. Catalysis 32, 315 (1974); 34, 162 (1974)
7. M.L. Cohen, S.G. Louie: Ann. Rev. Phys. Chem. 35, 537 (1984)
8. W.N. Delgass, G.L. Haller, R. Kellerman, J.H. Lunsford: Spectroscopy in Heterogeneous Catalysis (Academic, New York 1979)
9. H. Ibach, D.L. Mills: Electron Energy Loss Spectroscopy and Surface Vibrations (Academic, New York 1982)
10. M. Simonetta, A. Gavezzotti: Adv. in Quantum Chem. 12, 103 (1980)

11. R.L. Whetten, D.M. Cox, D.J. Trevor, A. Kaldor: Surf. Sci. 156, 8 (1985)
12. W. Weltner, Jr., R.J. Van Zee: Ann. Rev. Phys. Chem. 35, 29 (1984)
13. D.A. Case: Ann. Rev. Phys. Chem. 33, 151 (1982)
14. J.W.D. Connolly: In Semiempirical Methods of Electronic Structure Calculation, Part A: Techniques, ed. by G.A. Segal (Plenum, New York 1977)
15. J.A. Pople, D.L. Beveridge: Approximate Molecular Orbital Theory (McGraw-Hill, New York 1970)
16. G. Blyholder, J. Head, F. Ruette: Theoret. Chim. Acta 60, 429 (1982)
17. R.P. Messmer: In Chemistry and Physics of Solid Surfaces IV, ed. by R. Vanselow, R. Howe (Springer, Berlin, Heidelberg 1982) p.315
18. R.C. Baetzold, J.F. Hamilton: Prog. Solid State Chem. 15, 1 (1983)
19. Proc. 3rd Intern. Meeting on Small Particles and Inorganic Clusters, Berlin 1984, Surf. Sci. 156, 1-1071 (1985)
20. S.J. Riley, E.K. Parks, K. Kiu, S.C. Richtsmeier, G.C. Nieman, L.G. Pobo: Am. Chem. Soc. 190th National Meeting, Chicago (1985), Div. of Colloid and Surface Science, paper no. 127
21. R.C. Baetzold: J. Chem. Phys. 55, 4363 (1971)
22. H. Basch: J. Am. Chem. Soc. 103, 4657 (1981)
23. A.B. Anderson: J. Chem. Phys. 68, 1744 (1978)
24. H. Basch, M.D. Newton, J.W. Moskowitz: J. Chem. Phys. 73, 4492 (1980)
25. R.C. Baetzold: J. Chem. Phys. 55, 4363 (1971)
26. G. Blyholder: Surf. Sci. 42, 249 (1974)
27. J. Demuynck, M. Rhomer, A. Strich, A. Veillard: J. Chem. Phys. 75, 3443 (1981)
28. R.C. Baetzold: J. Phys. Chem. 80, 1504 (1976)
29. C. Bachmann, J. Demuynck, A. Veillard: Faraday Symp. Chem. Soc. 14, 170 (1980)
30. A. Herrmann, E. Schumacher, L. Wöste: J. Chem. Phys. 68, 2327 (1978)
31. E.A. Rohlfing, D.M. Cox, A. Kaldor, K.H. Johnson: J. Chem. Phys. 81, 3846 (1984)
32. J.D. Head, K.A.R. Mitchell: Mol. Phys. 35, 1681 (1978)
33. R.P. Messmer, S.K. Knudson, K.H. Johnson, J.B. Diamond, C.Y. Yang: Phys. Rev. B 13, 1396 (1976)
34. R.C. Baetzold: Surf. Sci. 106, 243 (1981)
35. J.D. Head, K.A.R. Mitchell: Mol. Phys. 35, 1681 (1978)
36. M.G. Mason, R.C. Baetzold: J. Chem. Phys. 64, 271 (1976)
37. R.C. Baetzold, M.G. Mason, J.F. Hamilton: J. Chem. Phys. 72, 366, 6820 (1980)
38. R.C. Baetzold: Inorg. Chem. 20, 118 (1981)
39. T.H. Upton, W.A. Goddard III: J. Am. Chem. Soc. 100, 5659 (1978)
40. A. Van der Avoird, H. DeGraaf, R. Berns: Chem. Phys. Lett. 48, 407 (1977)
41. C. Pisani, F. Ricca: Surf. Sci. 92, 481 (1980)
42. D.J.M. Fassaert, H. Verbeek, A. Van der Avoird: Surf. Sci. 29, 501 (1972)
43. G. Blyholder: J. Chem. Phys. 62, 3193 (1975)
44. T.H. Upton, W.A. Goddard III: In Chemistry and Physics of Solid Surfaces III, ed. by R. Vanselow, W. England (CRC, Boca Raton, FL 1982) p.127
45. C.F. Melius, J.W. Moskowitz, A.P. Mortola, M.B. Baillie, M.A. Ratner: Surf. Sci. 59, 279 (1976)
46. R.P. Messmer, D.R. Salahub, K.H. Johnson, C.Y. Yang: Chem. Phys. Lett. 51, 84 (1977).
47. Per. E.M. Siegbahn, M.R.A. Blomberg, C.W. Bauschlicher, Jr.: J. Chem. Phys. 81, 2103 (1984)
48. G. Pacchioni, J. Koatecky: Surf. Sci. 154, 126 (1985)

49. E. Miyoshi, Y. Sakai, S. Mori: Surf. Sci. 158, 667 (1985)
50. F. Ruette, G. Blyholder, J.D. Head: Surf. Sci. 137, 491 (1984)
51. K. Christmann, R.J. Behm, G. Ertl, M. van Hove, W. Weinberg: J. Chem. Phys. 70, 4168 (1979)
52. S. Andersson: Chem. Phys. Lett. 55, 185 (1978)
53. G. Blyholder, J. Head, F. Ruette: Surf. Sci. 131, 403 (1983)
54. J.E. Demuth: Surf. Sci. 65, 365 (1977)
55. F. Himpsel, D.E. Eastman: Phys. Rev. Lett. 41, 507 (1978)
56. B.N. Cox, C.W. Bauschlicher: Surf. Sci. 108, 483 (1981)
57. P.S. Bagus, C.W. Bauschlicher, Jr., C.J. Nelin, B.C. Laskowski, M. Seel: J. Chem. Phys. 81, 3594 (1984)
58. F. Ruette, A. Hernandez, E.V. Ludena: Surf. Sci. 151, 103 (1985)
59. D.E. Peebles, H.C. Peebles, D.N. Belton, J.M. White: Surf. Sci. 134, 46 (1983)
60. F. Ruette, E.V. Ludena, A. Hernandez, G. Castro: Surf. Sci. 167, 393 (1986).
61. F. Ruette, A. Hernandez, E.V. Ludena: Int. J. Quantum Chem., in press
62. K. Christmann, O. Schober, G. Ertl, M. Neumann: J. Chem. Phys. 60, 4528 (1974)
63. S.P. Walch: Surf. Sci. 143, 188 (1984)
64. F. Ruette, G. Blyholder: unpublished
65. E.W. Plummer, C.T. Chen, W.K. Ford, W. Eberhardt, R.P. Messmer, H.J. Freund: Surf. Sci. 158, 58 (1985)
66. G. Blyholder: J. Vac. Sci. Technol. 11, 865 (1974)
67. G. Blyholder: J. Phys. Chem. 79, 756 (1975)
68 H. Jörg, N. Rösch: Surf. Sci. 163, L627 (1985)
69. S.P. Mehandru, A.B. Anderson: Surf. Sci. 169, L281 (1986)
70. J.M. MacLaren, J.B. Pendry, D.D. Vnedensky, R.W. Joyner: Surf. Sci. 162, 322 (1985)
71. H. Kobayashi, H. Teranae, T. Yamabe, M. Yamaguchi: Surf. Sci. 141, 580 (1984)
72. D.B. Kang, A.B. Anderson: Surf. Sci. 155, 639 (1985)

Numerical Analysis in Self-Consistent Theories of Metal Surfaces

E.E. Mola and J.L. Vicente

Instituto de Investigaciones Fisicoquímicas Teóricas y Aplicadas (INIFTA), Casilla de Correo 16, Sucursal 4, 1900 La Plata, Argentina

1. Introduction

In metal surface calculations the rapid decrease of electron density near the surface, and the loss of translational symmetry in this region produce additional difficulties. In order to get self-consistency (to find the particle distributions from the picture of interactions in the system) and analytical response, we recently presented a metal film model /1/ where an asymptotic infinite square barrier potential is introduced. We managed to get numerical results out of this model by performing a careful numerical analysis of the equations to be solved.

2. Mathematical Formulation of the Problem

Let $\theta(x)$ be the step function (associated with a uniform background), $Q[\rho]$ a continuous functional (it represents the exchange and correlation potential) on the functions $\rho(x)$ (electron density) for $-b \leq x \leq b$ and let us find $\phi(x)$ (the electrostatic potential) and $\psi(x)$ (one-particle state) that satisfy the Poisson equation

$$\phi''(x) - \frac{\rho_o}{2a}\,\theta(a - |x|) + \rho(x) = 0 \quad ; \quad \phi(\pm b) = 0 \tag{1}$$

and $\psi(x)$ obtained from the Schrödinger type equation

$$\psi''(x) - \{\phi(x) + Q[\rho(x)]\}\psi(x) = \varepsilon\psi(x) \; ; \; \psi(\pm b) = 0 \quad . \tag{2}$$

Equation (2) will have a sequence of eigenvalues ε_m corresponding to solutions $\psi_m(x)$. We can define

$$\rho(x) = \sum_{m=1}^{M} (E - \varepsilon_m)\psi_m^2(x) \quad , \tag{3}$$

where M (the number of states) and E (the Fermi level) are determined by

$$\varepsilon_M < E \leq \varepsilon_{M+1} \qquad \text{and} \qquad \int_{-b}^{b} \rho(x)dx = \rho_o \tag{4}$$

If we assume that there exists a continuous function $\rho(x) \geq 0$ that satisfies (1) to (4), the problem will be how to find such $\rho(x)$. In order to do it we can use a straight iteration, starting with a trial density $\rho_1(x)$ under the condition given by (4-right), which will allow us to find $\phi_1(x)$ from

$$\phi_1(x) = \int_{-b}^{x} (x-\xi)\left[\frac{\rho_o}{2a}\,\theta(a - |\xi|) - \rho(\xi)\right]d\xi + \sigma(x) \; ; \; \sigma''(x) = 0 \quad . \tag{5}$$

By solving (2), a Sturm-Liouville problem that can be transformed /2/ into

$$\psi(x) - \frac{1}{\varepsilon}\int_{-b}^{b} K(x,\xi)\psi(\xi)d\xi = 0 \; ; \; K(x,\xi) \text{ Green's Function} \quad , \tag{6}$$

we obtain a new electron density $\rho_2(x)$ from (3), under the condition (4). As (1) to (4) can only be solved approximately starting with an arbitrary function $\rho_1(x)$, it will be seen that this is equivalent to solving a discrete problem whose limit approaches the solution. Next we shall develop two schemes to find such electron density $\rho(x)$:

A) Numerical Discretization on a Lattice, (NDL). A rather crude approximation is to replace the derivatives by a difference scheme. We denote ρ^j the approximation to the $\rho(x)$ at the point x_j. If there exists the limit of ρ^j when the mesh size $h = (x_j - x_{j-1})$ approaches zero, and if this limit is $\rho(x_j)$, there will be a convergent numerical solution /3/.

B) Expansion on a Finite Number of Basis Functions, (EFB). An alternative scheme is to consider an orthogonal and complete system of functions $\{w_k(x)\}$ $k = 0,1,\ldots$ that satisfies the boundary condition on (1) and (2) (eigenfunctions of an infinite square barrier potential between $\pm b$). Approximating $\psi(x)$ by an expansion $\tilde{\psi}(x)$ on the space generated by $\{w_k(x)\}$, $0 \leq k \leq J$, the NDL scheme on the net $\{x_j\}$, $0 \leq j < \infty$, becomes a system of linear equations whose solutions give the coefficients of the expansion. In this way, by using as $\{w_k\}$ the trigonometric system an equivalence is set up with a NDL ($h \to 0$ associated to $J \to \infty$). In the space generated by $\{w_k\}$, $0 \leq k \leq J$ an iterative method can be used. If $K(x,\xi)$ satisfies the Fredholm's theorem conditions (6) can be solved in this space /4/.

By considering the problem symmetry, one only needs to solve it in the interval $0 \leq x \leq b$, adding the conditions $\phi'(0) = 0$ and $\psi'(0) = 0$ to (1) and (2) respectively. Then from

$$\phi'(x) = \int_0^x \left[\frac{\rho_0}{2a}\,\theta(a - |\xi|) - \rho(\xi)\right]d\xi$$

and (4) we have $\phi'(b) = 0$ and $\sigma \equiv 0$. According to (1) and $\rho(\xi) \geqq 0$, if $a < x < b$

$$\phi'(x) = \tfrac{1}{2}\rho_0 - \int_0^x \rho(\xi)d\xi \;\; \geqq 0,$$

on the other hand, the boundary conditions give $\psi(b) = 0$ i.e. $\rho(b) = 0$ and taking into account the discretization error when $\int\rho(\xi)d\xi$ in $[\alpha,\beta]$ is approximated by $I(\rho;\alpha,\beta)$, we have $\delta\rho_0 = I(\rho;0,b) - \frac{1}{2}\rho_0 \neq 0$ that means $\tilde{\phi}'(b) \neq 0$ (see Fig. 1). If $h \to 0$ we get $\tilde{\phi}'(b) \to 0$ but this convergence can be slow and we want to accelerate it.

3. Analysis of the Solution

3.1. One way to accelerate the convergence when $h \to 0$, is by taking mesh sizes $h_\nu = h/2^\nu$ (ν=0,1,2...), and by calling z_ν the ρ_n^j or the ρ expansion coefficient

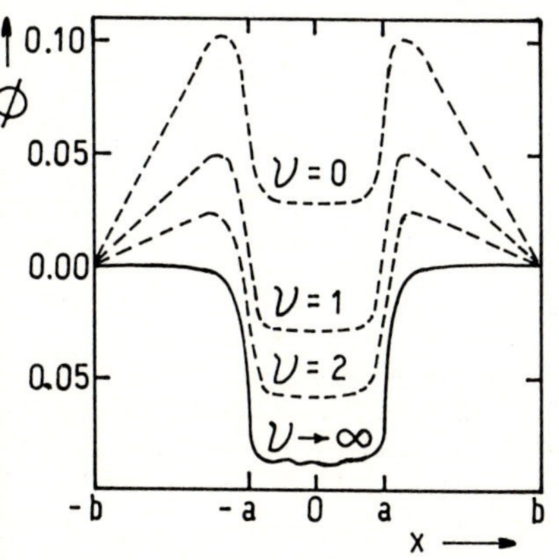

Figure 1. Dashed lines represent the solutions of (1) with mesh size h_ν. Full line represents the extrapolated solution (Note that $\phi'(b) \to 0$ when $h_\nu \to 0$).

Table I. Parameters of (1) and (3) as a function of the mesh size h_ν in 3.1

ν	0	1	2	3	$\to\infty$
E	0.0729	-0.0248	-0.0824	-0.0116	-0.1417
$\phi(0)$	0.1256	0.0279	-0.0296	-0.0588	-0.0889

in NDL or EFB schemes respectively; we can use

$$z'_\nu = z_\nu - (z_\nu - z_{\nu-1})^2/(z_\nu - 2z_{\nu-1} + z_{\nu-2}) \tag{7}$$

as extrapolation, where z'_ν converges faster than z_ν (See Table I).

3.2. An alternative way to find a $\rho(x)$, with $\phi'(b) \cong 0$ for a fixed h, rises from the asymptotic behaviour of the error in $\rho_n(x)$ as $n \to \infty$. Equations (1) and (2) are stable on the parameter ρ_0. Then we can change in the iteration scheme ρ_0 by $\rho_0 + \delta\rho_0$ ($\delta\rho_0$ represents the error in the charge introduced by the discretization $\phi'(b)\neq 0$) and a new $\rho + \delta\rho$, close to the older ρ, is obtained according to the Malkin's theorem /5/ (See Table II).

Table II. Parameters of (1) and (3) obtained by the methods 3.1, 3.2 and 3.3

Method	3.1	3.2	3.3
E	-0.1417	-0.1419	-0.1433
$\phi(0)$	-0.0889	-0.0875	-0.0904

3.3. Instead of the previous external perturbational methods, we can use an internal one changing in each step only a small $\sigma(x)$ in (6), by adding to $\phi_n(b) = 0$ the condition $\sigma'_n(b) = I(0,b;\rho_{n-1}) - \frac{1}{2}\rho_0$, n=2,3,... Therefore the evaluation of the left-hand side of (5) (see Table II) is obtained.

Finally the rate of convergence (instead of the previous asymptotic convergence) can be accelerated by using a new sequence ρ'_n as

$$\rho'_{n+1} = \rho'_n - (L[\rho'_n] - \rho'_n)^2/[L^2[\rho'_n] - 2L[\rho'_n] + \rho'_n] \ , \ \rho'_1 = \rho_1 \ , \tag{8}$$

where the iterative scheme was written as $\rho_{n+1} = L[\rho_n]$ (See Fig. 2).

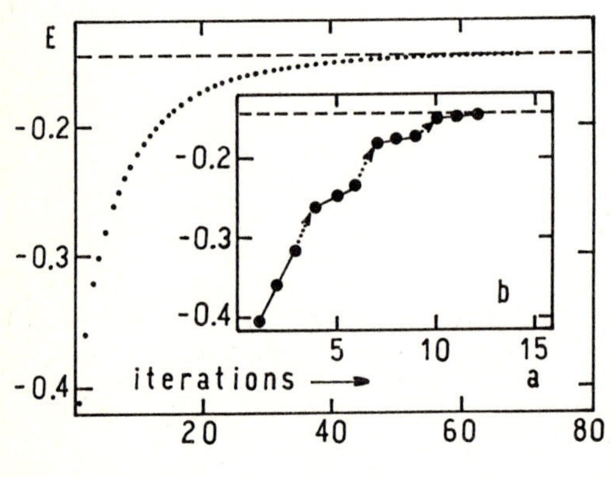

Figure 2. Parameter E (Fermi level) vs the number of iterations (a) Direct iteration, (b) Iteration accelerated by the algorithm (8).

4. Conclusions

Two different iterative schemes were presented to solve problems that appear in self-consistent solution of many-body phenomena at surfaces. One of them (EFB) has the advantage of allowing an analytical solution of the physical quantities involved in the calculation. The methods 3.1 and 3.2 were applied to study /1/ the electronic behaviour of a thin metal film and 3.3 to describe /6/ a theory of water adsorption on metal surfaces.

Acknowledgments

This work was financially supported by the CONICET, R. Argentina. INIFTA is an Institute depending on the Facultad de Ciencias Exactas, UNLP.

References

1. E.E. Mola and J.L. Vicente, J.Chem.Phys. 84, 2876 (1986)
2. J. Dieudonne: in Foundations of Modern Analysis (Academic, New York 1957)
3. R. Richtmyer: in Difference Methods for Initial Value Problems (Interscience, New York 1957)
4. A. Kolmogorov and S. Fomin: in Elements of the Theory of Functions and Functional Analysis (Graylock, Rochester, New York 1957)
5. L. Elsgoltz: in Differential Equations and Variational Calculus (Mir, Moscow 1969)
6. E.E. Mola and J.L. Vicente, Surface Sci., in press

Self-Consistent Calculation of Atomic Chemisorption on Metals

C.P. de Melo[1], *M.C. dos Santos*[2], *and B. Kirtman*[3]

[1]Departamento de Física, Universidade Federal de Pernambuco, 50000 Recife PE, Brasil
[2]Departamento de Química Fundamental, Universidade Federal de Pernambuco, 50000 Recife PE, Brasil
[3]Department of Chemistry, University of California, Santa Barbara, CA 93106, USA

1. Introduction

One of the most long standing theoretical challenges in the field of electronic structure calculations is how to treat in a self-consistent way the problem of a semi-infinite system with locally broken symmetry. Impurities in solids, defects in polymers, and the chemisorption of an atom or molecule on a metallic surface belong to this class of problems. Usual embedding techniques for the chemisorption problem, for example, /1-4/ include self-consistency in only a limited way: usually the transfer of electrons between the "surface complex" in the vicinity of the adsorbate and the rest of the system is not correctly accounted for, nor is charge conserved in a global sense for the whole system.

In recent years, the Hartree-Fock Local Space Approximation (LSA) method /5/, which was specially designed for the self-consistent calculation of localized electronic interactions in molecules and solids, has been applied to investigate the chemisorption phenomenon /5/, to study the interaction between finite molecular fragments /6/, and to examine the question of the localizability of hydrogen bonds in large systems /7/. A parameter in the H-F LSA method is the size of the "local space", the region where the perturbations introduced by the adsorbate are most felt.

In the present work we have examined the behavior of relevant chemisorption properties, such as charge on the adsorbate and spin distribution on the substrate, as a function of increasing local space sizes. The model considered here consists of a hydrogen atom interacting with a one-dimensional chain of transition metal atoms within the Anderson-Newns (AN) approximation /8,9/. Due to the crudeness of the model, we do not insist that the results agree with the experiment. Even so, the predicted energies for the adsorbate-induced surface states are in good agreement with the observed resonances in the photoemission spectra of hydrogen adsorbed on metal surfaces /10,11/.

2. Theory

The general features of the Hartree-Fock LSA method have been presented previously /5,6/. As a more complete version of this work will be presented elsewhere /12/, we will give here only a brief outline of the LSA treatment of the present problem.

We start with the non-interacting subsystems A (for adsorbate; in this case, the hydrogen atom) and M (for metal), which are described by their first-order density matrices, R^A and R^M, respectively. Since in the AN model the atomic orbitals are assumed to be orthogonal, the initial density matrix for the entire system, R_0, is just the direct sum R^A+R^M. Each sub-system is considered to be isolated; therefore, the unperturbed Hamiltonian matrices h^A and h^M, can be written as a sum of occupied and unoccupied blocks- e.g. $h^A=R^Ah^AR^A+U^Ah^AU^A$ - where we have introduced the projector into the virtual space U=1-R. As a consequence the total zeroth-order Hamiltonian is also block-diagonal.

After the coupling of the adsorbate to the metal surface is switched on, a mixing of occupied and virtual states will occur. At the end of the self-consistent cycle the HF solution will be reached when the Hamiltonian matrix is again diagonal into the new occupied and unoccupied spaces for the combined system.

During the whole procedure charge must be conserved, and the representability of the density matrix in terms of a single determinant wavefunction preserved. This is accomplished by imposing the constraint of idempotency of the first-order HF density matrix R.

McWEENY /13/ has shown that any self-consistent change in an idempotent density matrix R_n, can be written as

$$R_{n+1} = R_n+\Delta R^{(n)} = (R_n+v)(1+v^+v)^{-1}(R_n+v^+), \tag{1}$$

in which $v=U_nXR_n$ and X is an arbitrary symmetric matrix. Our <u>local space approximation</u> consists in supposing X to be confined to the <u>local space</u> Q comprising the set of atomic orbitals which span the region where the effect due to the interaction is most felt. In this way, $X=QXQ=X_Q$, where Q is the projector into the local space.

A particular feature of the LSA method is to relate all quantities to zeroth order density matrices. Although the cumulative departure from the original R_0 has a non-local character, it can be expressed as

$$\Delta R_T = R_0\tau_Q^R R_0 + R_0\tau_Q^{RU} U_0 + U_0\tau_Q^{UR} R_0 + U_0\tau_Q^U U_0 \ , \tag{2}$$

where the auxiliary matrices τ_Q have the dimensions of the local space. At the same time, use of the cyclic properties of the trace allows the chemisorption energy

$$\Delta E = \mathrm{tr}\{[h(R_0) + h(R)]\Delta R_T\} \tag{3}$$

to be computed within the local space. Therefore, although the LSA allows the free flow of charge outside the local space (cf.(2)), the Hartree-Fock problem is naturally projected into the local space, reducing the quantum chemistry calculation to molecular dimensions.

3. Model Calculation

The spin-polarized AN Hamiltonian can be written as

$$h = \sum_\sigma |\phi_{a\sigma}\rangle\{\varepsilon_a+JR_{a-\sigma,a-\sigma}\}\langle\phi_{a\sigma}| + \sum_{n\sigma} |\psi_{n\sigma}\rangle\varepsilon_{n\sigma}\langle\psi_{n\sigma}|$$
$$+ \sum_\sigma [|\phi_{a\sigma}\rangle V_{as}\langle\phi_{o\sigma}| + |\phi_{o\sigma}\rangle V_{as}\langle\phi_{a\sigma}|] \ , \tag{4}$$

where $\phi_{a\sigma}$ is the hydrogen atomic orbital of spin σ, $\psi_{k\sigma}$ is a metal eigenfunction, and $\phi_{o\sigma}$ is the atomic orbital of the "surface" substrate atom which couples to the adsorbate. ε_a=-13.6eV is the ionization potential of hydrogen, and Coulomb repulsion is included only for electrons of opposite spin on the adsorbate; J=12.9eV is the difference between the ionization potential and the electron affinity of the hydrogen atom. We parametrize the semi-infinite chain to represent a tungsten surface: the metal band is assumed to be half-occupied with a total d-band width of 10eV, and the substrate is initially in a non-magnetic state with the Fermi level at -4.6eV /14/.

To obtain the initial density matrix for the system, the semi-infinite chain is considered to result from slicing an infinite chain in half: use of transfer functions can then be made to relate the elements of the Green's function matrix on the site representation /15/. A final integration of the individual Green's function elements over the occupied band of energy produces the desired density matrix for the chain. The hydrogen atom contributes with an extra electron of definite spin. After convergence the corresponding net spin density is distributed along the chain: chemisorption induces spin polarization of the substrate.

The coupling V_{as} between the hydrogen atom and the substrate was chosen equal to -4156eV such that, for the largest size of the local space used (N_Q=8), the calculated binding energy agree with the experimental value of 3.0eV. As values in Table 1 show, charge is transferred to the surface complex from the surrounding metal chain. For N_Q=8, the net charge of the local space is 0.068e. The spin redistribution along the chain is even more vivid: the local space net magnetic moment is only 0.516e, instead of the exact asymptotic value of 1.0 to be reached when the perturbation is completely healed. These results highlight the importance of conserving charge in a global sense. After projecting the converged density matrix on an increasing number of atoms along the chain, we found that after the first 30(60) atoms the value of the net charge was reduced to 0.014e (0.007e), while the net magnetic moment had increased to 0.908e (0.955e). The pattern of alternating increase and depletion of charge reminiscent of Friedel oscillations was present along the chain. Results for smaller sizes of the local space show the same general behavior: for N_Q=8 the atomic magnetic moments are converged to within a little over 20% of their asymptotic values whereas the atomic charges are converged to within about 1%.

Direct diagonalization of the density matrix projected into the local space gives us a rigorous criterion for constructing surface-complex localized states: eigenfunctions associated to eigenvalues equal to unity are entirely confined to the local space. The energies corresponding to these adsorbate-induced surface states can be obtained after diagonalizing the Hamiltonian operator projected into these localized states. Two states having significant contribution of the hydrogen atomic orbital were found. Their extrapolated energies are 1.7 and 4.8eV below the Fermi level, in good agreement to the highest resonance peaks seen by spectroscopic surface studies /10-11/.

Table 1: Atomic charges and magnetic moments for atoms in local space. The adsorbate-surface coupling is V_{as}=-4156eV; the Fermi level of the metal is at -4.6eV; and the total d-band width is 10eV.

Atom	H	M_1	M_2	M_3	M_4	M_5	M_6	M_7
charge	1.179	0.906	1.031	0.974	1.004	0.988	0.994	0.992
magnetic moment	0.335	-0.166	0.163	-0.044	0.118	-0.012	0.107	0.015

4. Discussions

Results found in the present work show the importance of self-consistency effects upon the charge and spin densities on atoms in the surface complex. The analysis of the net charge and net magnetic moment of the substrate show that one has to go a long way into the chain before the perturbation introduced by the adsorbate can be considered to be healed. To stress this latter point we have used the same Hamiltonian approximation to perform cluster (of size N_Q) and similar calculations where self-consistency is included only for the adsorbate and the "surface" metal atom /16/. Both sets of results reveal the spin density distribution to be more sensitive to the correct boundary conditions than the charge on the chain, in agreement to the notion that spin effects have a longer range than charge disturbances.

The simple model calculation presented here shows the potential usefulness of the LSA method for the treatment of the chemisorption problem. At the present moment, we are using the method to investigate the interaction between a pair of adsorbed hydrogen atoms mediated by the substrate. The oscillating behavior of the charge and spin disturbances brought about by the presence of the adsorbate should lead to a mesh of atractive and repulsive sites for the binding of the second atom.

5. Acknowledgments

The work was supported in part by the Brazilian agencies FINEP, CNPq, and CAPES, and by a bilateral grant of the CNPq and National Science Foundation.

6. References

1 T.B. Grimley, C. Pisani: J. Phys. C7, 2831 (1974)
2 C. Pisani: Phys. Rev. B17, 3143 (1978)
3 J.L. Moran-Lopez, L.M. Falicov : Phys. Rev. B26, 2560 (1982)
4 M. Sell: Int. J. Quant. Chem. 24, 753 (1984)
5 B. Kirtman, C.P. de Melo: J. Chem. Phys. 75, 4592 (1981)
6 B. Kirtman: J. Phys. Chem. 86, 1059 (1982)
7 B. Kirtman, C.P. de Melo: to appear in Int. J. Quant. Chem.
8 D.M. Edwards, D.M. Newns: Phys. Lett. 24A, 236 (1967)
9 D.M. Newns: Phys. Rev. 17B, 1123 (1969)
10 E.W. Plummer: In *Interactions on Metal Surfaces*, ed. by R. Gomes, Topics Appl. Phys., Vol.4 (Springer, Berlin, Heidelberg 1975)
11 B. Feuerbacher, R.F. Willis: Phys. Rev. Lett. 36, 1339 (1976)
12 C.P. de Melo, M.C. dos Santos, B. Kirtman: to be submitted
13 R. McWeeny: Rev. Mod. Phys. 32, 335 (1960)
14 M. Baldo, F. Flores, A. Martin-Rodero, G. Piccitto, R. Pucci: Surf. Sci. 128, 237 (1983)
15 H.S. Brandi, C.P. de Melo : Solid State Commun. 44, 37 (1982)
16 J.L. Moran-Lopez, L.M. Falicov: J. Vac. Sci. Technol. 20, 831 (1982)

Self-Consistent Model of Hydrogen Chemisorption on Nickel

F. Aguilera-Granja[1], *J.L. Morán-López*[1], *and F. Mejía-Lira*[2]

[1]Centro de Investigación y de Estudios Avanzados del IPN, Departamento de Física, A.P. 14–740, 07000 México, D.F., México
[2]Instituto de Física, Universidad Autónoma de San Luis Potosi, A.P. H-1017, San Luis Potosi, 78150 SLP, México

1. Introduction

Hydrogen chemisorption on transition metals has been extensively studied due to its catalytic importance. Among the substrates used in heterogeneous catalysis there are several like Pt,Ni,Rh, crystallizing in fcc structure. Very often the exposed surface is the closed-packed (111). Experiments /1/2/ /3/ and calculations /4/ show that chemisorbed atomic hydrogen builds tetrahedra with the atoms of the (111) surface. We present a theory for the chemisorption of atomic hydrogen on the close-packed surface of fcc crystals. This theory is based on an Anderson-Newns Hamiltonian /5/6/. We mimic the geometrical structure of the surface and the substrate by means of Husimi cacti (Fig 1). The Husimi cacti (HC) have been shown as a satisfactory model for the fcc crystal structure /7/. We calculate self-consistently the magnitude of the magnetic moments at the adatom and at the surface atoms and give results for the chemisorption energy.

2. Model

The substrate is characterized by the Hamiltonian

$$H = -\sum_{i} \alpha C_i^{\dagger} C_i - \sum_{ij} t_{ij} C_i^{\dagger} C_j + h.c, \tag{2.1}$$

where α is the atomic level in each site, t_{ij} is the transition probability (hopping) between the sites i and j, $C_j^{\dagger}$ and C_j are the usual operators of creation and annihilation, respectively. This Hamiltonian has been solved for a clean (111) fcc surface modelled by HC /7/. In that case VERGES and YNDURAIN /7/ obtained from the solution of the Dyson equation, (E-H)G=1, the local density of states

$$n_i = -\frac{1}{\pi} \operatorname{Im} G_{ii}(E), \tag{2.2}$$

where G_{ii} is the Green function in an arbitrary site i.

In order to study the chemisorption on a paramagnetic substrate /8/ we use the Hamitonian

$$H_a = -\sum_{\sigma} \left[\varepsilon_a + U\langle n_{a\bar{\sigma}}\rangle\right] n_{a\sigma} - t_a C_{a\sigma}^{\dagger} C_{o\sigma} + hc + H_s, \tag{2.3}$$

where ε_a is the atomic level before chemisorption, t_a is the hopping integral for electronic transitions between adatom and substrate, U is the Coulomb integral, $\langle n_{a\sigma}\rangle$ is the average number of electrons of spin σ at the adatom, $C_{a\sigma}^{\dagger}$ is the creation operator of electrons at the adatom and C_o an annihilation operator at a site of the surface directly connected to the adatom.

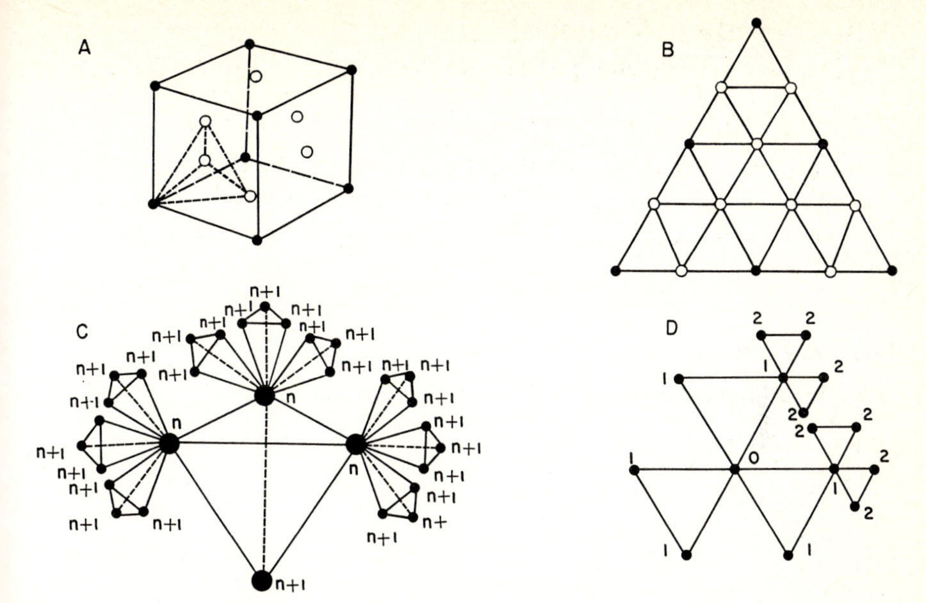

Figure 1 A) Structure of an fcc lattice; B) Structure of the (111) surface; C) 3D Husimi cacti used to simulate an fcc lattice; D) 2D Husimi cacti used to simulate the (111) surface

We used the following values for the parameters /6/. For the bandwidth and the work function of the substrate 3.8 eV and 4.5 eV, respectively; for the number of electrons of the d-band we took 9.3 electrons. The atomic level and Coulomb integral of the hydrogen atom were taken as -13.6 eV and 12.9 eV, respectively.

3. Selfconsistency

The solutions of the Dyson equation with (2.3) yields two spin-dependent Green functions for the adatom:

$$G^{\sigma}_{aa} = G^{\sigma}_{aa}(\varepsilon_a + U \langle n_{a\bar{\sigma}} \rangle) \quad \text{and} \quad G^{\bar{\sigma}}_{aa} = G^{\bar{\sigma}}_{aa}(\varepsilon_a + U \langle n_{\sigma} \rangle). \tag{3.1}$$

Following Anderson /5/ we solve the self-consistent equations

$$n_{a\sigma} = N\,(n_{a\bar{\sigma}}) \qquad \text{and} \qquad n_{a\bar{\sigma}} = N\,(n_{a\sigma}), \tag{3.2}$$

where /9/

$$n_{a\sigma} = \int_{-\infty}^{E_F} \rho^{\sigma}_{aa}\,(E, n_{a\bar{\sigma}})\, dE + (n_{a\sigma})_L \tag{3.3}$$

and

$$\rho^{\sigma}_{aa}\,(E, n_{a\bar{\sigma}}) = -\frac{1}{\pi}\,\mathrm{Im}\, G^{\sigma}_{aa}\,. \tag{3.4}$$

The last term of (3.3) is the contribution from the localized states at energies lower than the lowest edge of the band. The value of the Fermi energy E_F is fixed from the bulk solution.

Once we know the self-consistent populations $\langle n_a\uparrow\rangle$ and $\langle n_a\downarrow\rangle$ we calculate the magnetic moments at the adatom

$$\mu_a = \langle n_a\uparrow\rangle - \langle n_a\downarrow\rangle \tag{3.5}$$

and the induced ones at the surface sites of adsorption

$$\mu_{ind} = \int_{-\infty}^{E_F} (\rho_{sup}^{(\uparrow)} - \rho_{sup}^{(\downarrow)})dE, \tag{3.6}$$

where

$$\rho_{sup}^{\sigma} = -\frac{1}{\pi} \operatorname{Im} G_{oo} (E,n_{\bar{\sigma}}). \tag{3.7}$$

The chemisorption energy is given by

$$E = \sum_{\sigma} \int_{-\infty}^{E_F} \rho_{aa}^{\sigma}(E,n_{\bar{\sigma}})EdE + \int_{-\infty}^{E_F} \rho_{sup}^{\bar{\sigma}}(E,n_{\sigma})EdE - (\varepsilon_a + U\langle n_a\uparrow\rangle\langle n_a\downarrow\rangle) \tag{3.8}$$

$$- 2\int_{-\infty}^{E_F} \rho_{sup}(E)EdE \; .$$

4. Results

Figures 2a and 2b show how the electronic population and magnetic moments of the adatom depend on the hopping t_a. All energies are measured in units of the half-bandwidth of the substrate. The magnetic moment vanishes at t_a=1.12/10/11/. We observe a charge transfer from the surface to the adatom as reported by NEWNS /6/. The number of electrons n_a at the adatom has a maximum of 1.14 electrons at t_a=0.8.

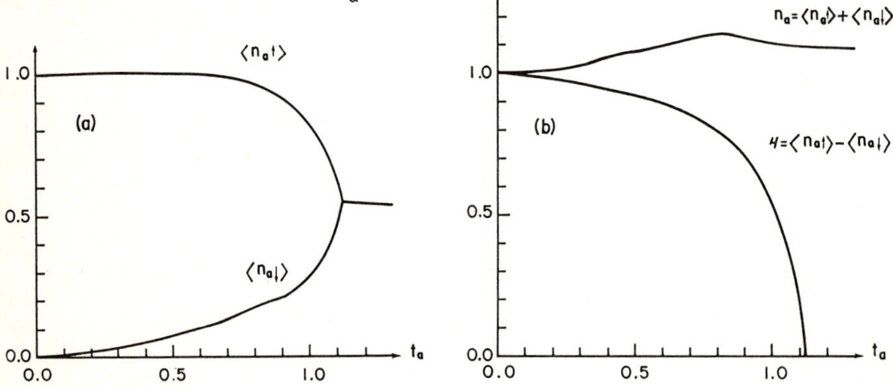

Figure 2 (a) Self-consistent populations $\langle n_a\uparrow\rangle$ and $\langle n_a\downarrow\rangle$ at the adatom; (b) total electric charge n_a and magnetic moment μ of the adatom

The magnetic moment of the adatom induces a moment in the surface atoms on which the chemisorption takes place (Fig 3). This induced magnetic moment is antiferromagnetic with respect to that of the adatom and appears at t_a~0.66 and vanishes at t_a=1.12, having a maximum at t_a=1.01.

Figure 4 shows our results for the chemisorption energy E as a function of the hopping t_a.

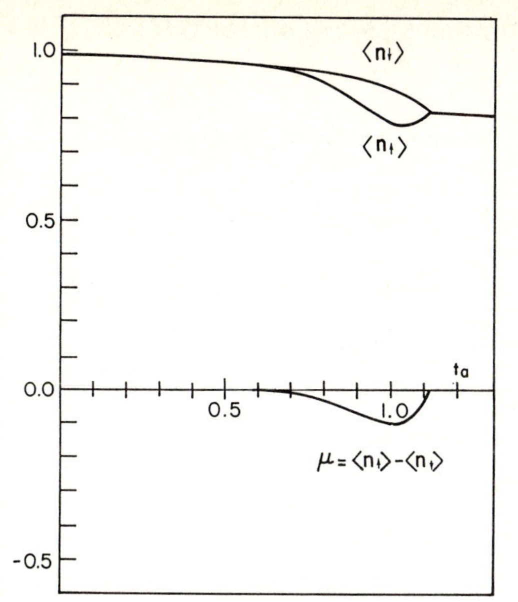

Figure 3 Self-consistent electronic populations <n↑> and <n↓>, and magnetic moment μ of a surface atom directly connected with the adatom. t_a is the hopping integral

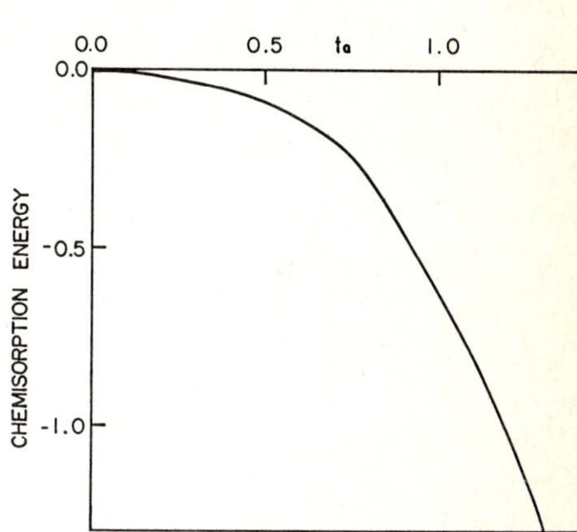

Figure 4 Chemisorption energy as a function of the hopping t_a. Both energy and hopping are in units of the half-bandwidth of the substrate

Acknowledgments

This work was supported in part by DGICSA-SEP (MEXICO) and by Programa Regional de Desarrollo Científico of OAS.

References

1. K. Christmann, R.J. Behm, G. Ertl, M.A. Van Hove and W.H. Weinberg, J. Chem. Phys. 70, 4168 (1979)
2. M.A. Van Hove, G. Ertl, K. Christmann, R.J. Behm and W.H. Weinberg, Solid State Commun 28, 373 (1978)
3. J. Behm, K. Christmann and G. Ertl, Solid State Commun 25, 763 (1978)
4. P. Nordlander, S. Holloway and J.K. Nørskov, Surf. Sci. 136, 59 (1984)
5. P.W. Anderson, Phys. Rev. 124, 41 (1961)
6. D.M. Newns, Phys. Rev. 178, 1123 (1968)
7. J.A. Verges and F. Yndurain, J. Phys. F8, 873 (1978)
8. F. Mejía-Lira, and J.L. Morán-López: In Electronic Structure and Properties of Hydrogen in Metals, ed. by J. Jena and C.B. Satterthwaite (Plenum, New York, London, 1983) p. 641
9. J. Heinrichs, Phys. Rev. B16, 4365 (1977)
10. J.L. Morán-López and L.M. Falicov, J. Vac. Sci. Technol. 20, 31 (1982)
11. J.L. Morán-López and L.M. Falicov, Phys. Rev. B26, 2560 (1982)

Order-Disorder Transitions of Hydrogen on Ni(111)

F. Aguilera-Granja[1] *and J.L. Morán-López*[2]

[1]Instituto de Física, Universidad Autónoma de San Luis Potosí, San Luis Potosí, 78000 SLP, México
[2]Centro de Investigación y de Estudios Avanzados del IPN, Departamento de Física, A.P. 14–740, 07000 México, D.F., México

The order-disorder transition of H chemisorbed at Ni(111) has been measured by BEHM, CHRISTMANN and ERTL /1/. They reported that the maximum transition temperature (270 K) occurs at a coverage $\Theta=0.5$, where Θ is defined as the ratio of the number of adsorbed particles to surface atoms of the substrate. Furthermore, they found that the phase diagram is not symmetric about $\Theta=0.5$, interpreting the LEED pattern as an ordered phase with (2x1) structure.

KITTLER and BENNEMANN /2/ have shown that the observed asymmetric phase diagram can be obtained from a model including three-body interactions between the H-atoms.

Later other authors /3,4/ interpreted those experimental results in terms of an ordered phase with a (2x2) superstructure in a honeycomb lattice, which should be symmetric about $n=0.5$ rather than about $\Theta=0.5$, where n is defined as the ratio of the number of adsorbed atoms to that of the adsorption sites, i.e. $n=\Theta/2$.

Here we address a different aspect of the problem, namely, since Ni is magnetic at temperatures higher than the maximum order-disorder transition temperature, and taking into account that recent experiments and calculations /5,6/ indicate that the surface magnetization is of considerable magnitude, it is expected that magnetism plays an important role in determining the phase diagram of H-Ni(111).

To study the magnetic effects on the order-disorder phase diagram, we assume that the H chemisorption takes place in a triangular lattice and we take into account chemical two-and three-body interactions for H-atoms (nearest neighbors) and only nearest-neighbor pair magnetic interactions.

To describe the spatial ordering, we subdivide the triangular lattice into three interpenetrating sublattices, α, β, and γ. We use as a basic cluster a triangle, and we denote the configuration probabilities by $t^{ijk}_{\alpha\beta\gamma}$, where $i,j,k=0,\uparrow,\downarrow$. The free energy can be written in terms of these probabilities as

$$U/2N = \sum \varepsilon^{ijk}_{\alpha\beta\gamma}\, t^{ijk}_{\alpha\beta\gamma}\,, \tag{1}$$

where the energy terms $\varepsilon^{ijk}_{\alpha\beta\gamma}$ are given by

$$\varepsilon^{\uparrow\uparrow\uparrow}_{\alpha\beta\gamma} = V_{\alpha\beta} + V_{\beta\gamma} + V_{\alpha\gamma} + W - 3J, \tag{2}$$

$$\varepsilon^{0\uparrow\downarrow}_{\alpha\beta\gamma} = V_{\beta\gamma} + J; \quad \varepsilon^{00\uparrow}_{\alpha\beta\gamma} = 0, \quad \text{etc.} \tag{3}$$

where $V_{\mu\nu}$ are the chemical repulsions, W is the three-body interaction and J is the magnetic coupling that can be ferromagnetic (J>0) or antiferromagnetic (J<0).

The configurational entropy is given by the expression /7/,

$$S/kN = \sum_{l,k} (Y^{lk}_{\alpha\beta} + Y^{lk}_{\beta\gamma} + Y^{lk}_{\alpha\gamma}) - 2\sum_{l,k,j} T^{lkj}_{\alpha\beta\gamma} - 1/3\sum_{l,\nu} X^{l}_{\nu} \quad , \tag{4}$$

where

$$Y^{lk}_{\nu\mu} = y^{lk}_{\nu\mu}(\ln[2^{l'+k'} y^{lk}_{\nu\mu}]-1) \quad , \tag{5}$$

$$T^{lkj}_{\alpha\beta\gamma} = t^{lkj}_{\alpha\beta\gamma}(\ln[2^{l'+k'+j'} t^{lkj}_{\alpha\beta\gamma}]-1) \quad , \tag{6}$$

$$X^{l}_{\nu} = x^{l}_{\nu}(\ln[2^{l'} x^{l}_{\nu}]-1) \quad , \tag{7}$$

and

$$y^{lk}_{\alpha\beta} = \sum_{j} t^{lkj}_{\alpha\beta\gamma}, \quad x^{l}_{\nu} = \sum_{k} y^{lk}_{\nu\mu}. \tag{8}$$

Furthermore, the probabilities are normalized

$$\sum_{lkj} t^{lkj}_{\alpha\beta\gamma} = 1. \tag{9}$$

In the above expressions

$$l' = \begin{cases} 1 & l=\uparrow \text{ or } \downarrow \\ 0 & l=0 \end{cases} . \tag{10}$$

An additional constraint to the probabilities is the coverage:

$$\sum_{lkj} g(l'k'j')t^{lkj}_{\alpha\beta\gamma} = 4\Theta, \tag{11}$$

where $g(l'k'j')=l'+k'+j'$. The equilibrium values for the probabilities are obtained by minimizing the free energy F=U -TS, with respect to all of them. The solution is obtained by means of the natural interaction method /7/.

To follow the order-disorder transition, one defines the spatial long-range order parameters

$$\eta_1 = x_\alpha - x_\beta \qquad\qquad \eta_2 = x_\gamma - x_\beta, \tag{12}$$

with $x_\nu = x^{\uparrow}_\nu + x^{\downarrow}_\nu$. The magnetic transition is described by the parameters

$$\xi_\nu = x^{\uparrow}_\nu - x^{\uparrow}_\nu \qquad \nu = \alpha,\beta,\gamma. \tag{13}$$

We show in Figs. 1 and 2 our results for two cases. In Fig. 1, we have assumed that the magnetic interactions are ferromagnetic. The values of the parameters used are given by the set ($V_{\alpha\gamma}$, $V_{\alpha\beta}$, W, J). Our values for the parameters have been chosen to resemble those used by KITTLER and BENNEMANN /2/. The dashed line is the Curie temperature and the solid line is the order-disorder transformation. The effect of magnetism in this case is to decrease the ordering temperature at high coverages, since the ferromagnetic

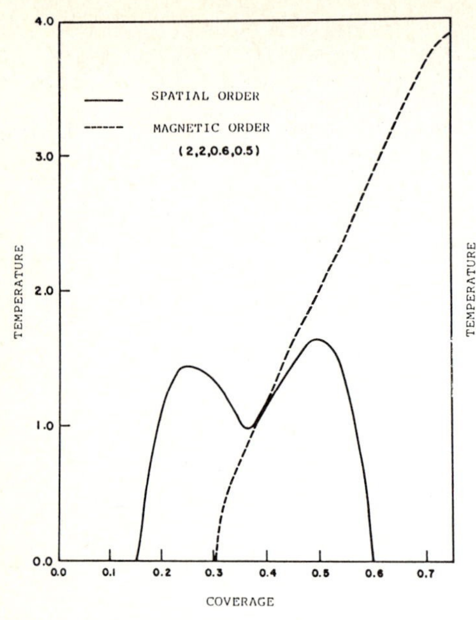

Fig. 1 Phase diagram for the ferromagnetic triangular lattice. The parameters used are given in the set in brackets.

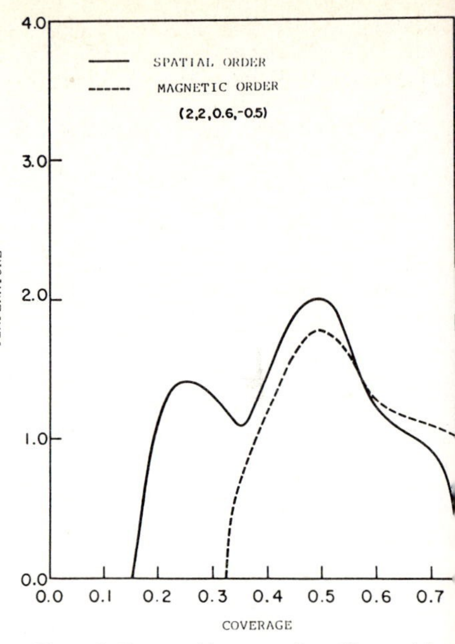

Fig. 2 Phase diagram for the antiferromagnetic triangular lattice. The parameters used are given in the set in brackets.

coupling favors the presence of a pair of hydrogen atoms, supressing thereby the spatial ordering.

A more interesting situation is obtained when the magnetic interactions are assumed antiferromagnetic. Again, the low coverage region does not get modified by the presence of magnetism, but the high coverage region gets completely changed. The antiferromagnetic interactions induce an ordered phase up to full coverage. It is worth noticing that the perfect antiferromagnet corresponds to $\Theta=2/3$ (in the units given in the figure that coverage corresponds to 0.5). Higher coverages gives rise to frustration, lowering thereby the Néel temperature.

Comparing our results with the measured phase diagram we find that the inclusion of magnetism does not improve the agreement between experiment and theory. This might be an indication that the the triangular phase proposed originally /1/ is not the correct one. However, similar studies in the honeycomb lattice are necessary to untangle the type of ordering and the role that magnetism plays in this system.

Acknowledgments

This work was supported in part by Dirección de Investigación Científica y Superación Académica de la Secretaría de Educación Pública and by Programa Regional de Desarrollo Científico y Tecnológico de la O.E.A.

References

1. J. Behm, K. Christmann and G. Ertl: Solid State Commun. 25, 763 (1978)
2. R.C. Kittler and K.H. Bennemann, Solid State Commun. 32, 403 (1979)
3. E. Domany, M. Schick and J.S. Walker: Solid State Commun., 30, 331 (1979)
4. K. Nagai, Y. Ohno and T Nakamura, Phys. Rev. B30, 1641 (1984)
5. C. Rau and S. Eichner, Phys. Rev. Lett., 47, 939 (1981)
6. E. Wimmer, A.J. Freeman and H. Krakauer, Phys. Rev. B30, 313 (1984)
7. R. Kikuchi, Phys. Rev. 81, 988 (1951)

Theoretical Study of Hydrogen Interaction on a Nickel Surface as a Function of Coverage

F. Ruette[1], *A.J. Hernández*[1,2], *and M. Sánchez*[1,3]

[1]Centro de Química, IVIC, Apdo. 21827, Caracas 1020-A, Venezuela
[2]Departamento de Química, Universidad Simón Bolívar, Apdo. 80659, Caracas 1080-A, Venezuela
[3]Departamento de Química, IUT R.C., Apdo. 1074, Caracas 101, Venezuela

1. INTRODUCTION

Although numerous studies on chemisorption of hydrogen on Ni single faces have been performed there are few theoretical studies concerning the change of physical properties as a function of the coverage. In particular, this work deals with the changes in the work function, heat of adsorption and magnetic moment as the surface coverage increases. It is a continuation of previous contributions to the study of hydrogen interaction with a nickel cluster as a model of a surface /1-4/. Our results give a good correlation between experimental results and the corresponding calculated theoretical quantities.

2. CALCULATIONAL DETAILS

The method used in these calculations is the MINDO/SR /5/ whose parameters are presented elsewhere /6/. The geometry of the nickel cluster (Ni_{18}) is shown in Fig. 1.

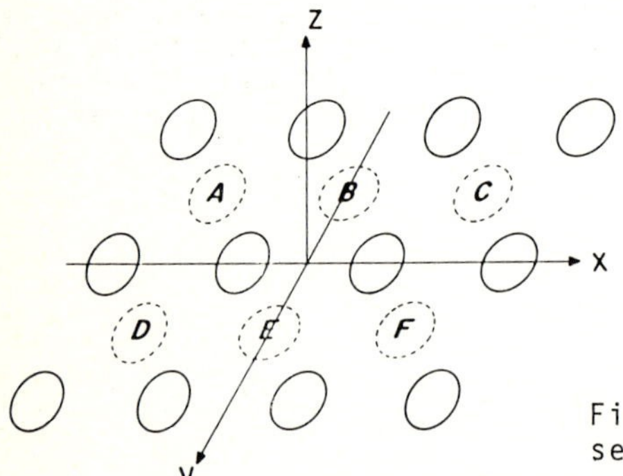

Fig. 1 Ni_{18} nickel cluster and selected adsorption sites

The solid line represents the top layer of atoms and the dashed line the bottom ones. The position where the hydrogens will chemisorb are labelled by the latin letters A,B,C, etc. According to theoretical /7/ and experimental /8/ results only the fourfold adsorption sites were used here.

3. RESULTS AND DISCUSSION

In order to study the coverage effects on a nickel surface, calculations of $Ni_{18}H_n$ (n=1,....6) clusters were performed. These calculations started with one hydrogen on the most stable four-fold adsorption site, and then progressively placed more hydrogens on the surface in the most stable remaining four-fold center sites until saturation on the cluster surface was reached. In Table 1, the adsorption sites, total hydrogen charges, heats of adsorption and multiplicities of the most stable hydrogen configuration for each $Ni_{18}H_n$ cluster are displayed.

Table 1. Properties of the $Ni_{18}H_n$ Systems (n=0,...5)

Hydrogen Adsorption Sites	Cluster	Total Hydrogen Charge	Heat of Adsorption kcal/mol	Multiplicity
-	Ni_{18}	-	-	21
B	$Ni_{18}H$	-0.050	-13.6	20
B,D	$Ni_{18}H_2$	-0.137	-34.5	19
B,D,F	$Ni_{18}H_3$	-0.249	-28.4	18
B,D,E,F	$Ni_{18}H_4$	-0.234	-16.7	19
A,B,D,E,F	$Ni_{18}H_5$	-0.256	-9.9	18
A,B,C,D,E,F	$Ni_{18}H_6$	-0.188	+12.7	17

3.1 Heat of Adsorption

The heat of adsorption values (ΔBE) presented in Table 1 are plotted vs. the number of adsorbed hydrogens as shown in Fig. 2.

Experimental results reported by CHRISTMANN et al. /9/ on the isosteric heat of adsorption indicate that for the roughest surface (110), at very low coverage, the heat of adsorption increases with the coverage. At higher coverage, the heat of adsorption becomes constant until it reaches a point from where it decreases rapidly. Our results on Fig. 1 reveal that in fact at the beginning of the hydrogen chemisorption ΔBE augments with the coverage but at higher coverage it diminishes very rapidly.

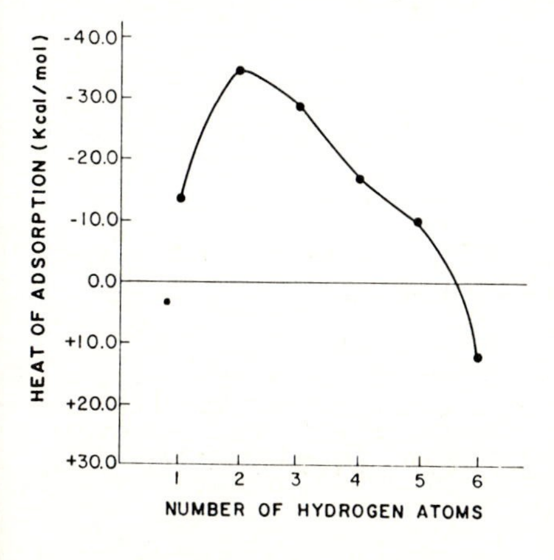

Fig. 2 Heat of adsorption as a function of the number of adsorbed hydrogens

The drastic ΔBE change with the coverage may be due to cluster size effects.

3.2 Work Function

The work function change ΔBE can be calculated by a classical model /10/ through the Helmoltz's equation, where ΔWF is directly proportional to the total amount of charge transferred from the surface to hydrogen. A plot of the electronic charge vs. the number of adsorbed hydrogens would give information on how the work function changes as the coverage varies (see Fig. 3).

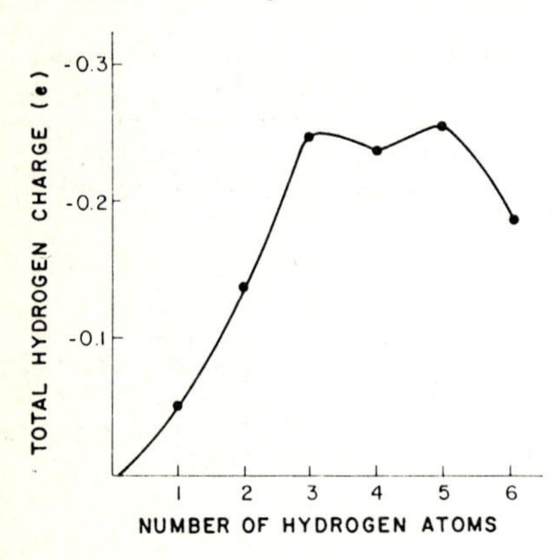

Fig. 3 Total hydrogen charge vs. the number of adsorbed hydrogens

Here it is observed that ΔWF changes almost linearly at small coverage. At coverage greater than $\theta=0.5$, ΔWF remains almost constant, and decreases abruptly to a θ value close to 1. This pattern is in agreement with an experimental work by CHRISTMANN /11/.

3.3 Magnetic Moment

The Ni_{18} cluster has a multiplicity of 21 that corresponds to a magnetic moment of 1.1μB. This result is in agreement with previous calculations of a Ni_{14} cluster /4/ where the magnetic moment was 1.2μB. The experimental value for the bulk is 0.56μB /12/.

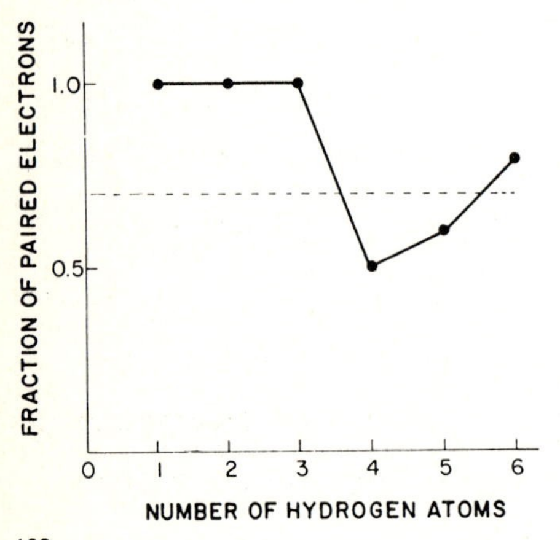

Fig. 4 Fraction of paired electrons vs. the number of adsorbed hydrogens

However, there is experimental and theoretical evidence /13,14/ that the magnetic moment for small clusters is enhanced with respect to the bulk. The fraction of paired electrons is plotted vs. the adsorbed number of hydrogens in Fig. 4 in order to analyze the hydrogen chemisorption effect on the magnetic moment.

The results indicate that at saturation coverage ($\theta \simeq 1$) the fraction of paired electron is about 0.8 which is very close to the experimental value of 0.7 /15/.

4. CONCLUSION

The results of this work demonstrate that a very small model of the surface (Ni_{18} cluster) and a simple method (MINDO/SR) may be used to give a reasonable prediction of the behavior of the change of some surface properties as function of coverage. Nevertheless the changes obtained from these calculations are more pronounced than those observed in actual surfaces.

ACKNOWLEDGEMENTS

The authors would like to express their gratitude to Drs. Claudio Mendoza, Paul H. Bulka, Miguel Luna and Javier González of IBM Venezuela Scientific Center for assistance with computer programs. A computer time grant on the 3081 computer at IBM Venezuela is gratefully acknowledged.

REFERENCES

1. F. Ruette, A. Hernández and E.V. Ludeña: Surface Sci. 151, 103 (1985)
2. F. Ruette, A. Hernández and E.V. Ludeña: Surface Sci. 167, 393 (1986)
3. F. Ruette, A. Hernández and E.V. Ludeña: Int. J. Quantum Chem. 29, 1351 (1986)
4. F. Ruette and G. Blyholder: Theoret. Chim. Acta (Submitted)
5. G. Blyholder, J. Head and F. Ruette: Theoret. Chim. Acta 60, 429 (1982)
6. F. Ruette, G. Blyholder and J. Head: J. Chem. Phys. 80, 2042 (1984)
7. F. Ruette, G. Blyholder and J. Head: Surface Sci. 137, 491 (1984)
8. I. Stensgaard and F. Jakobsen: Phys. Rev. Letters 54, 711 (1985)
9. K. Christmann, O. Schober, G. Ertl and M. Neumann: J. Chem. Phys. 60, 4528 (1974)
10. J. Hölzl and F.K. Schulte: In Springer Tracts Mod. Phys. Ed. G. Höhler, Vol. 85 (Springer-Verlag, Berlin 1979) p. 1
11. K. Christmann: Z. Naturforsch. 34a, 22 (1979)
12. H. Darmon, R. Heer and J.P. Meyer: J. Appl. Phys. 39, 669 (1968)
13. P.A. Montano, H. Purdum, G.K. Shenoy, T.I. Morrison and W. Schulze: Surface Sci. 56, 228 (1985)
14. K. Lee, J. Callaway, K. Kwong, R. Tang and A. Ziegler: Phys. Rev. B 31, 1796 (1985)
15. R.E. Dietz and P.W. Selwood: J. Chem. Phys. 35, 270 (1961)

Theoretical Study of the Iron-Hydrogen Interaction

M. Sánchez[1], *F. Ruette*[2], *and A.J. Hernández*[3]

[1]Instituto Universitario de Tecnología (R.C.), Apdo. 40347, Caracas-Venezuela
[2]Centro de Química IVIC, Apdo. 21827, Caracas 1020-A, Venezuela
[3]Depto. de Química, Universidad Simón Bolívar, Apdo. 80659, Caracas 1080-A, Venezuela

1. INTRODUCTION

Because of its natural abundance and physico-chemical properties iron is very important in metallurgy and plays a relevant role in industrial catalytic hydrogenation processes /1/. However the literature dealing with theoretical calculations and spectroscopic data on iron dihydrides is scarce and there are rather few works concerning the basic chemistry of H_2 dissociation on transition-metal surfaces /2-9/.

In the present work we report MINDO/SR calculations for dihydrogen bound to an iron atom in the range of geometrical possibilities going from the symmetrically-bonded "molecular" coordination mode to the linear "dissociated" H-Fe-H structure. The treatment of this simple system is considered as a preliminary step in the description of the properties of H_2 over a (100) iron surface.

2. METHOD OF CALCULATION

The calculations were done with a semiempirical SCF method which is a modification of MINDO/3 /10/ referred to as MINDO/SR /11/. The modifications include symmetry, selective orbital occupancy and the generation of all singly and doubly-excited configurations out of the Hückel configuration used to obtain an initial density matrix in the SCF cycle. We have introduced this last feature in order to perform a search for the most stable electronic states at any given molecular geometry. This state is maintained along the dissociation reaction coordinate using the calculated Fock matrix at any given point as the initial guess for the next point /12/. The parameters used in this work are those used by Blyholder et al. /13/ in the description of the interaction of atomic hydrogen with iron, where a Fe($3d^7 4s^1$, quintet) ground state was predicted.

3. RESULTS AND DISCUSSION

Calculations with optimization of geometry for a variety of configurations and multiplicities were carried out, using symmetry-adapted wave functions in C_{2v} symmetry. Table 1 shows the electronic configuration, bond distances, charge density, and total

Table 1. Bonding properties for the most stable dissociation states of FeH_2

State	Configuration	Bond Distances (Å) R(Fe-H)	R(H-H)	Charge H_2(e)	Total Energy (a.u.)
1^5B_1	$\alpha: 1a_1 2a_1 1b_2 1a_2 1b_1 3a_1 4a_1$ $\beta: 1a_1 3a_1 1b_1$	2.09	0.77	+0.077	-20.7290
1^3B_1	$\alpha: 1a_1 1b_2 1b_1 1a_2 2a_1 3a_1$ $\beta: 1a_1 2a_1 3a_1 2b_1 1$	1.66	1.14	-0.141	-20.7230
2^5B_1	$\alpha: 1a_1 1b_2 1a_2 2a_1 1b_1 3a_1 2b_1$ $\beta: 1b_1 3a_1 2b_1$	1.50	3.00	-0.175	-20.8005

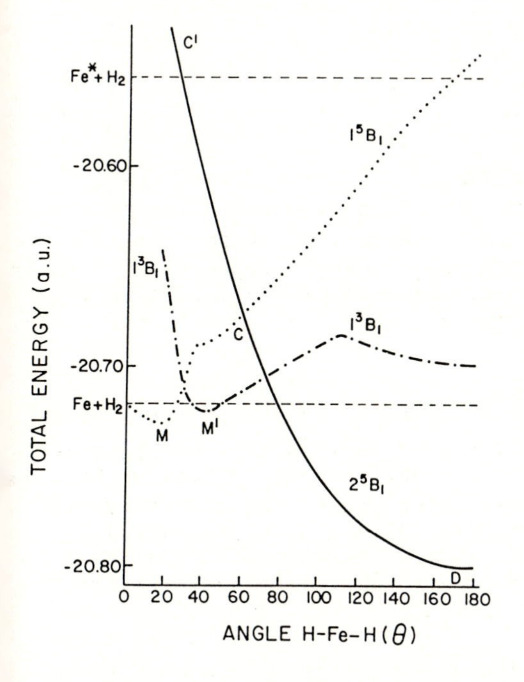

Fig. 1
Dissociation curves for the most stable of FeH_2

energy of the most stable states depicted in Fig. 1 at their optimum dissociation angle Θ (points M,M' and D).

The crossing point between the physisorbed 1^5B_1 state and the chemisorbed 2^5B_1 state (point C in Fig. 1) is located 29 Kcal/mol above the energy of the isolated $Fe(3d^7 4s^1)$ + $H_2(\sigma_g{}^2)$ system, i.e. favoring the H_2 desorption. Experimental results by Ozin et al. /6/ suggested the dissociation of dihydrogen on a previously photoexcited Fe* $(3d^6 4s^1 4p^1)$ atom. Our calculations indicated that the energy of Fe* + H_2 is very close to point C´ in Fig. 1 (representing the vertical excitation from the physisorbed 1^5B_1 state to the chemisorbed 2^5B_1 state), i.e. it favours

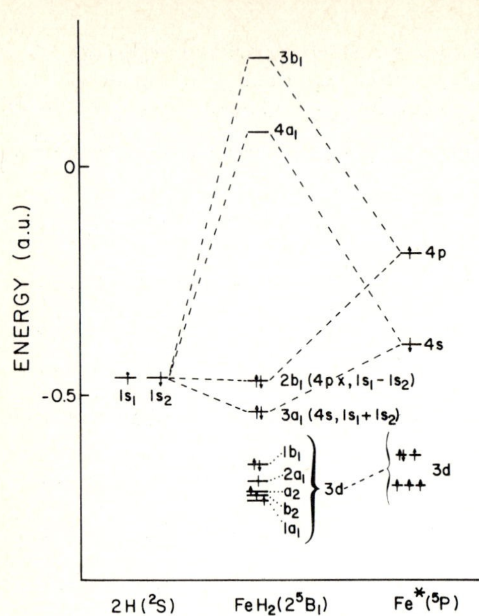

Fig. 2
Molecular orbital diagram for the 2^5B_1 state of H-Fe-H.

the dissociation path M⟶ C´⟶ D proposed by Ozin. The molecular orbital diagram depicted in Fig. 2 shows unambiguously that the chemisorbed $FeH_2(2^5B_1)$ molecular orbitals are correlated to the atomic orbitals of H and Fe*.

The most stable linear geometry of the chemisorbed state found is mainly due to the dissociation behavior of the bonding $2b_1$ orbital, as indicated in previous calculations reported for NiH_2 /4,14/. The matrix isolation experiments of Ozin et al. /6/ indicated a nonlinear chemisorbed state, although this geometry may differ from the one found in the gaseous state /15/.

4. CONCLUSIONS

The MINDO/SR calculation of FeH_2 showed that it was not possible to dissociate H_2 on only one iron atom, i.e., a chemisorption path below the energy of the isolated Fe + H_2 system was not found. The excitation Fe($3d^7 4s^1 4p^0$, quintet) ⟶ Fe*($3d^6 4s^1 4p^1$, quintet) provided a dissociation path in agreement with experimental results. The final $^5B_1$ dissociated state is found to be linear and has a physisorbed state as a precursor.

ACKNOWLEDGEMENTS

The authors express their gratitude to IBM of Venezuela for the computer time granted.

REFERENCES

1. G. Ertl: Catal. Rev. Sci. Eng. 21, 201 (1980)
2. M.R.A. Blomberg and P.E.M. Siegbahn: J. Chem. Phys. 78, 5682 (1983)
3. A. Youakim: Ph. D. Dissertation Southern Illinois University (1981)
4. F. Ruette, G. Blyholder and J. Head: J. Chem. Phys. 80, 2042 (1984)
5. M.P. Guse, R.J. Blint and A.B. Kunz: Int. J. Quantum Chem. 11, 725 (1977)
6. A.G. Ozin and J.G. McGraffrey: J. Phys. Chem. 88, 645 (1984)
7. F. Ruette, A. Hernández and E.V. Ludeña: Surface Sci. 151, 103 (1985)
8. F. Ruette, A. Hernández and E.V. Ludeña: Int. J. Quantum Chem. 29, 1351 (1986)
9. F. Ruette, E.V. Ludeña, A. Hernández and G. Castro: Surface Sci. 167, 393 (1986)
10. G. Blyholder, J. Head and F. Ruette: Theoret. Chim. Acta 60, 429 (1982)
11. R.C. Bingham, M.J.S. Dewar and D. H. Lo: J. Am. Chem. Soc. 97, 1285 (1985)
12. F. Ruette and F.V. Ludeña: J. Catalysis 61, 266 (1981)
13. G. Blyholder, J. Head and F. Ruette: Surface Sci. 131, 403 (1983)
14. J. Demuynck and H.F. Schaefer: J. Chem. Phys. 72, 311 (1980)
15. E.F. Hayes, A.K.Q. Siu and D.W. Kisker: J. Chem. Phys. 59, 4587 (1973)

A CNDO Study of Surface Hydroxyl Groups of γ-Alumina

*L.J. Rodríguez, F. Ruette, E.V. Ludeña, and A.J. Hernández**

Instituto Venezolano de Investigaciones Científicas, IVIC, Apdo. 21827, Caracas 1020-A, Venezuela
*Departamento de Química, Universidad Simón Bolívar, Apdo. 80659, Caracas 1080, Venezuela

1. INTRODUCTION

The wide use of alumina as a support in various types of catalysts has made it important to obtain a detailed knowledge of its surface structure. The surface of aluminas is covered by hydroxyl groups and their number and distribution depend on the temperature of the sample. There exist various crystallographic types of aluminas, among which, the η- and γ-Al_2O_3 are those usually employed in industrial catalysis.

In the present work, the CNDO semi-empirical molecular orbital method was applied in order to calculate the vibrational frequencies of the hydroxyl groups on the γ-alumina surface. Representative models were advanced for the different hydroxyl groups present on the low index planes which are considered as those most likely exposed in the crystallites. For all these models the stretching vibrational frequencies of hydroxyl groups were calculated. As a result of the present calculation we have obtained a series of OH stretching frequencies which are much higher than the experimental one but which lead to a very satisfactory coincidence when scaled with respect to the latter frequencies. This coincidence allows us to tentatively assign the positions of the hydroxyl groups responsible for the observed bands.

2. METHOD AND PARAMETRIZATION

In the present work we have performed molecular orbital calculations using the CNDO/2 method /1/ for several possible structures corresponding to different types of interactions of OH with the γ-Al_2O_3 surface. This method has been widely applied to the study of surfaces /2/. The search for the optimum geometry was accomplished using an energy gradient subroutine and the stretching vibrational OH frequencies were computed for each one of these structures.

As is well known, semi-empirical methods overestimate the force constant and thus lead to poor predictions of the stretching frequencies. In the present situation, however, as the problem at hand is one of assigning a series of frequencies for the same compound when the OH position has varied, all that was required were the relative values of these frequencies and in this case we expected the errors to be minimized. We used the standard CNDO/2 values for the spd orbital exponents and for all other parameters /1,3,4/. The calculations were carried out with an IBM 3081 computer. A GEOMO program /5/, modified by MAYER /6/, was employed. This program was adapted by SCHMIDLING /7/ using PULAY's formalism /8/ for calculating vibrational frequencies. The optimal frequencies were calculated by determining the curvature of the potential well leading to the most stable structures, according to the energy criterion, for the cases considered.

3. RESULTS AND DISCUSSION

The models used for the calculations are depicted in Fig. 1. They were constructed according to KNOZINGER's scheme /9/ in order to simulate the chemical environment present in the aluminas after dehydration.

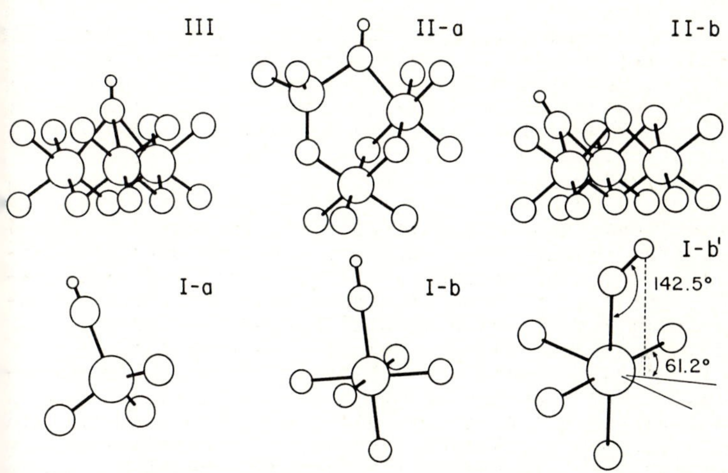

Fig. 1: Models used for the calculation. Plots were made using the PLUTO program (Sam Motherwell, University Chemistry Laboratory, Cambridge, England).

The experimental and calculated frequencies for all of the models are given in Table 1. In this Table, scaled frequencies are also included. These were obtained by matching the extreme calculated frequencies with the corresponding mean values of ex-

Table 1. Comparison of calculated and experimental frequencies for the six OH groups.

GROUP	FREQUENCIES $/cm^{-1}/$ EXPERIMENTAL /a/	CALCULATED	SCALED	CALCULATED OH CHARGE
IIA	3705	5688	3705	-0.1608
III	----	5692	3711	-0.1561
IA	3732	5709	3735	-0.2094
IB'	3742	5714	3743	-0.2959
IB	3770	5737	3774	-0.3317
IIB	3792	5750	3792	-0.1666

(a) Ref. 9

perimental ones and by interpolating the remaining frequencies. This scaling was necessary in view of the fact that the CNDO method overestimates bond orders and yield values which are far too large, for the stretching frequencies.

Clearly, the KNOZINGER structures account for all the possibl ones within the layer. But there still are, however, others whic may occur at the borders or at "steps" on an imperfect surface. At these sites, the coordination of these structures would be incomplete and consequently the OH group may move outward giving rise to distorted configurations. Although we have not analyzed in detail all the possible distorted structures located at the border or "steps", it is easily seen that an incomplete III-type structure would be akin to the II-type structures and that an incomplete II-type structure would give rise, in turn, to a I-type one. The I-type structures could be distorted at these sites due to an incomplete oxygen atom shell encountered about an Al catio For this reason we have investigated structure IB', which is ver similar to IB but which lacks an oxygen atom. Optimization of angles and distances led to the distorted configuration IB' depict in Fig. 1. The calculated OH frequency was 5714 cm^{-1} which when scaled yields the value 3743 cm^{-1}, which as can be seen from Table 1, is in very close agreement with the observed frequency of 3742 cm^{-1}. The calculated frequencies of 3711 cm^{-1} for the III and IIA structures, respectively, are close enough to be classified as only one band. With this provision, it is clear from Table 1 that there exists an excellent accordance between those calculated frequencies and the experimental ones.

In the last column of Table 1 we list the calculated charge o the hydroxyl group for each one of the models, in order to asses whether there exists a relationship between total OH charges and the stretching frequencies. We can observe that for all models considered, except for IIB, there is indeed a correlation betwee increasing negative total charge and frequency. Several calculations were performed for different types of structures corresponding to an OH in a IIB position. The simplest one consisted of an OH group linked to two Al^{+3} cations and the most complicated one was a cluster of 17 atoms. All our calculations led to similar results. Thus, we may conclude that although in some cases the total charge may be related to the stretching frequenc there are other factors which also have a decisive influence. It is important to notice, however, that our calculated total charges do not coincide with KNOZINGER and RATNASAMY assignments

ACKNOWLEDGEMENTS

The authors would like to express their gratitude to Drs. Claud Mendoza, Pablo Bulka, Miguel Luna and Javier González of the IBM Venezuela Scientific Center for their technical assistance. A computer time grant on the 3081 computer at the IBM Venezuela is gratefully acknowledged. L.R. would like to thank "Consejo Nacio nal de Investigaciones Científicas y Tecnológicas", CONICIT, for financial support.

1. J.A. Pople and D.L. Beveridge: In Approximate Molecular Orbi Theory. (McGraw-Hill, New York, 1970)

2. a) R.D. Baetzold: Adv. in Catal. 25, 1 (1976); b) F. Ruette and E.V. Ludeña: J. Catal. 67, 266 (1981); c) W.J. Mortier and P. Geerlings: J. Phys. Chem. 84, 1982 (1980); d) W. Grabowski, M. Misono and Y. Yoneda: J. Catal. 61, 103 (1980); e) S. Beran, J. Dubsky, V. Bosacek and P. Jiru: React. Kinet. Catal. Lett. 13(2), 151 (1980); f) S. Beran and P. Jiru: React. Kinet. Catal. Lett. 17(1), 47 (1981); g) S. Beran: J. Mol. Catal. 10, 177 (1981); h) S. Beran, P. Juri and L. Kebelkova: J. Mol. Catal. 12, 341 (1981) and 16, 299 (1982)
3. J.A. Pople, D.P. Santry and G.A. Segal: J. Chem. Phys. 43, S129 (1965)
4. J.A. Pople and G.A. Segal: J. Chem. Phys. 44, 3289 (1966)
5. D. Rinaldi: "GEOMO" Program System, QCPE 290
6. I. Mayer and M. Révész: Comput. Chem. 6(3), 153 (1982)
7. D.G. Schmidling: J. Mol. Struct. 25, 313 (1975)
8. P. Pulay: Mol. Phys. 17, 197 (1969)
9. H. Knözinger and P. Ratnasamy: Catal. Rev. Sci. Eng. 17(1), 31 (1978)

Theoretical Study of the Oxygen Atom Interaction with an FeSi(100) Surface

L.J. Rodríguez[1], *F. Ruette*[1], *E.V. Ludeña*[1], *G.R. Castro*[1,2], *and A. Hernández*[1,3]

[1]Centro de Química, Instituto Venezolano de Investigaciones Científicas, IVIC, Apdo. 21827, Caracas 1020-A, Venezuela
[2]Departamento de Física, Universidad Central de Venezuela, Apdo. 2120, Caracas, Venezuela
[3]Departamento de Química, Universidad Simón Bolivar, Apdo. 80659, Caracas 1080-A, Venezuela

1. INTRODUCTION

Transition metal silicides have currently been the object of considerable interest due to their important applications in semi-conductor technology /1/. The oxidation behavior of metal/Si alloys has be studied by means of photoelectron spectroscopy techniques in order tc attain a detailed understanding of the nature of oxygen-alloy inter action /2/.

In the present work, MINDO/SR /3/ calculations of the interacti of atomic oxygen with the FeSi(100) surface represented by a 12-irc 12-silicon cluster are reported. The atomic parameters for Si, Fe and O and the Fe-Si bonding parameters have been discussed in a pre vious work /4/. The Si-O (α = 0.70137, β = 0.52852) and Fe-O (α = 1.75 β = 1.06944) bonding parameters were chosen to reproduce the equilibrium bond lengths and bond energies of the SiO and FeO molecules /5/.

The B-20 cubic structure of iron silicide has been reported el where /6/. For the present calculations we have modeled the (100) face of a single FeSi crystal by the central-iron $Fe_{12}Si_{12}$ cluster de picted in Fig. 1a and by the central-silicon $Si_{12}Fe_{12}$ cluster depict ed in Fig. 1b, where only the top layer atoms are drawn. Because of the complicated structure of the models chosen we present in Fig. 2a-2c the zOy, zOx and yOx projections, respectively, of the central-iron $Fe_{12}Si_{12}$ cluster. Calculations were performed for a large number of adsorption sites of which only the most stable ones are include in Fig. 1. The 24-atoms cluster was assumed to be large enough to repr ent several types of oxygen adsorption sites on a FeSi(100) surface, as well as small enough to enable such calculations to be carried out in reasonable amount of time.

2. CALCULATIONS

A summary of the most stable adsorption energies, charges and bon distances for the oxygen-cluster interactions is given in Table 1. For these calculations, the Fe and Si positions were taken to be those of the bulk material /4c/ and the X-O bond distances (with X = Fe or Si) and the total spin multiplicity of the cluster were optimized.

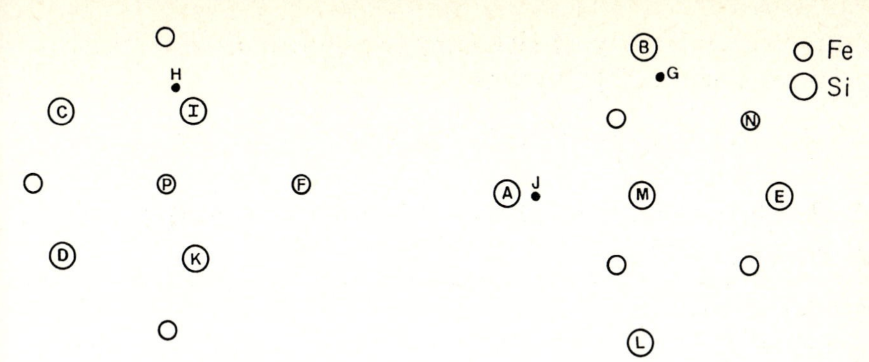

Fig. 1a. Top layer atoms of central-iron Fe_{12} Si_{12} cluster

Fig. 1b. Top layer atoms of central-silicon Si_{12} Fe_{12} cluster

TABLE 1. STABLE OXYGEN ADSORPTION SITES

SITES	ADSORPTION ENERGY (kcal/mol)*	CHARGE O ATOM	O CLOSEST NEIGHBOR X	BOND DISTANCE X-O (Å)
A	-104.22	6.52	Si	1.58
B	-97.29	6.54	Si	1.58
C	-94.78	6.54	Si	1.60
D	-58.92	6.52	Si	1.60
E	-42.81	6.52	Si	1.68
F	-40.39	6.18	Fe	1.62
G	-39.04	6.40	Si	1.78
H	-36.57	6.51	Si	1.68
I	-26.94	6.48	Si	1.70
J	-25.37	6.37	Si	1.77
K	-18.86	6.50	Si	1.70
L	-18.68	6.49	Si	1.68
M	-15.54	6.60	Si	1.68
N	-15.41	6.24	Fe	1.90
O	+25.78	6.20	Fe	2.16

(*) With respect to the isolated cluster and oxygen atom

It is important to point out the following features:

a) The most stable adsorption sites shown in Table 1 are "on top" sites located directly above a surface cluster atom. Only the sites G, H and J are in bridge positions closer to an edge Si atom.
b) The first most stable adsorption sites (A-E in Table 1) are all located at "on-top" Si positions belonging to the edge of the cluster.
c) The site F is located directly above a Fe atom on the edge of the Fe Si cluster, and it is about 64 Kcal/mol less stable than the site A.
d) The oxygen atom has a negative charge in all sites shown in Table 1 and this charge is larger when oxygen is closer to a Si atom.
e) In general, the bond distance between oxygen and its closest neighbor Si atom is shorter than in the analogous cases involving a Fe atom in the same cluster.

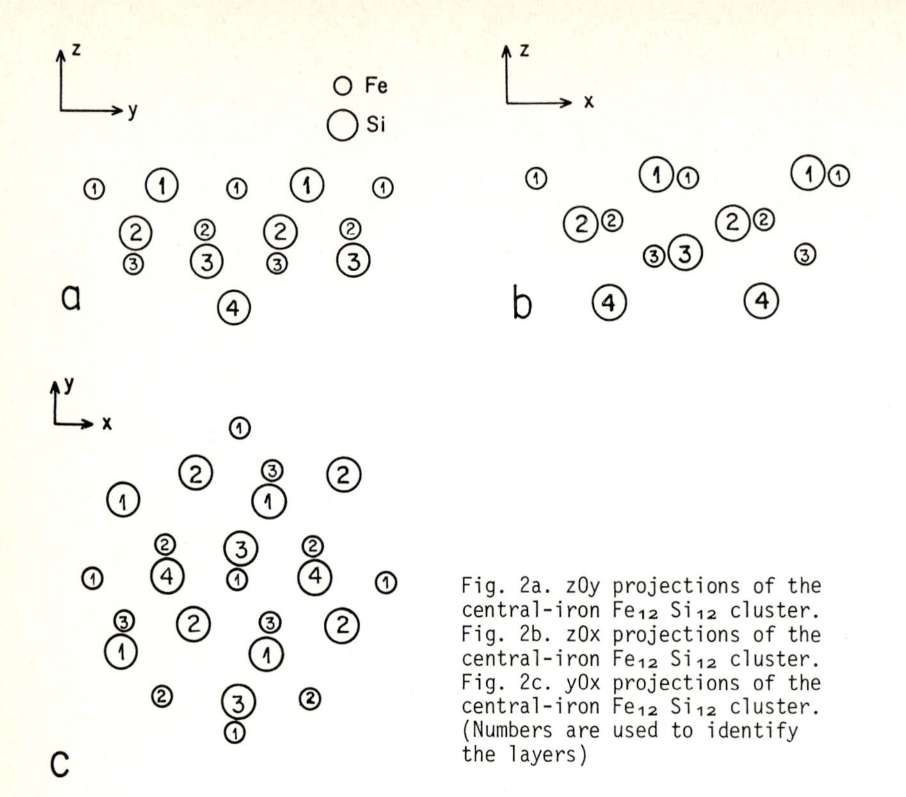

Fig. 2a. zOy projections of the central-iron Fe_{12} Si_{12} cluster.
Fig. 2b. zOx projections of the central-iron Fe_{12} Si_{12} cluster.
Fig. 2c. yOx projections of the central-iron Fe_{12} Si_{12} cluster.
(Numbers are used to identify the layers)

The larger thermodynamics stability calculated for the adatom on top of the edge atom, although in agreement with previous calculations on the adsorption of hydrogen on Ni surfaces /7/, may not be a representative situation found in real surfaces. The P and M adsorption sites represent more accurately the coordination expected. Our results indicate that the oxygen adsorption is stable on site M on top of the central Si atom, while it is unstable on site P on top of the central Fe atom.

It is interesting to notice that in addition to our prediction that oxygen atoms preferably chemisorb at Si sites, it has been experimentally observed that in the early stages of the subsequent oxidation process it is the Si that oxidizes, rather than Fe, in accordance with thermodynamics predictions.

ACKNOWLEDGEMENTS

The authors would like to express their gratitude to Drs. Claudio Mendoza, Pablo Bulka, Miguel Luna and Javier Gonzalez of the IBM Venezuela Scientific Center for assistance with computer programs. A computer time grant on the 3081 computer at IBM Venezuela is gratefully acknowledged. L. R. Would like to thank the Consejo Nacional de Investigaciones Científicas y Tecnológicas, CONICIT, for financial support.

REFERENCES

1. J.M.Poate and R.C. Dynes: IEEE Spectrum 38, February, 1986
2. G.R. Castro, J.E. Hulse, J. Kuppers and A. Rodriguez Gonzalez-Elipe: Surf. Science 117,621 (1982)
3. G. Blyholder, J. Head and F. Ruette: Theoret.Chim.Acta 60,429(1982)
4. a) R.C. Binghan, J.S. Dewar and D.H. Lo: JACS 97(6),1285 (1975). b) G. Blyholder, J. Head and F. Ruette: Surf.Science 131,403(1983). c) L.J. Rodriguez, F. Ruette, E.V. Ludena, G.R. Castro and A.J. Hernandez: Surf.Science (in press)
5. G. Herzberg:"SPECTRA OF DIATOMIC MOLECULES", Van Nostrand Reinhold, USA, 1950
6. L. Pauling and M. Soldate: Acta Crys. 1,212(1948)
7. F. Ruette, A. Hernandez and E.V. Ludeña: Surf.Science 151,103(1985).

Water Adsorption on Metal Surfaces. Theory of the Work Function Changes

E.E. Mola and J.L. Vicente

Instituto de Investigaciones Fisicoquímicas Teóricas y Aplicadas (INIFTA), Casilla de Correo 16, Sucursal 4, 1900 La Plata, Argentina

A self-consistent model for the interaction of a monolayer of water with a metal surface is presented. The model describes the lowering of work function values experimentally measured when water is adsorbed on different metals at low temperatures. It also shows that multilayer water adsorption has no further effect on the electronic structure of the metal.

1. Introduction

Several attempts have been made to study theoretically the interaction of water molecules with a metal surface. Badiali et al. /1/ employed the continuous film model to describe the adlayer of water, following the viewpoint proposed by Lang /2/ to describe the adsorption of a layer of alkali atoms. In that paper the metal electron density is described by a very simple trial function containing a couple of unknown parameters. The Friedel-type oscillations which must exist at the surface are not included, arguing that for metals of relatively large electron density these oscillations are of very small amplitude.

In order to describe the interaction of an adlayer of water with a metal surface we present a model where the electron density profile is obtained by solving self-consistently a Schrödinger equation, avoiding in this way the use of approximate trial density functions. The electronic distribution obtained exhibits the correct profile with the Friedel oscillations. This model calculation shows how the amplitude of these oscillations affects the spill-over of the electrons into the adlayer.

2. Experimental Studies

Bujor and coworkers /3/ studied water vapour adsorption on the Cu(110) face by Auger Electron Spectroscopy (AES) and work function change measurement. L. Bange et al. /4/, using ultraviolet photoelectron spectroscopy, found molecular adsorption of water on a clean Cu(110) surface. Mariani and Horn /5/ studied the orientation of water molecules on Cu(110) by angle-resolved photoemission with polarized light. Klaua and Madey /6/, using ESDIAD (Electron Stimulated Desorption Ion Angular Distribution) and TDS (Thermal Desorption Spectroscopy) studied the adsorption of water on Ag(001) and Ag(111). Netzer and Madey /7/ also studied the adsorption of water on Al(111) by ESDIAD, LEED, AES and thermal desorption. A similar desorption temperature has been observed by one of the authors /8/ when H_2O is thermally desorbed from polycrystalline gold films. Heras et al. /9/ reported some results on the adsorption and decomposition of water on various metal films (Fe, Co, Ni, Cu, Pt and Au).

From all these studies it clearly emerges that, at low temperatures, water is molecularly adsorbed, causing a large decrease in work function on all these metals.

3. Model Description

To study the interaction between a metal and an adsorbate one can consider the adlayer as a continuous film covering the surface /1/. This viewpoint can be useful to describe water vapour adsorption on clean metal surfaces when surface saturation is reached, or, to describe the interaction of water molecules with metal surfaces in electrochemical systems. Therefore, it seems reasonable to consider that, in the absence of specific adsorption, the coverage by water can be represented by a continuous film. Its response to charge is characterized by a dielectric constant, K, as a function of the field strength, following the Booth model /10/, that accounts for the reduction due to the saturation effect. The metal is represented by a slab, whose dimensions are very large compared with the thickness L. The planar uniform-background is used, the positive metal ions are smeared out into a constant positive charge density $n_+(\underline{r})$ -jellium- with $n_+(\underline{r}) = (4/3\pi r_s^3)^{-1}$ if $|z| < \frac{1}{2}$ L and $n_+(\underline{r}) = 0$ elsewhere (z is the coordinate perpendicular to the faces and r_s is the Wigner-Seitz radius). More details of the model employed to describe the metal were recently published /11/. The ground-state electron density $n(\underline{r})$ of the system of M interacting electrons in an external potential can be found by solving self-consistently the one-particle Schrödinger-type equation

$$[-\tfrac{1}{2}\nabla^2 + \phi(\underline{r}) + v_{xc}(\underline{r})]\psi_j(\underline{r}) = \varepsilon_j\psi_j(\underline{r}) \tag{1}$$

and

$$n(\underline{r}) = \sum_{j=1}^{M} |\psi_j(\underline{r})|^2 , \tag{2}$$

where the ψ_j are the M lowest-lying orthonormal solutions of (2). ϕ is the electrostatic potential and will satisfy the Poisson type equation

$$\frac{d^2\phi}{dz^2} = -(4\pi/K)[n_+(z) - n(z)] , \tag{3}$$

where K, as a function of field strength, is obtained, inside the dielectric region, from the Booth model /10/, see Fig. 1. The charge density n_+, and therefore the associated density n, depend solely on z. v_{xc} is the exchange-correlation potential written as

$$v_{xc}(z) = -6\alpha[3n(z)/\pi]^{1/3} \tag{4}$$

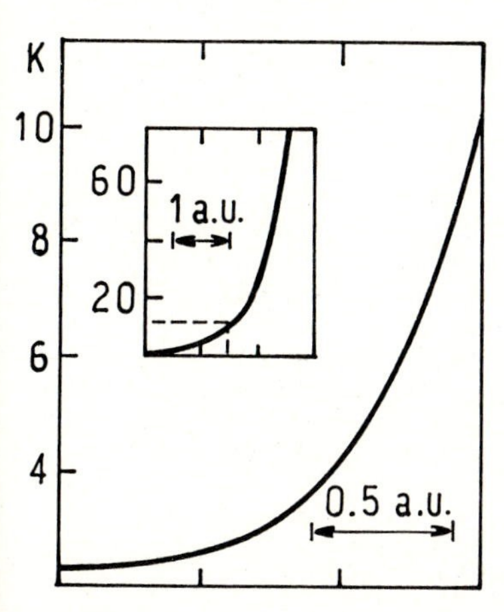

Figure 1. Dielectric constant K vs distance to the jellium edge. The field is first calculated from the electrostatic potential and then K is obtained through the Kirkwood method at 77°K

with a parameter α. The equations are solved in the following way: (i) An initial electron density n(z) and K ≡ 1 (vacuum condition) is chosen; (ii) from (3) and (4) we obtain ϕ and v_{xc}; (iii) the new n(z) and K(z) is calculated and then steps (ii) and (iii) are repeated until self-consistency is reached.

4. Comparison between Experimental and Numerical Results

Work function values were derived for a clean surface and for a surface covered by water molecules: the presence of water reduces the work function of the metal. From Table I we learn that for a given metal electron density, the adlayer influence on the work function increases in those metals where the exchange and correlations effects are smaller. This may be due to the fact that the greater the exchange and correlation effects the smaller is the spill-over of the electrons outside the jellium edge. Figure 2 shows that there is an increase of the electron density in the region, mainly due to the reduction of Coulomb interactions.

Table I. Clean surface work function (Φ) and the decrease ($\Delta\Phi$) vs parameter α

α	0.22	0.24	0.26	0.28
Φ (eV)	3.732	4.171	4.640	5.093
$\Delta\Phi$(eV)	-0.611	-0.569	-0.528	-0.490

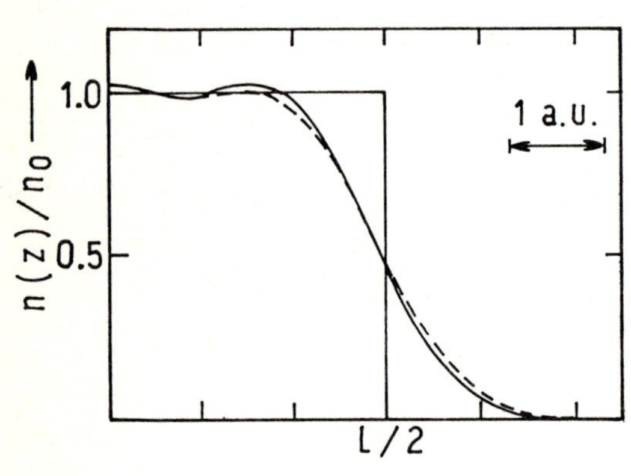

Figure 2. Electron density profile for the bare surface and for the surface in contact with water. (——) bare surface; (---) surface in contact with water

It is a very well-known fact from experimental studies that the work function decrease upon adsorption of water at low temperatures (77°K) ceases when a complete monolayer of water molecules is adsorbed on the metal surface. Further adsorption of water has no effect on the work function value. The model shows that a dielectric layer thicker than the electron density spill-over (i.e. a monolayer of admolecules of water) should not cause further decrease of the work function. In other words, this fact confirms the well-known metal "blindness" to multilayer water adsorption. Uniform values of the dielectric constant K, as those proposed by Bockris et al. /12/, Trasatti /13/ and Mola et al. /14/, capable of reproducing experimental results in connection with double layer problems, should be compared with a weighted average of K(z) with $n(z)/n_+$ as a normalized weight function.

Table II. Experimental and theoretical changes of work function at 77°K in eV

	Al	Zn	Cu	Au	Ag	Hg
$\Delta\Phi_{exp}$	-1.0 /15/	--	-0.73 /9/	-0.60 /8,9/	--	--
$\Delta\Phi_{theor}$	-1.19	-1.03	-0.78	-0.54	-0.56	-0.82

The self-consistent treatment of the electrostatic potential makes the model suitable for use in more general problems such as, (i) the influence of charged planes in the vicinity of the metal surface on its electronic structure and Fermi level and (ii) the contribution of the metal to the double layer capacitance in a metal-electrolyte interface.

Acknowledgments

This work was financially supported by the CONICET, R. Argentina. INIFTA is an Institute depending on the Facultad de Ciencias Exactas, UNLP.

References

1. J. Badiali, M. Rosinberg and J. Goodisman, J.Electroanal.Chem. 130, 31 (1981)
2. N. Lang, Phys.Rev. B4, 4234 (1971)
3. Phy Sy Uy, J. Bardolle and M. Bujor, Surface Sci. 129, 219 (1983)
4. K. Bange, D. Grider and J.K. Sass, Surface Sci. 126, 437 (1983)
5. C. Mariani and K. Horn, Surface Sci. 126, 279 (1983)
6. M. Klaua and T. Madey, Surface Sci. 136, L42 (1984)
7. F. Netzer and T. Madey, Surface Sci. 127, L102 (1983)
8. E. Mola, Thesis, Universidad Nacional de La Plata, Buenos Aires (1973)
9. J. Heras and E. Albano: in Proc.4th Inter.Conf. on Solid Surfaces, Cannes, 1980, p. 263; and J. Heras and L. Viscido, Appl.Surface Sci. 4, 238 (1980)
10. F. Booth, J.Chem.Phys. 19, 391 (1951)
11. E. Mola and J. Vicente, J.Chem.Phys. 84, 2876 (1986)
12. T. Andersen and J. Bockris, Electrochim.Acta 9, 347 (1964)
13. S. Trasatti, J.Electroanal.Chem. 150, 1 (1983)
14. E. Mola and J. Vicente, Surface Sci., in press
15. R. Eastment and C. Mee, J.Phys. F3, 1738 (1973)

Theoretical Study of the Electronic Structure of FeSi

L.J. Rodríguez[1], *F. Ruette*[1], *E.V. Ludeña*[1], *G.R. Castro*[1,2], *and A.J. Hernández*[1,3]

[1]Centro de Química, Instituto Venezolano de Investigaciones Científicas, IVIC, Apdo. 21827, Caracas 1020-A, Venezuela
[2]Departamento de Física, Universidad Central de Venezuela, Apdo. 2120, Caracas, Venezuela
[3]Departamento de Química, Universidad Simón Bolívar, Apdo. 80659, Caracas 1080-A, Venezuela

1.- INTRODUCTION

Transition metal silicides prepared by depositing metal films on silicon have been broadly studied in the past ten years using photoelectron spectroscopy. Even though much experimental and theoretical work has been devoted to the study of silicides of Pt, Pd, Au, Ag /1/ and to a lesser degree to the silicides of Ni, Co and Fe /2/, there is still much uncertainty concerning the detailed nature of the metal-silicon interaction.

In the present work we report calculations on the electronic structure of a 7-iron-7-silicon atom cluster using the MINDO/SR method which represnts a modification of the QCPE-290 algorithm described by RINALDI /3/ to include transition metal atoms /4/. The atomic parameters of Fe were reported in a previous work dealing with the band structure of iron /5/. The Fe-Si bonding parameters α_{Fe-Si} =3.00175 and β_{Fe-Si} =0.67739 were chosen to reproduce the equilibrium bond length and bond energy of the diatomic FeSi molecule /6/. The density of state (DOS) for the sp- and d-bands were evaluated from the MINDO/SR eigenvalues by means of the Gaussian approximation described by SIMONETTA and GAVEZZOTTI /7/.

Iron silicide has a B-20 cubic structure with four Fe atoms in positions (X,X,X; X+1/2, 1/2-X,$\bar{X}$;...) and four Si atoms in equivalent ones in the unit cell (a=4.489 Å; X_{Fe} =0.137 and X_{Si} =0.842) /8/. Here, every atom is bonded to seven different ones in the first coordination sphere and to six atoms of the same kind in the second one. Fig. 1a depicts the 14 atom cluster used in this work to represent the bulk of FeSi, where the central atom can be taken either as iron (Fe_7Si_7) or as silicon (Si_7Fe_7). The DOS obtained from these two bulk models of FeSi are compared with those computed from the Si_{17} cluster model depicted in Fig. 1b.

2. CALCULATIONS

The calculated bond orders and diatomic energies with respect to the central Si atom and corresponding orbital occupancies of all atoms represented in the central-silicon (Si_7Fe_7) bulk cluster are given in Table 1.

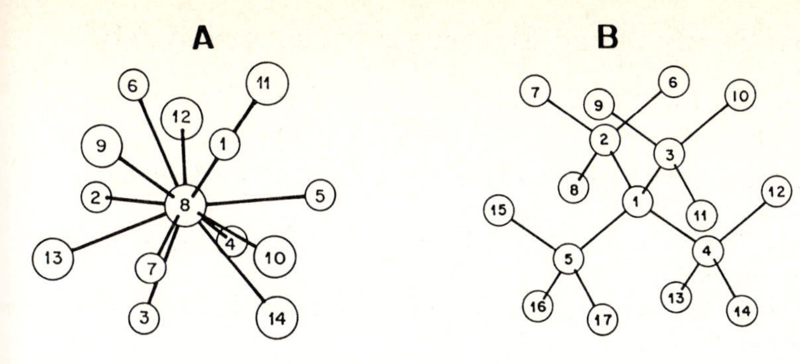

Fig. 1. Models used for the calculation: a) Si_7Fe_7 and b) Si_{17} Plots were made using the PLUTO program (Sam Motherwell, University Chemistry Laboratory, Cambridge, England)

Table 1. BONDING PROPERTIES AND ORBITAL OCCUPANCIES IN Si_7Fe_7

Si-X*	Bond orders sp-sp	sp-d	Total	Diatomic Energies	Orbital sp	Occupancies d	total
8-1	0.9210	0.0044	0.9254	-0.2427	1.6606	6.9881	8.6487
8-2	0.9532	0.0039	0.9571	-0.2119	1.4388	6.9950	8.4338
8-3	0.9532	0.0039	0.9571	-0.2119	1.4388	6.9950	8.4338
8-4	0.9532	0.0039	0.9571	-0.2119	1.4388	6.9950	8.4388
8-5	0.6679	0.0022	0.6701	-0.2108	1.6437	6.9940	8.6377
8-6	0.6679	0.0022	0.6701	-0.2108	1.6437	6.9940	8.6377
8-7	0.6679	0.0022	0.6701	-0.2108	1.6437	6.9940	8.6377
8(Si)	-	-	-	-	2.7440	-	2.7440
8-9	0.4460	-	0.4460	-0.0070	3.4874	-	3.4874
8-10	0.4460	-	0.4460	-0.0070	3.4874	-	3.4874
8-11	0.4460	-	0.4460	-0.0070	3.4874	-	3.4874
8-12	0.4516	-	0.4516	-0.0409	3.6435	-	3.6435
8-13	0.4516	-	0.4516	-0.0409	3.6435	-	3.6435
8-14	0.4516	-	0.4516	-0.0409	3.6435	-	3.6435

* (X= 1-7, first neighbor Fe-atoms; X= 9-14, second neighbor Si-atoms)

The diatomic energy entries in this Table show unambiguously that the central Si atom is mainly bonded to seven first-neighbor Fe-atoms. The bond orders displayed in Table 1 reveal the Si-Fe bonding in Si_7Fe_7 is of the sp-sp type with negligible participation of the Fe(3d) orbitals. The orbital populations give an average $Fe\left[(4s4p)^{1.5}\ (3d)^{7.0}\right]$ valence electronic configuration which confirms the negligible participation of 3d orbital and shows electronic transfer from Si(sp) to Fe(sp). A 20% expansion of the 3d orbitals did not change the above results significantly i.e., any participation of the 3d orbitals of Fe may be discarded as opposed to the experimental evidence accumulated in the Pt and Pd silicides /9/.

The calculated total DOS of Si_7Fe_7 and the corresponding sp and d contributions are depicted in Fig. 2a. The d-band is well localized around -8.0 eV compared to the sp-band which spans a / larger

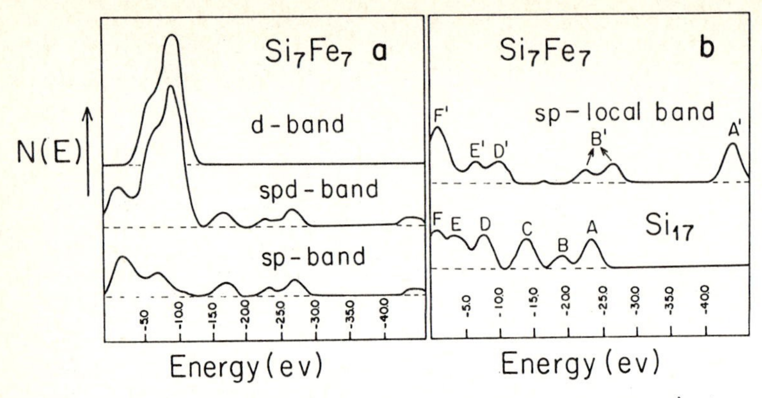

Fig. 2. Total DOS calculated from MINDO/SR of: a). d-, spd-, and sp-bands in Si_7Fe_7 and b) central-silicon local band for Si_7Fe_7 and Si_{17}

energy region and overlaps the d-band in the neighborhood of the Fermi level. A molecular orbital analysis of the band structure in Fig. 2a. reveals that the d- and sp-bands are noninteracting, as given by the bonding properties displayed in Table 1.

The sp-DOS localized on the central Si atom of Si_7Fe_7 is depicted in Fig. 2b, where the sp-band confined on the central Si atom of Si_{17} is also included for comparison. Several interesting features in the local sp-band are described as follows: a) the peak marked as A in Si_{17} is stabilized by a Si(3s)-Fe(4s) bonding interaction in Si_7Fe_7 giving rise to the peak marked as A; the antibonding peak C is destabilized and transferred above the Fermi level, b) The peaks marked B, D, E, F in Si and B', D', E', F' in Si_7Fe_7 are all equivalent with respect to the sp-sp interaction between the central Si atoms and first neighbors. The slight stabilization observed for the primed peaks is due to the smaller electronic population for Si in Si_7Fe_7 as compared to Si_{17} (see Table 1) which reduces the electron-electron repulsion inside the sp-band, c) the splitting of the peak marked as B' is due to the change in coordination of the central Si atom in going from Si_{17} to Si_7Fe_7 , d) the peaks marked as A, B, A' and B' are essentially of s character while the peaks marked as D, E, F, D', E', and F' are mainly pure p-bands. This is in agreement with the experimentally observed slight change of LVV Auger line of Si in going from pure silicon to iron silicide /10/. Similar results were also found for the Fe_7Si_7 cluster with a central Fe-atom.

In conclusion, the nature of the iron-silicon interaction in iron silicide is of the sp-sp type with negligible participation of the Fe(3d) orbitals. A small charge transfer from Si to the metal atom, is also observed. This is in agreement with previous results on SiPt /9/.

ACKNOWLEDGEMENTS

The authors would like to express their gratitude to Drs. Claudio Mendoza, Pablo Bulka, Miguel Luna and Javier Gonzalez of the IBM Venezuela Scientific Center for assistance with computer programs. A computer time grant on 3081 computer at IBM Venezuela is gratefully acknowledged. L.R. would like to thank "Consejo Nacional de Investigaciones Cientificas y Tecnologicas", CONICIT, for financial support.

REFERENCES

1. a) L. Braicovich, I. Abbati, J.N. Miller, I. Lindan, S. Schwarz, P.R. Skeath, C.Y. Su and W.E. Spicer: J. Vac. Sci. Technol. 17 (5), 1005 (1980); b) J.N. Miller, S.A. Schwarz, I. Lindan, W.E. Spicer, B. De Michelis, I. Abbati and L. Braicovich: J. Vac. Sci. Technol. 17(5), 920 (1980); c) P.S. Ho, G.W. Rubloff, J.E. Lewis, V.L. Moruzzi and A.R. Williams: Phys. Rev. B 22(10), 4784 (1980)
2. a) G.R. Castro, J.E. Hulse, J. Kuppers and A. Rodriguez Gonzalez-Elipe: Surface Sci. 117, 621 (1982); b) B. Egert and G. Panzner: Phys. Rev. B 29(4), 2091 (1984); c) G.R. Castro and I. Escalona: In Lectures on Surface Physics ed. by M. Cardona and G.R. Castro, (Springer, Berlin, Heidelberg 1986).
3. D. Rinaldi: "GEOMO" Program System, QCPE 290
4. G. Blyholder, J. Head and F. Ruette: Theoret. Chim. Acta, 60, 429 (1982)
5. G. Blyholder, J. Head and F. Ruette: Surface Sci. 131, 403 (1983)
6. K.A. Gingerich: In Current Topics in Materials Science, ed. by E. Kalsdis, Vol. 6 (North-Holland Publishing Co. 1980)
7. M. Simonetta and A. Gavezzotti: In Advances in Quantum Chemistry, Vol. 12 Ed. by P. Lowdin (Academic Press, New York, 1980) p. 103
8. L. Pauling and M Soldate: Acta Cryst. 1, 212 (1984)
9. I. Abbati, L. Braicovich, B. De Michelis, O. Bisi, R. Rovetta: Solid State Comm. 37, 119 (1981)
10. Ref. 2-b and 2-c

The Effect of Particle Diffusion on the Insulator-Conductor Transition in Adsorbed Monolayers

H.O. Mártin[1], *E.V. Albano*[2], *and A. Maltz*[3]

[1]Laboratorio de Física Teórica, Facultad Ciencias Exactas, UNLP, C.C. 67, 1900 La Plata, Argentina
[2]Instituto de Investigaciones Fisicoquímicas Teóricas y Aplicadas (INIFTA), Facultad Ciencias Exactas, UNLP, C.C. 16, Sucursal 4, 1900 La Plata, Argentina
[3]Departamento de Matemáticas, Facultad Ciencias Exactas, UNLP, C.C. 172, 1900 La Plata, Argentina

1.- Introduction

Let us briefly review insulator-conductor transition experiments. A fraction Φ of metal atoms is deposited onto an insulator or semiconductor substrate between two electrodes. Initially, Φ is small and the film is made up of many physically disconnected islands. When Φ increases the isolated islands grow in size and some of them get, for the first time, in mutual contact forming a continuous film. At this critical concentration Φ_c, the electrical resistivity between the electrodes abruptly decreases /1/.

Theoretically, the insulator-conductor transition has been analysed with the aid of the Standard Site Percolation Model (henceforth SSPM) where the condensed metal atoms on the substrate are represented by occupied sites on a two-dimensional lattice. Each site is present (absent) with probability P (1-P) independent of whether the other sites are occupied or empty. A cluster is a group of occupied sites connected by nearest-neighbour (n-n) distance. For $P > P_c$ (P_c is the critical probability) one percolating infinite cluster is formed in the infinite lattice (see e.g. /2/). In the SSPM, $\Phi_c = P_c$ but this model does not consider the mobility of the atoms as well as the lateral interactions between them. Since the properties of very thin films are known to be dependent on the diffusion of interactive particles, we shall study two models (Models A and B, defined below) where this effect is taken into account.

2.- The Models

In the models the surface where the diffusion of particles takes place is represented by a regular two-dimensional lattice. Each site of the array can be either occupied by only one particle or empty. When n-n lateral interactions between particles are taken into account, the probability per unit of time $P_{ij}(z')$ that a given i particle, with z' n-n occupied sites, jumps to one given j empty site is /3/

$$P_{ij}(z') = \mu \exp[-(E + z'\omega)/kT] \quad , \tag{1}$$

where E, k, T and ω are the activation energy of diffusion of an isolated particle, the Boltzmann constant, the temperature and the interaction energy ($\omega > 0$ attractive, $\omega < 0$ repulsive) between two n-n particles respectively. The pre-exponential μ is usually assumed to be a constant.

Considering (1), the rules for the models A and B are: i) Initially, for both models, the lattice is randomly covered with probability Φ as in the SSPM. This can be thought of as the condensation of incident atoms into a monolayer at very low temperature. ii) After that, the diffusion starts. In the Monte Carlo simulation, at each time step, the jumping probability P_{Aij} and P_{Bij} for the models A and B, respectively, are:

$$P_{Aij}(z,z')=\begin{cases}(zN)^{-1}\exp(-z'\omega/kT) & , \ \omega > 0 \\ (zN)^{-1}\exp[(z-1-z')\omega/kT] & , \ \omega < 0\end{cases} \qquad (2)$$

$$P_{Bij}(z,z')=\begin{cases}(zN)^{-1} & , \ z' = 0 \\ (zN)^{-1}\exp(-\tilde{\omega}/kT) & , \ z' \geq 1, \ \tilde{\omega} > 0 \ ,\end{cases} \qquad (3)$$

where N is the number of particles on the lattice and z the coordination number of the lattice. The normalization factor $(Nz)^{-1}$ means that each time step, an i particle and one of its z n-n sites are randomly chosen. If the selected n-n site is already occupied, the i particle does not move. Due to the interactions, the largest probability P_{Aij} results when z' = 0 (z'=z-1) for the attractive (repulsive) case. As we will work with fixed values of ω/kT, (2) is equivalent to (1) after an appropriate change of time scale. Note that in the model B only the attractive case is studied and that $\tilde{\omega}$ is an effective interaction energy which does not depend on the number z' of n-n occupied sites (see (1)). Thus, the model B is a simplified version of model A, but it includes the most important effects when only the attractive interaction is considered. The main difference of the models from the "static" SSPM (see Introduction) is to introduce the mobility of particles with n-n interactions. The behaviour of the models has been studied using Monte Carlo simulations on LxL square lattices ($L \leq 251$) with periodic boundary conditions. When the system has reached the steady state, instantaneous configurations (IC) are analysed every 2N effective movements (EM) (one EM takes place when one particle actually jumps from one site to another one). The data for each value of Φ, L, ω/kT, ($\tilde{\omega}/kT$) have been obtained by averaging over 30-360 IC, starting with 2-36 different initial configurations.

3.- Results

3.1.- The Critical Concentration

On an LxL lattice, a percolating cluster is a cluster which has either its length or its width (or both) equal to L. Let F_L be the fraction of percolating clusters of many configurations. Fig. 1 shows F_{LA} (F_{LS}) for the model A (SSPM) versus Φ (P). Due to the attractive (repulsive) interaction, F_{LA} becomes shifted from F_{LS} to the left (right).

In the thermodynamic limit ($L = \infty$), let us denote by Φ_c the critical concentration for the diffusion models (i.e. only for $\Phi > \Phi_c$ an infinite percolating cluster appears). On the other hand, in finite lattices, one can define the L-dependent threshold Φ_L demanding that $F_L(\Phi_L) = 0.9$. Using the finite-size scaling arguments one can obtain Φ_c extrapolating Φ_L versus $L^{-1/\nu}$, where ν is the correlation length exponent ($\nu = 4/3$ as in the SSPM). Fig. 2 shows the results obtained working with $31 \leq L \leq 251$.

Note that for model A, Φ_c decreases continuously from $\Phi_c \cong 0.661$ to $\Phi_c \cong 0.512$ when ω/kT increases from -2.0 to 2.0. For model B, Φ_c against $\tilde{\omega}/kT$ has also a similar behaviour but the difference with P_c is smaller than in model A. Our results clearly suggest that the insulator-conductor threshold should

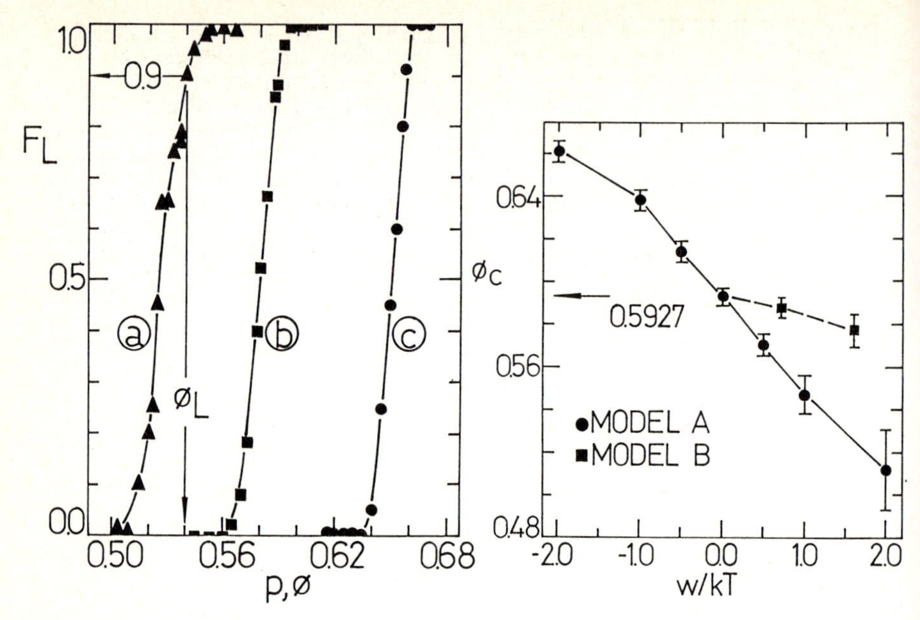

Figure 1. The fraction F_L ($L = 201$) of percolating clusters. (a) and (c) F_{LA} for model A against Φ for $\omega/kT = 1.0$ and $\omega/kT = -2.0$, respectively. (b) F_{LS} for the SSPM versus P (these results agree with those of F_{LA} for $\omega/kT = 0$). The method employed in the determination of Φ_L (see the text) is shown in (a).

Figure 2. The critical concentratio Φ_c for the infinite system versus ω/kT ($\tilde{\omega}/kT$) for model A (model B). The error bars take the uncertainties in the determination of each Φ_L into account. The arrow indicates the best available value of P_c for the SSPM on a square lattice.

depend on both the interaction between condensed particles and the temperatur Due to this behaviour, it should be expected that for strongly repulsive systems the competition between the available room on the lattice and the dissociative movement of particles would cause the stabilization of Φ_c at a certain maximum value Φ_{cMax}. For attractive systems the fact that Φ_c decrease when ω/kT is increased suggests the competition between the available number of particles on the lattice and the associative effect among clusters which are bridged by sticking particles. This situation should cause the stabilization of Φ_c at a certain minimum value Φ_{cMin}.

3.2.- The Fractal Dimension and the Critical Exponents

In an infinite cluster, the number M of particles inside a square of linear size l (centred in one point belonging to the cluster) scales as $M \propto l^D$, wher D is the fractal dimension of that cluster. Fig. 3 shows the ln-ln plot of M vs l for both the SSPM and model B with $\exp(-\tilde{\omega}/kT) = 0.1$, 0.5 and 1.0, evaluated for the largest cluster at probability P (concentration Φ) slightly below its critical value P_c (Φ_c). From the slopes of Fig. 3 we obtain $D = 1.90 \pm 0.02$ for $0.1 \leq \exp(-\tilde{\omega}/kT) \leq 1.0$ as well as for the SSPM. These results agree within the error bars with the well-known exact value $D = 91/48 \cong 1.896$ for the SSPM.

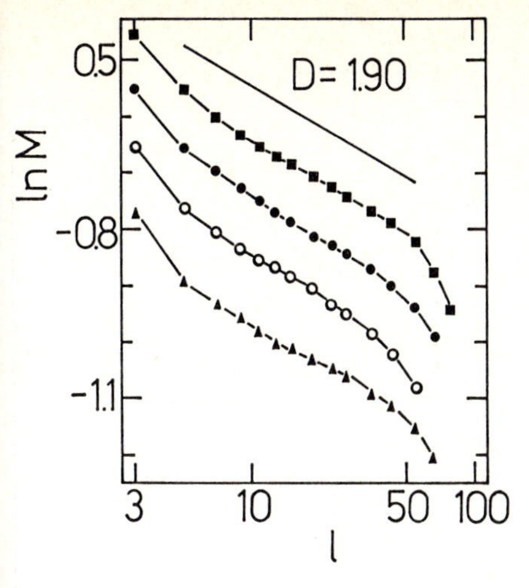

Figure 3. Ln-ln plot of M against l (L = 201)■for the SSPM with P = 0.583; ●,o,▲, for model B with $\tilde{\omega}/kT$ = 0.0, 0.6931, 2.3026 and with Φ = 0.583, 0.573 and 0.560, respectively. Each point has been averaged over 225-360 instantaneous equilibrium configurations using 5-36 different initial configurations. For the sake of clarity the results have been shifted in the ordinate axis.

For model B we have also computed the correlation length ν and the percolation probability β exponents (also we have preliminary results of ν and β for model A). All these results agree with those of the SSPM. As all the standard percolation exponents (such as ν, β, γ,...) are mutually related (there are only two free exponents), our results strongly suggest that all these exponents are the same as those of the SSPM.

4.- Conclusions

From the obtained results we conclude that:

1) The diffusion of particles with n-n lateral interactions does change the insulator-conductor monolayer transition threshold from that of the SSPM. This critical concentration depends on the particular details of the models. We expect that the change could be measured by varying the annealing temperature of very thin adsorbed conductor films on insulator substrates.

2) The diffusion does not change the universality class since the studied models and the SSPM have the same critical exponents.

Acknowledgments

Two of us (H.O.M. and E.V.A.) acknowledge the financial support of the CONICET (Argentina).

References

1. E. Harnik, S. Kovnovich and T. Chernobelskaya: Thin Solid Films, 126, 155 (1985)
2. D. Stauffer: Phys.Rep. 54, 1 (1979)
3. D. King: J.Vac.Sci.Technol. 17, 241 (1980)

The Surface Ferromagnetic Phase in Semi-Infinite Heisenberg Ferromagnets

S. Selzer and N. Majlis

Instituto de Física, Universidade Federal Fluminense,
Outeiro de S. João Batista s/no., 24210 Niterói, RJ, Brasil

We consider in this paper the conditions under which one can expect to find surface ferromagnetic behaviour in a Heisenberg semi-infinite crystalline ferromagnet, at temperatures higher than the transition temperature T_c^b from the ferro- to the paramagnetic bulk phase. This problem was studied in a previous paper by SELZER et al. |1| under the random phase approximation (RPA). That paper (hereafter called I) described the effect of surface anisotropy of the exchange interaction at the surface of a semi-infinite system. It was proved therein that a correspondence can be established, for $T > T_c^b$, between two systems: a) the semi-infinite one already mentioned; b) a film of a few planes of spins. System b) is not identical to a thin self-supported film because of the lack of symmetry resulting from the fact that the thin film is surrounded on one side by vacuum and on the other by the rest of the spins of the system. As a consequence, the parameters of the Hamiltonian on both sides of the film are different (Fig. 1).

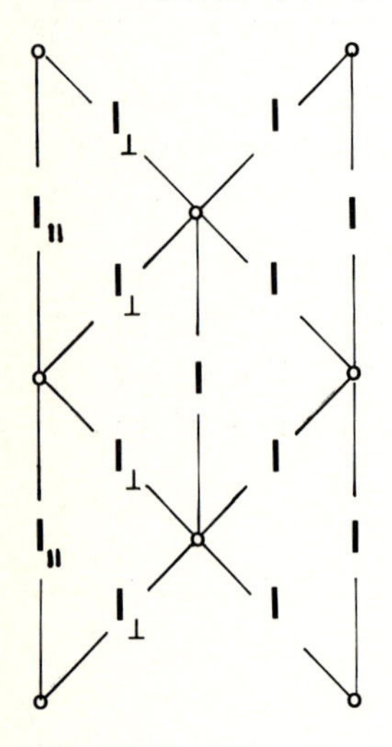

Fig. 1 - Exchange parameters in the first three planes of a semi-infinite Heisenberg ferromagnet. The left plane is the surface plane

The results which were obtained in RPA for the magnetization of the first n planes (n = 2 or 3) of the semi-infinite system, coincide within a few percent with those obtained in the same approximation for a film of n planes.

The Hamiltonian considered was:

$$H = \sum_{<ij>} h_{ij} ,$$

where $<ij>$ are all different nearest neighbour pairs, and

$$h_{ij} = I_{ij}(S_i^x S_j^x + S_i^y S_j^y + \eta_{ij} S_i^z S_j^z) \quad .$$

The parameters in (2) are: $I_{ij} = I$ if (i,j) are both bulk spins; $I_{ij} = I_{//}$ if i and j are on the surface plane; $I_{ij} = I$ if one spin of the pair (i,j) is on the surface and the other is on the nearest plane; $\eta_{ij} = \eta > 1$ if (i,j) are both on the surface; otherwise, $\eta_{ij} = 1$ (see Fig. 1). The z axis coincides with the direction of the magnetization.

We considered in I the effect of varying η upon the magnetization and the transition temperature T_c^s, defined as the ferro- to paramagnetic transition temperature of the film (system b) above). In I we assumed $I_{//} = I_{\perp} = I$.

In the present work we extend I by considering simultaneously the effect of the parameters $\varepsilon_{//} \equiv I_{//}/I$ and η upon the surface magnetization m_s and the "surface transition temperature" T_c^s. Based on the results of I, we assume that m_s obtained in RPA for a film of two planes will be approximately the same as that obtained also in RPA for the surface plane of a semi-infinite system with the same exchange parameters, for $T > T_c^b$. Therefore we calculated the magnetization of two planes with spins located on sites corresponding to points on (111) planes of an f.c.c. lattice. Extrapolating to $m_s = 0$ we obtained the transition temperature T_c^s. Actual calculations were performed for S = 7/2.

In Fig. 2 we show T_c^s of this film, in units such that $T_c^b = 8.3$, as a function of η, for fixed values of $\varepsilon_{//}$. We observe, as in I, that T_c^s varies linearly with η.

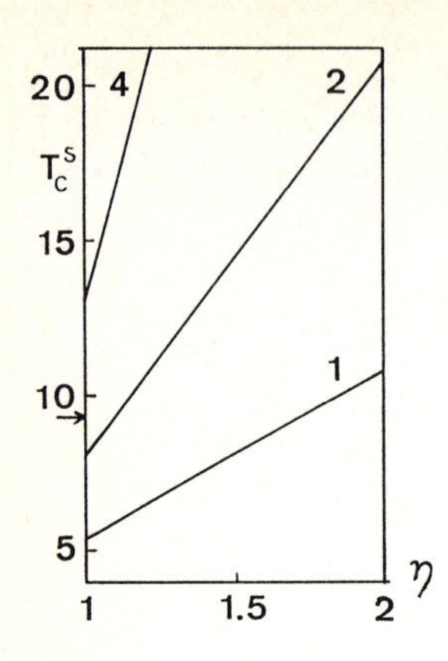

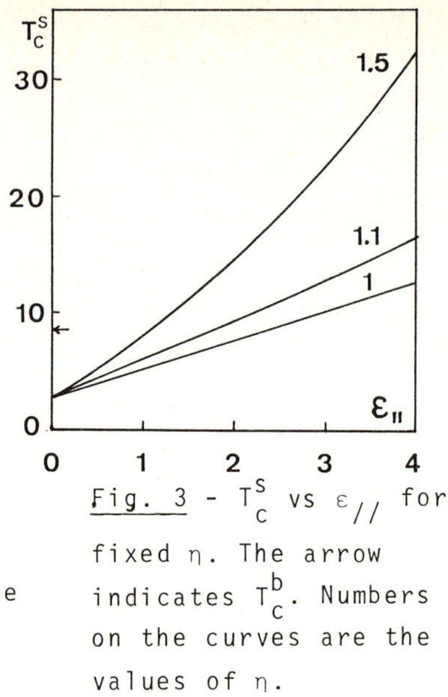

Fig. 2 - T_c^s vs η for fixed $\varepsilon_{//}$. The arrow shows T_c^b. Indicated on the lines are the corresponding values of $\varepsilon_{//}$.

Fig. 3 - T_c^s vs $\varepsilon_{//}$ for fixed η. The arrow indicates T_c^b. Numbers on the curves are the values of η.

MARIZ et al. |2| applied techniques of renormalization group in real space to hierarchical lattices of Heisenberg spins with S = 1/2 with a Hamiltonian equivalent to (1) and (2), except for a change of notation. The agreement of both types of calculation is remarkable.

It is worthwhile to mention that, at first sight, finding a surface critical temperature $T_c^s > T_c^b$, without introducing anisotropy ($\eta = 1$), seems to contradict the fact that a two-dimensional array of spins with isotropic Heisenberg ferromagnetic interactions has $T_c^s = 0$. However, "surface" means usually, as in our case, a system of several atomic planes, since the interactions $I_\perp$ are always present. Such a system exhibits $T_c^s > 0$.

In recent experiments by WELLER et al. |3| a surface magnetic phase was found in Gd, confirming previous results of measurements by RAU et al. |4|. We expect to apply our calculations to Gd. in future work.

Acknowledgements - This work was performed with financial support of Financiadora de Estudos e Projetos (FINEP), Conselho Nacional de Desenvolvimento Científico e Tecnológico (CNPq) and

Coordenação de Aperfeiçoamento de Pessoal de Nível Superior (CAPES) of Brazil. We thank Dr. Constantino Tsallis for sending us a preprint of MARIZ et al. |2| previous to publication.

References

1. S. Selzer, N. Majlis: Phys. Rev. B 27, 544 (1983)
2. A.M. Mariz, U.M.S. Costa, C. Tsallis: preprint, 1986
3. D. Weller, S.F. Alvarado, W. Gudat, K. Schröder, M. Campagna: Phys. Rev. Lett. 54, 1555 (1985)
4. C. Rau: J. Mag. Mag. Mater. 31-34, 874 (1983)

Part III

Surface Spectroscopies

Theoretical Aspects of Electronic Spectroscopies at (Adsorbate Covered) Surfaces

G. Doyen, D. Drakova[+], *and F. von Trentini*

Institut für Physikalische Chemie, Universität München,
Theresienstr. 37, D-8000 München 2, Fed. Rep. of Germany

A model Hamiltonian is used to calculate the electronic structure of the ground state and the excited states independently in a self-consistent way (Δ SCF). The electronic transition operators are defined by the same Hamiltonian which is used to determine the wave functions. The method is used to calculate the spectral distribution of emitted electrons in the case of gas atoms and simple molecules interacting with alkali and transition metal surfaces. Relaxation effects and image force screening are discussed.

1. INTRODUCTION

The purpose of this contribution is to consider four electronic spectroscopies in the context of a model Hamiltonian for adsorption : photoemission, inverse photoemission, Auger deexcitation and ion neutralization. In photoemission (ultraviolet photoelectron spectroscopy, UPS) the incident probe is a photon and the detected signal consists of emitted electrons. Exhaustive treatments of various aspects of this spectroscopy are available /1/.

Unoccupied states between Fermi level and vacuum level, which are not accessible by photoemission, can be reached by means of the time reversed process, i.e., inverse photoemission /2/. The incident probe is an electron and the photons are detected. Inverse photoemission is an inefficient process. About one photon per 10^8 electrons is created. For this reason the technique has been developed only recently and relatively few results exist.

In Auger deexcitation AD (also called Penning ionization, PI, metastable de-excitation spectrosopy, MDS, or metastable quenching spectroscopy, MQS) electronically excited, metastable rare gas atoms are used as the incident probe. The emitted electrons are detected. The application of this method to surface problems is relatively new as well /3/. The prominent property is the extreme surface sensitivity.

At metal surfaces with large work function (> ca. 4 eV), metastable rare gas atoms will resonance ionize. One then switches to a different spectroscopy, i.e., ion neutralization spectroscopy, which was pioneered by Hagstrum several decades ago /4/. In this case the emitted electron originates from the metal and therefore the kinetic energy distribution is completely different from AD. The interpretation is more involved although the fundamental theory is quite similar.

2. THEORY OF ELECTRONIC SPECTROSCOPIES

2.1 The generalized Ehrenfest Theorem

Initial $|I\rangle$ and final states $|F\rangle$ are eigenstates of the Hamiltonian $H_{surf} + H_{probe} + V_{surf-probe}$. H_{surf} describes the interacting adsorbate - surface system, H_{probe} is the Hamiltonian of the incoming probing particle and $V_{surf-probe}$ stands for the interaction between the probe and the surface. $|I+\rangle$ describes the neutral ground state of the (adsorbate covered) surface in interaction with the field of the probing particle. $|F\rangle$ describes an ionic state. For photoemission, e.g., it is positively charged and contains the removed electron in an outgoing plane wave. In the case of inverse photoemission it would be negatively charged and contain an outgoing photon. It is assumed that the final state can be written as a product:

$$|F\rangle = |N-1\rangle|particle,out\rangle \text{ or } |F\rangle = |N+1\rangle|particle,out\rangle. \quad (1)$$

Here $|N-1\rangle$ and $|N+1\rangle$ are excited states of interacting electrons plus ion cores. In some cases the separability assumption is invoked also for the initial state:

$$|I+\rangle = |N\rangle|probe,in\rangle . \quad (2)$$

The approximation is used for direct and inverse photoemission. For Auger deexcitation and INS it is crucial to look for a better description of the initial state, because the interaction is much stronger than in the other spectroscopies. From Lippmann's generalization /5/ of Ehrenfest's theorem, the transition rate from $|I\rangle$ to $|F\rangle$ is

$$R_{fi} = 2\pi \; |\langle F|V_t|I+\rangle|^2 \; \delta(E_F - E_I) \quad ; \tag{3}$$

V_t is the transition - inducing potential. It describes the interaction of the emitted particle with the rest system.

2.2 The transition operator

V_t consists in general of three parts: the electron-photon interaction V_{e-p}, the electron-ion interaction V_{e-i} and the electron-electron interaction V_{e-e}:

$$V_{e-p} = -(i/2c) \sum_{i,j,s,q} a^+_{is} \langle i|A_q \nabla + \nabla A_q|j\rangle a_{js} \; (b_q + b^+_q) \quad ; \tag{4}$$

a^+_{is} and a_{js} are electron creation and destruction operators, the corresponding operators for the photon are b^+_q and b_q. A_q is the position-dependent vector potential which has to be evaluated in the presence of both the microscopic current density in the metal induced by the radiation, and the photocurrent /6/.

$$V_{e-i} = \sum_{i,j,s} \langle is|v_M - Q/r|js\rangle a^+_{is} a_{js} \quad ; \tag{5}$$

v_M is the effective one-particle potential of the metal and $-Q/r$ denotes the ion core potential of (adsorbed) gas particles.

$$V_{e-e} = \sum_{i,j,s,t} (isjs|ktlt) a^+_{is} a^+_{kt} a_{lt} a_{js} \quad ; \tag{6}$$

$(isjs|ktlt)$ is the common notation of a two-electron integral. All three operators can be responsible for the emission of an electron, if the necessary energy is provided by either an electronic deexcitation or the absorption of a photon. With the help of the theorem by GELL-MANN and GOLDBERGER /7/ two of them can be removed from the transition matrix element $\langle F|V_t|I\rangle$ by introducing the incoming wave state $|F-\rangle$ which then contains the effect of those two operators /8/. In the case of inverse photoemission, only V_{e-p} is transition inducing, because no outgoing electron is detected asymptotically.

2.3 Matrix elements and particle hole excitation spectra

Because of the product form (1) of the final state $|F\rangle$ the electron creation operator a^+_{is} furthest to the left in the above expressions of the transition operators can be directly applied. It yields a non-vanishing result only, if $|is\rangle$ coincides with the state of the emitted electron. In the case of photoemission one obtains e.g.:

$$\langle N-1|\langle \text{electron},ks|V_{e-p}|N\rangle|\text{photon},q\rangle = -(i/2c)\sum_j \langle ks|A_q\nabla+\nabla A_q|js\rangle * \langle N-1|a_{js}|N\rangle \quad (7)$$

If in the sum over j one optical matrix element $\langle ks|A_q\nabla+\nabla A_q|js\rangle$ dominates, the rate (3) is proportional to the single hole excitation spectrum

$$p^-_j = \sum_{N-1} |\langle N-1|a_{js}|N\rangle|^2 \delta(E_F-E_I) \quad (8)$$

The sum runs over all excited states of the ionized system. If the ionized final state is calculated in Koopmans´ approximation (also called frozen orbital approximation), the excitation spectrum would consist of a sum of δ-functions with the weights $|c_{jl}|^2$, where c_{jl} is the coefficient of the j-th basis orbital in the l-th Hartree-Fock orbital of the ground state which is supposed to be non-degenerate. If $|N-1\rangle$ is determined by a separate self-consistent Hartree-Fock calculation (as is done in the present investigation), p^-_j shows additional peaks (so called satellite structure). The weights have to be determined by evaluating the overlap of the many-body states $|N-1\rangle$ and $(a_{js}|N\rangle)$. The single hole excitation spectrum is often used in the literature to estimate the relative intensity observed in photoemission. The corresponding quantity relevant for inverse photoemission is the single particle excitation spectrum:

$$p^+_j = \sum_{N+1} |\langle N+1|a^+_{js}|N\rangle|^2 \delta(E_F-E_I) \quad (9)$$

In the case of two - electron Auger processes the single hole excitation spectrum is of importance as well for those processes which are induced by the one-particle potential V_{e-i}. In order to conserve the total energy in this case the ionization process has to be accompanied by an electronic deexcitation. This means that Koopmans' approximation yields no intensity for Auger processes mediated by V_{ei}. Therefore processes of this kind are termed relaxation-induced transitions (RIT).

For Auger processes mediated by the two-electron potential V_{e-e} the relevant spectral function describes the formation of two holes and the addition of one particle. These contributions to the Auger rate are called correlation - induced transitions (CIT).

3. THE ELECTRONIC MODEL HAMILTONIAN

3.1 Model of the gas particle - transition metal surface interaction

A model Hamiltonian operator for chemisorption systems contains terms describing the metal surface H_M, the adparticle H_A and their interaction H_{AM}:

$$H = H_A + H_M + H_{AM} \quad . \tag{10}$$

Based on this general form for the adsorption Hamiltonian, a model has been proposed for the study of gas adsorption on transition metal surfaces /9-11/. It has been successfully used for the study of many chemisorption systems /10/ and for the extraction of theoretical interaction potentials between noble gases and transition metal surfaces /11/, which compared favourably with 'experimental' potentials from atomic beam scattering experiments. The physical ideas behind the model will be reminded only briefly. Some recent developments are outlined in the next paragraph. For more details the reader is referred to refs. /9-11/.

Partial cancellation was assumed between large electrostatic interactions in the adsorption system. Only that part of adelectron - metal electron repulsion was retained in the Hamiltonian which is not cancelled by a corresponding attraction between adelectrons and metal ion cores. This is the 'renormalized' electron - electron repulsion. Electron - electron repulsion was included explicitly and consistently in the local region overlapped by the gas particle wave functions.

An attractive gas particle - core potential was introduced for the metal electrons in the region overlapped by the adorbitals. This is a Goeppert-Mayer-Sklar type (GMS) /12/ potential and it was assumed to cancel perfectly the repulsion between metal and gas electrons in the overlap region.

3.2 Treatment of diffuse virtual adorbitals

When attempts are made to use the model of DOYEN et al. /9-11/ for the study of adsorption accompanied by diffusion of the gas atoms inside the metal lattice, some problems are encountered. The reason lies in the non-orthogonality and overcompleteness of the basis. An extended basis set including diffuse empty adatom wave functions is certainly necessary for the description of gas particle-induced spectral features as studied by inverse photoemission and Auger deexcitation. For the asymptotic case $S_A = 1$ when the metal wave functions form a complete basis set and represent the adatom wave functions, a special parameterization of the hopping between them is necessary, which should merge smoothly to small overlap situations further outside the metal surface. Therefore the non - orthogonality between so called ´adsorbate projected metal wave functions for different adorbitals is accounted for by explicitly evaluating the overlap integrals between them as well as the hopping terms in the metal part of the Hamiltonian due to coupling of metal wave functions through the adorbitals. Physically adequate hopping between adatom and metal through the core potentials is introduced for high overlap situations. Divergences due to the restriction of electron-electron terms to the adparticle are removed by introducing electron-electron repulsion between metal electrons in the overlap region. They are estimated under the conditions of hybridizational and rotational invariance.

4. APPLICATIONS

4.1 Photoemission and inverse photoemission: CO/Cu(100) and H/Ni(111)

The parameterization of the model Hamiltonian consists of choosing the wave functions and the one- and two-electron integrals in such a way that the description of the separated system agrees with available experimental data. An effective one-electron description is used, which means that relaxation effects in the <u>separated</u> gas particle - surface are not handled <u>explicitly</u>, but are <u>implicitly</u> included in the choice of the one- and two-electron integrals. Relaxation processes obtained in this theory are due to the adsorbate - substrate interaction and are only part of the total relaxation effects as they would result from an ab-initio type calculation. As image force effects are included only in a <u>static</u> way, they do not contribute to relaxation, although they are important for screening by renormalizing the core and Coulomb integrals.

The Hamiltonian is diagonalized numerically on a computer. This is possible, if the continuous metal spectrum is discretized. The method has been described in detail /9 /. The overlap integrals are evaluated numerically exactly by constructing adsorbate projected metal states. For the d-band the situation is more complicated and the reader is referred to the discussion in ref. /9/. The ´T-B method´ outlined there has been used in the present work.

An example of relatively weak adsorption is CO on Cu(100). Included are eight occupied molecular spin orbitals (4σ, 5σ, 1π) and four empty ones (2π). It is generally accepted that the molecular axis is perpendicular to the metal surface and that the carbon atom is directed towards the surface, the oxygen atom standing away. The coupling of all molecular orbitals with both the 3d-band and the 4sp-band is included. This is an improvement over earlier calculations /10/. The most important couplings are in the order of decreasing strength: 2π to unoccupied sp-band; 2π to d-band; 2π to occupied sp-band; 5σ to unoccupied sp-band; 4σ to unoccupied sp-band; 4σ to occupied sp-band; 5σ to occupied sp-band; 1π - coupling. This order agrees roughly with previous model assumptions. A potential energy curve has been calculated for the on top position. The minimum is at a C-Cu distance of 2.3 A and corresponds to a well depth of 0.63 eV.

Non-image relaxation effects induced by the CO - metal interaction are found to be small due to the weak interaction of the occupied MO´s and the diffuse character of the 2 -orbital. The diffuseness implies small electron-electron repulsion from the

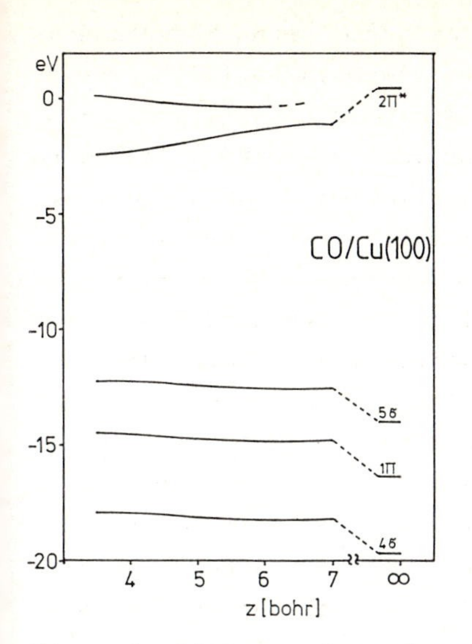

Fig. 1 Distance dependence of CO - derived orbitals and resonances

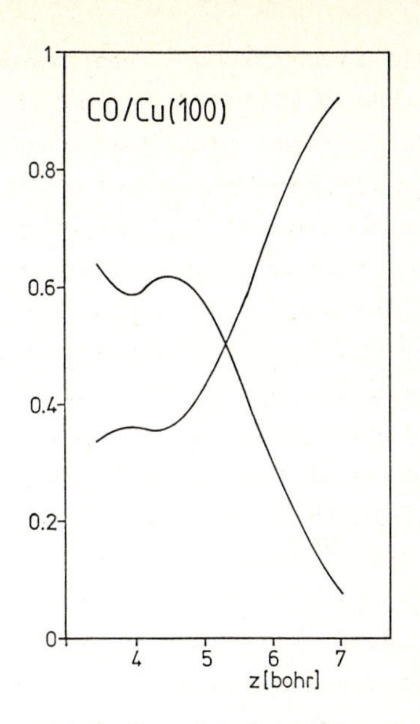

Fig. 2 Projection of the affinity resonances onto the 2π * - orbital

core, which is further reduced by the image renormalization. Non-image relaxations shifts turn out to be smaller than 0.05 eV. Therefore the Koopmans picture holds to a large degree and is sufficient for interpretation.

The results are displayed in fig. 1 and 2. Compared to the gas phase values a general upward shift of the ionization energies towards the Fermi level is observed. The largest contribution is the image shift which is 1.5 eV at the equilibrium position. The additional screening shift due to repulsion from the metal electrons is 0.12 eV.

An interesting feature is obtained for the 2π - derived affinity levels. Because the 2π -orbital interacts strongly with the empty sp-band density of states, two such resonances are observed which might be termed bonding and anti-bonding. One resonance is situated slightly below the vacuum level, whereas the other one (depending on the distance from the surface) is 1 to 2.5 eV deeper. The interesting aspect is that the character of these two resonances changes with distance from the surface. At far distances the higher one has predominantly metal character and the lower one is mainly 2π derived. Upon closer approach to the surface the characters of both resonances become more mixed, and at the equilibrium position they have interchanged their major affiliation, the higher one being now mainly 2π - derived and the lower one having the larger metal character. This is due to the attraction of the metal states by the CO-core potential. At nearer distances the 2π -projected metal state has considerable overlap with the 2π and consequently experiences a significant perturbation by the adcore (called GMS-attraction in our terminology). This moves the 2π -projected metal state down in energy so that the corresponding diagonal matrix element of the Hartree-Fock operator is below the value for the 2π - orbital itself. Interference (hybridization) effects lead then to mixing of the basis wave functions and further splitting of the two levels. This splitting of the 2π- derived affinity levels has been observed experimentally and explained in a qualitatively similar way /13/.

As an example of stronger chemisorption the interaction of atomic hydrogen with the 111-surface of nickel was investigated. For adsorption in the center position the H - Ni-atom separation at equilibrium was calculated to be 1.84 A. The predicted adsorption energy is 2.7 eV. Both values agree with the experimental findings.

Strong relaxation effects are predicted for the formation of negative hydrogen ions near the nickel surface. Upon filling the spin down 1s orbital the spin up 1s orbital shifts towards the Fermi level and empties partially. This implies that the electron-electron - induced hopping $(1s\uparrow 1s\uparrow | 1s\downarrow met\downarrow)$ decreases and consequently the effective self-consistent hopping matrix element $\langle 1s\downarrow | H_{sc} | met\downarrow \rangle$ becomes more negative (for positive overlaps). It also implies that the 1s down diagonal matrix element of the Hartree-Fock matrix decreases in energy by ca 3.5 eV. The quantity determining the admixture of metal wave functions in the 1s-orbital is

$$V^{eff}_{1s-met} = \langle 1s\downarrow | H_{sc} | met\downarrow \rangle - E_{1sv} \langle 1s | met \rangle \quad \cdot \qquad (11)$$

The latter of the two terms on the r.h.s. decreases more strongly than the former, but in such a way that the two terms nearly cancel. Hence the mixing between the 1s and the metal states changes drastically, i.e., the Hartree-Fock orbitals relax.

The Δ SCF - single particle - hole excitation spectrum including relaxation effects is displayed in fig. 3. In the case of forming a <u>positive</u> ion, relaxation effects are found to be much smaller. The relaxation shift is 0.1eV towards smaller ionization potentials. The important difference to the negative ion is in the width of the 1s induced resonance which is ca. 1eV for the positive ion and ca. 3.5 eV for the negative ion. As the energetic separation of the discretized levels is roughly 1eV, a complete 1s electron is removed in the case of positive ionization, whereas only part of a 1s electron is added in the case of negative ionization.

The width of the discretization is determined by the time the photoelectron (incoming or outgoing) spends in the surface

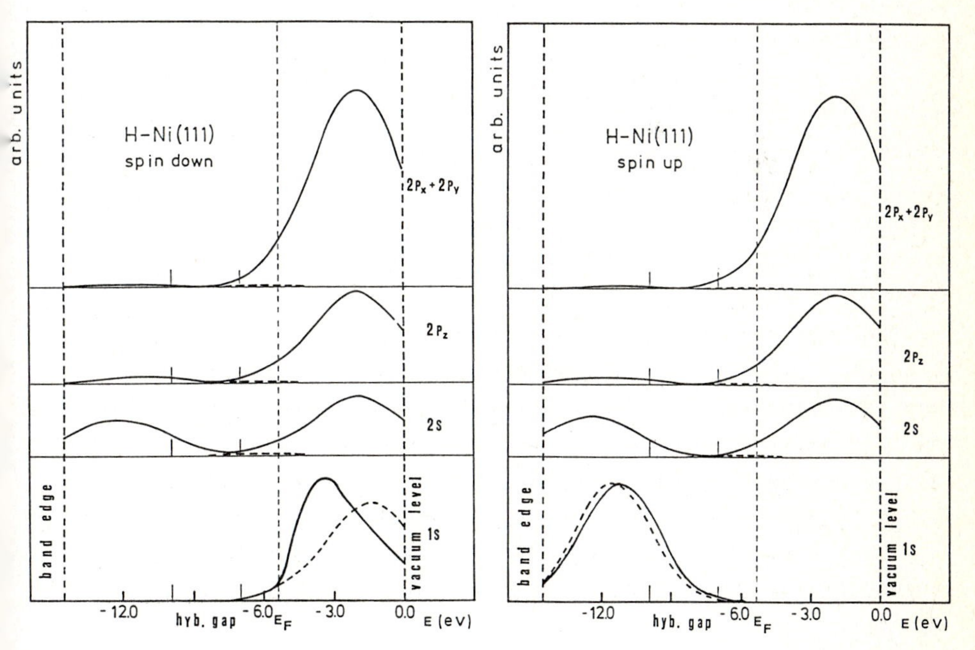

<u>Fig. 3</u> Spin unrestricted Δ SCF - single particle - hole excitation spectrum for hydrogen adsorbed on Ni (111). The broken curves for the 1s spin orbitals are the results in Koopmans approximation. The broken curves for the other orbitals indicate the weight in the d-band.

region. In the present calculation relaxation processes which are faster than ca. 10^{-15} seconds are included.

4.2 Auger deexcitation at a caesium surface

Electronically excited, metastable rare gas atoms hitting a solid surface relax to the ground state with a probability near unity /14/. The deexcitation energy is transferred to electrons which are emitted and can be analyzed with respect to their energy, momentum and angular distribution. This 'Metastable Deexcitation Spectroscopy' (MDS) can be used to obtain information about the electronic structure of the surface. The advantage over UPS is the extreme surface sensitivity, which is comparable to that of tunneling spectroscopies (FEED or STM). It is tempting to suppose that MDS involves essentially a tunneling process (of a substrate electron into the empty gas atom core state) and that the emission of the excited electron is a process localized on the gas atom which is independent of the specific nature of the surface. An important question is whether the deexcitaion process can be understood in perturbation theory, i.e., whether the substrate wave functions are unperturbed by the presence of the excited rare gas atom. In this case one would obtain direct information about the local surface density of states of the clean surface. Otherwise a reliable theory will be necessary to understand the experimental spectra.

The work reported here concentrates on investigating this problem. Alkali surfaces are ideal for gaining a first understanding as the work function is low (resonance ionization can be excluded, see below) and they can be theoretically treated as a Sommerfeld metal (free electron gas with a surface barrier). AD at alkali surfaces has been experimentally studied in detail /15

The theoretical task involves the evaluation of: (i) the potential energy surface of the initial state; (ii) the wave function of the initial state at each distance; (iii) the potential energy surfaces of all final states; (iv) the wave functions of all final states; (v) the matrix elements of the transition operators and the sum over all final states to obtain the rate; (vi) the Franck Condon factors for the core movement. Point (vi) has not been attacked in the present investigation.

Auger deexcitation of triplet helium at a caesium surface has been studied /16/. Two transition operators need attention V_{e-e} (CIT) and V_{e-i} (RIT) (cf. section II.2). It turns out that the RIT - matrix element dominates by several orders of magnitude. It can be approximately expressed by a formula which suggests a physical interpretation:

$$T^{RIT}_{fi} = \langle k | v_M - 2/r | 2s \rangle \langle 1s\downarrow, fin | met\downarrow, in \rangle \tag{12}$$

The first factor describes the emission of the spin-up electron from the 2s basis function into the continuum state. The second factor accounts for the transition of the spin-down electron from the initially occupied metal state into the 1s level of the final state. The first factor of (12) is approximately constant in the considered energy interval, whereas the overlap $\langle 1s\downarrow, fin | met\downarrow, in \rangle$ is roughly proportional to the 2s amplitude mixed into the metal wave function by the interaction. The diffuse 2s state does not shield the helium core very effectively in the initial state so that the metal functions expand into the region of the adparticle. This expansion is described by a 2s admixture. In the final state the helium core is shielded by two 1s electrons and the perturbation of the metal states by the helium potential is rather small. In particular the 2s admixture has vanished and the kinetic energy distribution reflects the shape of the occupied part of the filled 2s$\downarrow$ resonance. The conclusion from this theoretical investigation is therefore that perturbation theory is not sufficient to treat Auger deexcitation. AD does not probe the unperturbed local metal density of states, but contains rather information about the surface density of states as modified by the interaction with the excited metastable rare gas atom. The theoretically obtained kinetic energy distribution is compared to the experimental spectrum in fig. 4.

4.3 Ion neutralization at a Cu (100) surface

Metastable helium atoms incident on clean transition metal surfaces produce electronic spectra which are equivalent to those obtained by ion neutralization spectroscopy (INS). This can be explained, if one assumes that resonance ionization occurs from

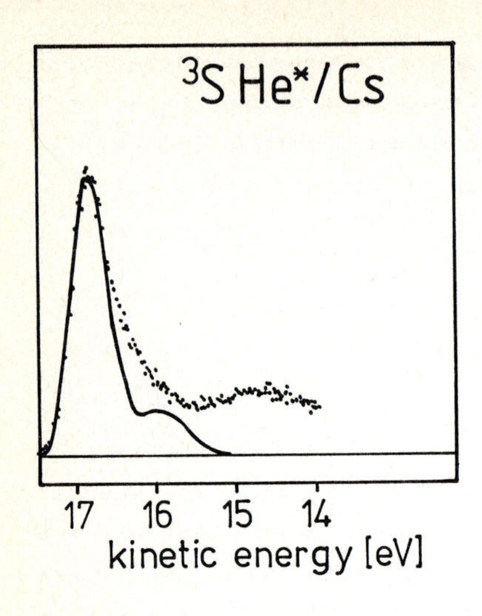

Fig. 4 Comparison of theory (full curve) and experiment /15/ for AD at a Caesium surface. The theoretical results obtained for the equilibrium position have been convoluted with a Gaussian of width 0.25 eV in order to simulate the limited resolution of the spectrometer and averaging over different deexcitation distances.

the asymptotically occupied 2s-orbital to the degenerate empty metal wave functions. This process is in fact much faster than the Auger process. It is energetically allowed, if the work function of the metal surface is larger than the ionization potential of the 2s-electron, which is 4.768 eV for the triplet helium. For alkali surfaces the work function is lower and resonance ionization is forbidden.

After resonance ionization there is a positive helium ion in front of the surface, which is then neutralized by an Auger process. The important difference to MDS is that now the emitted electron does not come from the helium atom but from the sp-band of the metal. This leads to spectra with much less structure which are often deconvoluted in order to extract some information.

The theory is essentially the same as for AD (MDS) . The RIT process is again found to dominate by orders of magnitude. Auger neutralization has been examined for the system He^+/Cu(110) /17/. The same Hamilton operator and transition operators (and in fact also the same computer code) have been used as in the case of Auger deexcitation. The only difference is in the asymptotic occupation of the orbitals. Delocalized surface states are found to have a strong influence on the kinetic energy distribution of the emitted electrons, whereas metal d-states give a negligible contribution. Qualitative agreement with the experimental spectra from a Cu (110) /18/ surface can be achieved only, if a metal surface state is included in the theory, which was experimentally observed by means of angular resolved photoemission /19/. To our knowledge these are the first calculations for INS including so called matrix element effects.

+ Permanent address: Faculty of Chemistry, University of Sofia, Sofia, Bulgaria

References

/1/ B. Feuerbacher, B. Fitton and R. F. Willis, editors: "Photoemission and the electronic properties of surfaces", Wiley, 1978

/2/ V. Dose, Prog. Surf. Sci. 13, 225 (1983); F. J. Himpsel and T. Fauster, J. Vac. Sci. Technol. A2, 815 (1984)

/3/ H. Conrad, G. Ertl, J. Küppers, S. W. Wang, K. Gerard and H. Haberland, Phys. Rev. Lett. 42, 1082 (1979); H. Conrad, G. Ertl, J. Küppers, W. Sesselmann and H. Haberland, Surf. Sci. 93, L75 (1980)

/4/ H. D. Hagstrum, Phys. Rev. 96, 336 (1954).

/5/ B. A. Lippmann, Phys. Rev. Letters 15, 11 (1965); Phys. Rev. Letters 16, 135 (1966)

/6/ N. W. Ashcroft, in ref. /1/, page 21

/7/ M. Gell-Mann and M. L. Goldberger, Phys. Rev. 91, 398 (1953)

/8/ G. Doyen and T. B. Grimley, in ref. /1/, page 137

/9/ G. Doyen, Surf. Sci. 59, 461 (1976); G. Doyen and G. Ertl, Surf. Sci. 65, 641 (1977)

/10/ G. Doyen and G. Ertl, Surf. Sci. 69, 157 (1977); G. Doyen and G. Ertl, J. Chem. Phys. 68, 5417 (1978); G. Doyen, Surf. Sci. 122, 505 (1982)

/11/ D. Drakova, G. Doyen and F. von Trentini, Phys. Rev. B32, 6399 (1985)

/12/ see e.g. M. J. S. Dewar, 'The Molecular Orbital Theory of Organic Chemistry', McGraw-Hill, 1969

/13/ J. Rogozik, V. Dose, K. C. Prince, A. M. Bradshaw, P. S. Bagus, K. Hermann and Ph. Avouris, Phys. Rev. B32, 4296 (1985)

/14/ H. Conrad, G. Doyen, G. Ertl, J. Küppers, W. Sesselmann and H. Haberland, Chem. Phys. Lett. 88, 281 (1982)

/15/ B. Woratschek, PhD - Thesis, Munich, 1986

/16/ F. von Trentini and G. Doyen, Surf. Sci. 162, 971 (1985)

/17/ F. von Trentini, PhD - Thesis, Munich, 1984

/18/ W. Sesselmann, PhD - Thesis, Munich, 1983

/19/ P. Heimann, J. Hermanson, H. Miosaga and H. Neddermeyer, Surf. Sci. 85, 263 (1979)

XPS Relaxation and AES Cross-Relaxation Energies in Small Metal Particles

N.J. Castellani and D.B. Leroy

Planta Piloto de Ingeniería Química (UNS-CONICET) and
Departamento de Física (UNS), C.C. 717, 8000 Bahía Blanca, Argentina

1 Introduction

It is well known that the core level electron binding energy in small metal particles is shifted with respect to the bulk metal, as can be observed by XPS measurements. The general trend is that the binding energy increases with decreasing particle size. On the other hand, MASON/1/ has shown that the Auger parameter decreases when the particle size decreases in the case of small supported particles. Both shifts have been related to initial and final state effects. This latter is due to the relaxation of valence electrons around electron holes created during the photoemission or the Auger process. In order to obtain the initial state contribution corresponding to chemical effects, which are of interest e.g. in catalysis, it is necessary to take off the experimental shift the relaxation contribution. The aim of this work is to calculate the relaxation energy (R_1) and the cross-relaxation energy (R_c) in the case of small metal particles.

2 Spherical Electron Gas Model

The calculation of screening for an external potential in a small metal particle has been developed by CINI/2/. The conduction electrons are treated as an electron gas confined to a finite volume and linear response is assumed. The screening charge density is decomposed in a discrete set of normal modes adapted to the geometry of the finite particle. In the present case, only spherical particles will be considered. The dynamical properties of the finite electron gas are taken into account through a size-dependent dielectric function which was earlier derived by CINI/3/. The screening energy of an external perturbing potential V^e may be written as:

$$E_s = \frac{1}{8\pi\,\Omega} \sum_q \frac{\varepsilon(\lambda_q) - 1}{\varepsilon(\lambda_q)} (\lambda_q \; V^e_q)^2 \qquad (1)$$

where Ω is the volume of the particle, (λ_q) the dielectric function, λ_q the generalized wavenumber of the mode, and V^e_q is the generalized Fourier transform of the external potential.

The influence of the core hole position on the screening energy may be studied considering a point charge at a distance d from the center of a particle of radius r/4/. Equation (1) can be solved analytically and the results of such a calculation are shown in Table 1. As may be seen, the screening energy decreases very slowly on moving the hole outwards from the center towards the surface, except very close to it where an abrupt drop occurs. Consequently only a central core hole position will be considered in further calculations.

In order to calculate relaxation energies, the external potential corresponding to the creation of a core hole remains to be specified. In a previous paper /5/, we have shown that the use of a Coulomb potential is not suitable and it is convenient to use pseudopotentials. That is done in the next sections.

d/r_0	Es / Ev	
	r_0 = 10 a.u.	r_0 = 40 a.u.
0.00	3.30	6.20
0.50	3.30	6.20
0.90	3.10	6.20
0.98	2.30	5.30
1.00	1.90	4.00

Table 1: Screening energy as a function of d/r_0 for two particle radii r_0

3 Relaxation Energy in the Single Core Hole Case

As it has been demonstrated in our previous paper /5/, the orthogonality of screening valence wave functions to the core wave functions may be considered through a pseudopotential for the core hole perturbation. In the present case, we have used an Ashcroft-like pseudopotential:

$$V(r) = -\frac{1}{r} \quad \text{for} \quad r > r_c \,,$$
$$V(r) = -\frac{1}{r_c} \quad \text{for} \quad r \leqslant r_c \,, \qquad (2)$$

where the cut-off radius r_c is the radius of the sphere which contains 95% of charge density of the highest core electron wave function of the same angular momentum quantum number 1 as the screeening electrons.

The pseudopotential linear-response electron-gas method is then applied to calculate the relaxation energy R_1 in small metal particles as a function of particle size, for three simple metals. An important reduction of R_1 is observed for particles smaller than 40 a.u. (Fig.1). Furthermore, convergence to the bulk limit is slow. This last value may be compared with the result obtained by the density functional theory (DFT) /6/. Here, the metal relaxation energy was calculated as the sum of two contributions: the intra-atomic and the extra-atomic ones, the second term being calculated assuming the excited atom model. As can be seen from Fig. 1, the values obtained by two radically different methods are in good agreement.

4 Auger Cross-Relaxation Energy

In a strictly linear response theory, the double core hole relaxation energy R_2, as has been shown elsewhere /7/, is four times the single relaxation energy R_1, where R_1 is calculated with the r_c defined in section 3. However, non-linear effects due to shrinking of the core in the presence of the spectator hole can be taken into account, even in linear response theory, through an adequate pseudopotential with a smaller cut-off radius. This new pseudopotential was defined as in section 3, but replacing r_c by r_c^*, the cut-off radius of an atom with a core hole.

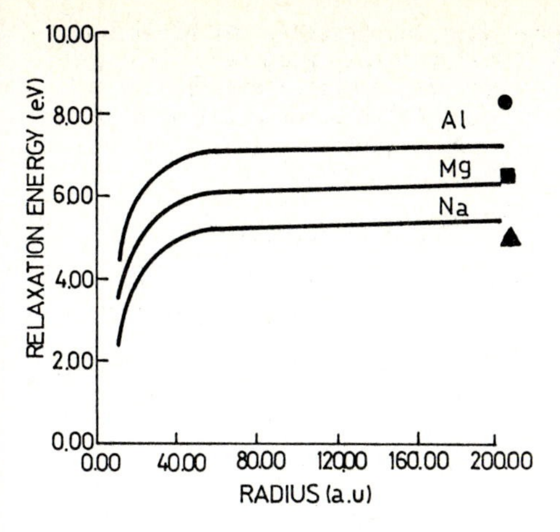

Fig. 1: Relaxation energy as a function of particle size. DFT results are also indicated: Al (●), Mg (■), Na (▲).

As the Auger cross-relaxation energy R_c is given by:

$$R_c = R_2 - 2 R_1 \tag{3}$$

and taking into account the above consideration, we obtain:

$$R_c = 2 R_1 (r_c) + \Delta R_c, \tag{4}$$

where ΔR_c is the deviation from the linear response rule:

$$\Delta R_c = 4 R_1 (r_c^*) - 4 R_1(r_c). \tag{5}$$

The results of the calculation of cross-relaxation energy as a function of particle size are given in Fig. 2 for the same three simple metals as in section 3. The same general trend is observed as in the R_1 case. Comparison with results obtained by DFT is also shown in Fig. 2. Although the order of magnitude is the same, some discrepancy always of the same sign is discernible, probably because non-linear effects are not completely included in our model.

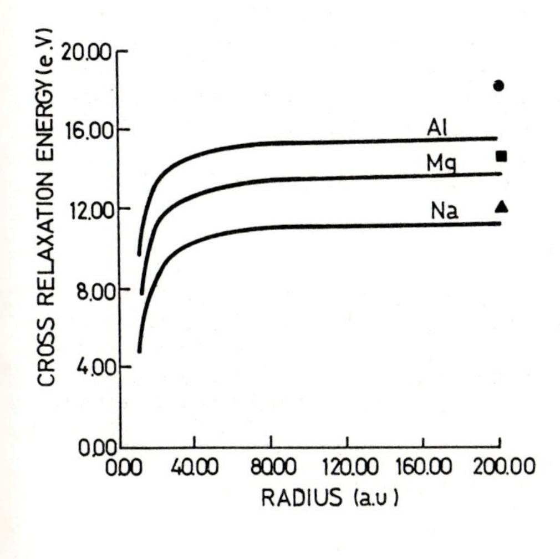

Fig. 2: Auger cross-relaxation energy as a function of particle size. DFT results are also shown.

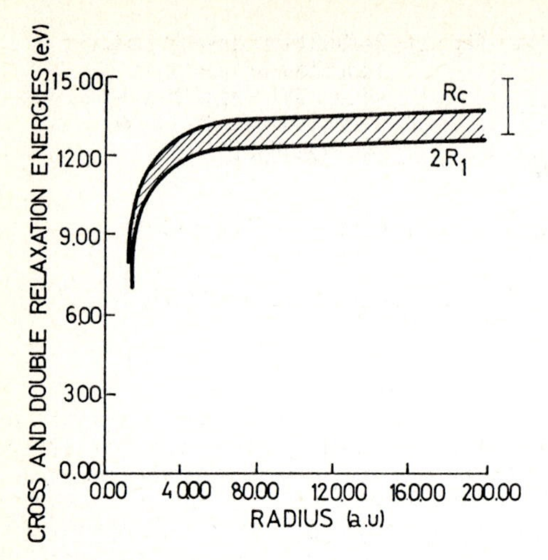

Fig. 3: Deviation R_C given by the shaded area as a function of particle size for magnesium. DFT result is also indicated.

The comparison between the $R_C = 2\ R_1$ rule and the Auger cross-relaxation energy calculated through a pseudopotential with a smaller cut-off radius r_c^* is shown in the case ot magnesium in Fig. 3. It appears that the deviation ΔR_C is not negligible and increases with the particle size. Deviations can also be evidenced by DFT calculations in the bulk limit (see Fig. 3). Both methods, completely different, demonstrate that non-linear effects are important.

5 Conclusions

The main conclusions of this work are:

- the screening energy is almost independent of the position of the core hole in the particle (except near the surface).
- the single hole relaxation energy R_1 and the Auger cross-relaxation energy R_C have been calculated in the range of small particles for Na, Mg and Al. R_1 and R_C evidence a pronounced reduction below 40 a.u. and convergence to the bulk limit is slow.
- comparison with our DFT calculations in the bulk limit show a good agreement, specially in the case of Na and Mg. ALMBLADH and BARTH /8/, using a spherical DFT model, have obtained similar results.

The good agreement between our DFT calculations, spherical DFT calculations and pseudopotential linear-response electron-gas model - three completely different methods - must be pointed out.

- ΔR_C variations are not negligible and depend on the particle size.

1. M.G. Mason: Phys.Rev. B 27, 748 (1983)
2. M. Cini: Surf.Sci. 62, 148 (1977)
3. M. Cini and P.Ascarelli: J.Phys. F4, 1998 (1974)
4. N.J. Castellani, D.B.Leroy and W. Lambrecht: Phys.Stat.Sol. b 129, K 69 (1985)
5. N.J. Castellani, D.B.Leroy and W. Lambrecht: Chem.Phys. 95, 459 (1985)
6. W. Lambrecht, N.J. Castellani and D.B. Leroy: J.Electr.Spectr.Rel.Phenom 37, 87 (1985)
7. C.D. Wagner and P. Biloen: Surf.Sci. 35, 82 (1973)
8. C.O. Almbladh and U. von Barth: Phys.Rev. B13, 3307 (1976)

Electron-Hole Pairs Bulk Production in Simple Metals

J. Giraldo[1], *P. Apell*[2], *and R. Monreal*[3]

[1]Departamento de Física, Universidad Nacional de Colombia, Bogotá, Colombia
[2]Institute of Theoretical Physics, Chalmers University of Technology, S-412 96 Göteborg, Sweden
[3]Departamento de Física de Sólidos, Universidad Autónoma de Madrid, E-28049 Madrid, Spain

The electron (e-h) pair excitation in the bulk of a metal has been considered in two recent studies. Whereas one concentrates on the bulk contribution to photoemission /1/, the other separates surface and bulk effects in the outcome of the electron energy loss (EEL) experiment /2/. These excitation mechanisms are found to contribute even more to the energy loss of the scattering electrons than the standard treatment. Since photoyield and EEL are a consequence of induced e-h pairs excitation, it is worthful to envisage this common feature within the same scheme. A linking for both phenomena may be found in the outstanding results of the optical experiment, the anomalous skin effect (ASE) /3/.

In this report we present results for the optical absorption due to e-h pairs excitation in the bulk of simple metals. The pure surface effect arises in the strong variations of the field within the very surface region of the solid. Interband transitions are essentially bulk excitations. It is a common practice to neglect the intraband volume term. It will be seen that this contribution to absorption can be handled within reasonably simple physical models. We limit ourselves to present an illustrative derivation of this contribution in free-electron like metals perturbed by either photons or electrons. For the purposes of this work the semiclassical infinite barrier (SCIB) model is used to describe the metal response. It is known to be quantitatively insufficient to describe the optical e-h pairs absorption in solids when regarded as a surface effect /4/. In spite of this shortcoming we have chosen that model for two reasons. Firstly, we are interested in the bulk contribution alone. In this case it can be expected to yield saisfactory results provided any interference effect between surface and bulk can be neglected /5/. Secondly, this model admits an analytical treatment to a certain extent, thereby making it more transparent for our discussion and interpretation. Finally, it should be mentioned that for very low frequencies it becomes dominant over the surface contribution.

The bulk absorption due to e-h pairs excitation can be described using the surface impedance Z /6/.For our purposes it is conveniently divided in transverse and longitudinal parts /7/:

$$Z_\tau = iq\{\int dp_1 [\sin^2\theta/\varepsilon - p_1^2/(q_1^2 - q^2\varepsilon_\tau)]/\pi q_1^2\}, \qquad (1)$$

$$Z_\ell = iq\{\int dp_1 (1/\varepsilon_\ell - 1/\varepsilon)\sin^2\theta/\pi q_1^2\}, \qquad (2)$$

where $q=\omega/c$ is the free space wave vector at frequency ω for a wave incident at an angle θ with the surface normal, and

$q_{\parallel}^2 = p^2 + k_{\parallel}^2$ ($k_{\parallel} = q\sin\theta$ is a conserved quantity). ε_ℓ and ε_τ are the longitudinal and transverse dielectric functions characterizing the metal response. Their bulk RPA forms are used in our calculations /8/. In the limit $q \to 0$, both ε_ℓ and ε_τ approach Drude form with relaxation τ, ω_p being the plasma frequency:

$$\varepsilon(\omega) = 1 - \omega_p^2/\omega(\omega + i/\tau). \qquad (3)$$

The longitudinal contribution is seen to be zero for a transverse EM field (s-polarized light). It becomes eventually the main contribution when the parallel component of momentum of the incident field increases. In order to distinguish them clearly both situations will be considered separately.

A. Transverse Contribution. The basic ingredient for the optical absorption is the real part of the surface impedance, as expressed by the well-known relationship $A = 4\cos\theta\,\mathrm{Re}(Z)/|\cos\theta + Z|^2$ /4/, where Re(Z) denotes the real part of Z. Under the assumption of normal incidence (transverse field), Re(Z) takes the form:

$$\mathrm{Re}(Z) = q^3 \int dq_1\, \mathrm{Im}(\varepsilon_\tau)/\pi\{[q_1^2 - q^2\,\mathrm{Re}(\varepsilon_\tau)]^2 + [q^2\,\mathrm{Im}(\varepsilon_\tau)]^2\}. \qquad (4)$$

In order to facilitate a division into the various optical regimes, we proceed to make some approximations based on the very small value of q compared to q characterizing e-h pairs. For that purpose we replace $\mathrm{Re}(\varepsilon_\tau)$ by $\varepsilon(\omega)$ given by (3) in the limit $\tau \to \infty$. It is convenient at this point to introduce three length scale parameters in the following way. The classical mean free path, simply related to the relaxation time through $\ell = V_F\tau$, where V_F is of the order of Fermi velocity, the electronic length, defined as the distance traveled by an electron during one period of oscillation of the EM field $\ell_0 = V_F/\omega$ and the classical optical length $\ell_e = c/\omega$, i.e., the penetration depth at low frequencies. In this way we compare two different contributions to the optical absorption.

i. Classical absorption. It is obtained using Drude form of the dielectric constant. It is found to be given by

$$\mathrm{Re}(Z_{cl}) = (1/2\omega_p\tau)\{2/[1 + (1 + (\alpha/r)^2)^{1/2}]\}^{1/2}, \qquad (5)$$

where $\alpha = (1/\omega_p\tau)(c/V_F)$ and $r = \ell_0/\ell_e$. It should be noticed that this expression does not contain any information about the e-h pair mechanism, since a local version of ε has been employed.

ii. Electron-hole pairs excitation. It is estimated through the following approximation to ε_τ /6/: $\varepsilon_\tau(q,\omega) = 1 - \omega_p^2/(\omega^2 - \beta^2 q^2)$. At low frequencies one readily gets:

$$\mathrm{Re}(Z_{eh}) = (1/\omega_p\tau)\beta(\ell/\ell_e)(\beta r)^2/[1 + (\beta r)^2]^2. \qquad (6)$$

The dispersion coefficient β is a number of the order of unity. V_F/c is about 10^{-2}. A polar form of $\varepsilon_\tau(q,\omega)$ above has been used:
$\mathrm{Im}(\varepsilon_\tau) = (\pi\omega_p^2/2\omega\beta)[\delta(\omega/\beta - q_1) + \delta(\omega/\beta + q_1)]$.

B. Longitudinal Field. In this case the e-h pairs production is better seen in EEL experiments. Now the imaginary part of the reflection coefficient $\rho(k_{\parallel},\omega)$ gives the spectral strength for the excitation of the solid /8/. For an electron very close to the surface retardation effects may be neglected. The $c \to \infty$ limit yields

$$\mathrm{Im}\,\rho(k_{\parallel},\omega) = \mathrm{Im}[(1-\tilde{Z})/(1+\tilde{Z})], \tag{7}$$

where $\tilde{Z}=\lim_{q\to 0} qZ/ik_{\parallel}$, and Z is actually Z_{ℓ} given by (2).

Instead of θ, for electrons it is more convenient to work with $k_{\parallel}$. This is roughly ω/V_i for an incoming electron with velocity V_i losing an energy ω to the solid. The length scales are the same as before, except for ℓ_e which must be replaced by $1/k_{\parallel}$. This is the characteristic length-scale for the penetration of the electron EM field.

The classical contribution follows a similar behavior to the previous case. The e-h pairs contribution to ρ may be written in terms of the surface dielectric function $\varepsilon_s(k_{\parallel},\omega)=k_{\parallel}\int dp_1/\pi q_1^2 \varepsilon_{\ell}$:

$$\mathrm{Im}(\rho_{eh}) = \mathrm{Im}[(1-\varepsilon_s)/(1+\varepsilon_s)] = 2\,\mathrm{Im}(-\varepsilon_s)/|1+\varepsilon_s|^2 . \tag{8}$$

In the spirit of our transverse approximation one retrieves for $\mathrm{Im}(\rho_{eh})$ the same form recently derived by PERSSON et al./2/. Their calculation is therefore equivalent to an approximate SCIB model. Taking the Lindhard expression for $\mathrm{Im}(\varepsilon_{\ell})$ and performing the integration over the e-h pairs spectrum, one gets a slightly different version for $\mathrm{Im}(\rho_{eh})$. Using the same notation as in /2/:

$$\mathrm{Im}(\rho_{eh}) = (\omega/\omega_p)\nu^3 G(\nu)/2, \tag{9}$$

where $\nu=\ell_e/\ell_0$ (compare with 1/r above) and $G(\nu)$ is now changed to

$$G(\nu) = \begin{cases} 8[1-(1+1/2\nu^2)(1-1/\nu^2)^{1/2}], & \nu>1, \\ 4[1-(\nu_0/\nu)^2]^{1/2}[2+(\nu_0/\nu)^2]A, & \nu_0<\nu<1, \\ 0, & \nu<\nu_0=h\omega/4E_F, \end{cases} \tag{10}$$

with $A=1+3(\nu_0^4)/8$. The main difference here is that there is no e-h pair excitation when $\nu<\nu_0$ ($k_{\parallel}>2k_F$). For $\nu<1$ G is very close to 8, the value given in /2/. For the experiments considered by these authors, $k_{\parallel}\ll k_F$. Therefore our result has not any bearing on their conclusions. However some recent studies should deserve a deeper analysis of that work /9/.

In what follows we present plots of some relevant expressions. It should be noticed that $r=\ell_0/\ell_e \propto 1/\omega$. Figure 1 exhibits the behavior of both Z_{eh} and Z_{cl} when e-h pairs play an important role ($\ell/\ell_e \gg 1$). The inset in Fig.1 was obtained with $\ell/\ell_e \ll 1$, where a classical description is expected to give reliable results. The different regions of the skin effect are explicitly indicated in both figures. Figure 2 establishes a comparison between the results obtained by using the Lindhard transverse dielectric function. In this figure β has been adjusted to get the maxima of both calculations at the same frequency. The factor characterizing the material (absent in the bulk dielectric function when assuming a homogeneous electron gas) has been chosen to predict the same value as the numerical calculation at $\ell_0=\ell_e$. Exact agreement cannot be claimed, but it is encouraging that the main features of the absorption coefficient within a region of tens to hundreds of meV can be obtained using the simple approach proposed in this paper. Regarding EEL experiments, we have seen that the SCIB model and its approximation only give the same results at very low $k_{\parallel}$ and intermediate frequencies below the plasma oscillation.

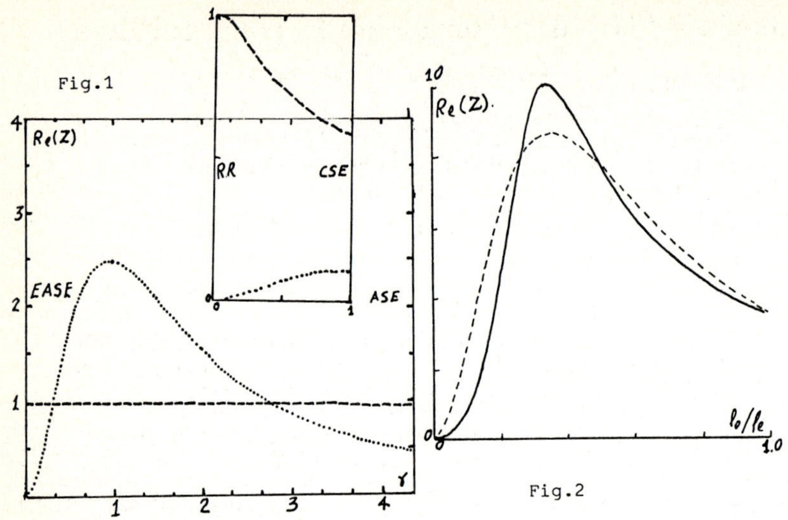

Fig.1. Schematic picture of the e-h pairs' contribution (dotted line) and the classical one (dashed) to the surface impedance in a semi-infinite metal. Re(Z) is plotted in units of $(1/2)\omega_p\tau$. The inset is for $\ell \ll \ell_e$ .The main figure shows the same results for $\ell \gg \ell_e$.RR stands for the relaxation region and CSE is the classical skin - effect region. ASE shows the region of the anomalous skin effect and EASE designs the extreme anomalous region. The approximate expressions (5) and (6) were used.

Fig.2. Results from a numerical calculation using the full Lindhard dielectric function (continuous) and our approximate approach given by (6). The results have been normalized to the maximum value obtained with the full ε_τ. β has been chosen as 2.77 to get the maxima at the same place.

References

1. B.C.Meyers and T.E.Feuchtwang: Phys. Rev. B27, 2030 (1983)
2. B.N.J.Persson and S.Andersson: Phys. Rev. B29, 4382 (1984)
3. G.E.H.Reuter and E.H.Sondheimer: Proc. Roy. Soc. London, A195, 336 (1948)
4. P.Apell: Physica Scripta 24, 795 (1981)
5. B.N.J.Persson and E.Zaremba: Phys. Rev. B31, 1863 (1985)
6. F.Garcia-Moliner and F.Flores: Introduction to the theory of solid surfaces, Cambridge University Press, New York (1979)
7. S.Lundqvist, P.Apell and J.Giraldo: Rev.Col.Física (to appear)
8. J.Lindhard: Kgl. Danske Videnskab. Selskab, Mat. Fys. Medd. 28, No.8, (1954)
9. A.Liebsch: Europhys. Lett. 1, 361 (1986)

Two-Dimensional Dipolar Complexes in Helium Films

U. de Freitas[1], *J.P. Rino*[2], *and N. Studart*[2]

[1]Departamento de Física e Ciência dos Materiais, Instituto de Física e Química de São Carlos, Universidade de São Paulo, 13560 São Carlos, SP, Brazil

[2]Departamento de Física, Universidade Federal de São Carlos, 13560 São Carlos, SP, Brazil

The existence of surface electrons localized above a liquid helium film which wets a solid substrate is well established both theoretically and experimentally /1-3/. These electronic states are induced by image forces coming from the film and substrate and by a potential barrier which prevents the electron from penetrating the liquid, resulting in a quantum well structure in the direction perpendicular to the surface. Along the parallel direction, the electrons behave like a two-dimensional gas interacting via a pair-potential obtained from the solution of Poisson's equation with appropriate boundary conditions. In the experiments performed on thick films or bulk helium, the highest attainable electron densities are in the range such that the Fermi energy is much smaller than k_BT. Therefore the electron gas can be described by a 2D classical plasma with a 3D interaction and plasma parameter defined as the ratio of the mean potential energy and the mean kinetic energy, $\Gamma=\langle V\rangle/\langle K\rangle=e^2/ak_BT$, with $\pi a^2 n=1$ and n being the particle density. The correlational properties of the 2D plasma have been studied by Rino et al. /4/ in the liquid phase and by Peeters /5/ in the solid phase. An interesting feature of this system is that the particle interaction can be changed through the variation of the distance between the electron layer and the substrate. For example, for a metal substrate and thin films, the strong screening of the electron-electron interaction by the substrate leads to a dipolar-type potential.

Recently, interest has been growing in a similar charged system which consists of surface electrons on the helium films located above positive ions trapped on a dielectric substrate. It is well known that positive ions can be easily injected into a helium film and be attracted to the solid substrate by image forces /6/. The system is stable and the minimum film thickness above which the ion cannot be detached from the substrate in the presence of the electron field is estimated as 20-70Å, depending on the kind of substrate /7/. The electron-ion complex has large dipole moments and forms a quasiparticle state that has been called a diplon /8/. The existence of this new surface state has been demonstrated by measuring the microwave absorption of the electrons /2/.

In this work, we investigate the properties of a system of diplons in the classical regime for different values of the film thickness and several kind of substrates.

The interaction energy between two diplons lying on a substrate with dielectric ε_s is

$$\Phi(r) = \frac{4e^2}{\varepsilon_s+1}\left(\frac{1}{r} - \frac{1}{(r^2+d^2)^{1/2}}\right) + \frac{e^2(\varepsilon_s-1)}{\varepsilon_s+1}\left(\frac{1}{r} - \frac{1}{(r^2+(2d)^2)^{1/2}}\right), \qquad (1)$$

where r is the distance between dipoles and d is the distance between the electron layer and the ion layer. In the above expression the dielectric constant of the liquid He was assumed to be one. For small interparticle distance ($r<<d$), we have essentially a 3D Coulombic potential $\Phi(r)=e^{\star 2}/r$, with a renormalized charge $e^{\star}=e\sqrt{(\varepsilon_s+3)/(\varepsilon_s+1)}$. If the diplons are far apart ($r>>d$) one has a bare dipolar interaction $\Phi(r)=\alpha e^2d^2/r^3$ with $\alpha=2\varepsilon_s(\varepsilon_s+1)^{-1}$. Our calculations require the Fourier Transform of the potential (Eq.1)

$$\Phi(q) = \frac{2\pi e^2}{q} F(qd) , \tag{2}$$

where $F(qd)=\delta_1(1-\exp(-qd))+\delta_2(1-\exp(-2qd))$, with $\delta_1=4/(\varepsilon_s+1)$ and $\delta_2=(\varepsilon_s-1)/(\varepsilon_s+1)$. In the long-wavelength limit, $2qd<<1$, $F(qd)=2qd$ and $\Phi(q)$ does not depend on q and ε_s and has the same form for both diplons and surface electrons above a helium film and metallic substrate, $\Phi(q)=4\pi e^2 d$ /4/. In the other limi $qd>>1$, one has the usual $\Phi(q)=2\pi e^{\star 2}/q$, which is similar to the system of electrons on bulk helium, but with different effective electron charge.

In order to evaluate the static properties and the collective excitations of the diplon gas, we use a mean-field approximation, the STLS method, where short range correlations are included in the system. In this scheme, the equation of motion of the one-particle distribution function is truncated und the assumption that the coupling of the particle to the medium is given throu the static pair correlation function $g(\vec{r})$. Setting $g(\vec{r})=1$, corresponds to the well-known classical RPA and implies that the coupling between the partic under consideration with the environment is neglected. Self-consistency is introduced by application of the fluctuation dissipation theorem linking $S(\vec{q})$ the Fourier transform of $g(\vec{r})$, to the density-density response function of the system /9/. The result in $\vec{k}$-space is a non-linear integral equation for the static Structure Factor

$$S(\vec{q}) = \frac{1}{1+\frac{n}{k_BT}\Phi(\vec{q})\left(1-G(\vec{q})\right)} , \tag{3}$$

where

$$G(\vec{q}) = -\frac{1}{n}\int \frac{\Phi(\vec{k})}{\Phi(\vec{q})}\frac{\vec{k}.\vec{q}}{q^2}\left(S(\vec{q}-\vec{k})-1\right)\frac{d\vec{k}}{(2\pi)^2} \tag{4}$$

is the local field correction, responsible for the inclusion of short-range correlations. The self-consistent solution of equations (3) and (4) is obtai by the iteration method. With a reasonable input $S(\vec{q})$, the local field $G(\vec{q})$ calculated and a new $S(\vec{q})$ is obtained. The whole procedure is repeated until self-consistence in $S(\vec{q})$ is achieved with the required precision, ε. In our se $\varepsilon=10^{-4}$. Our numerical calculations were performed for several plasma para ters, film thicknesses and sapphire, neon and metallic substrates. The resul of the self-consistent local field function as a function of the wave number are shown in Fig.1 for $\Gamma=1$ and several values of the film thickness which is in units of the core radius a. For large d, we get the expected behavior of $G(\vec{q})$ in both 2D and 3D usual Coulomb interactions. As d decreases, $G(\vec{q})$ tend to a constant value independent of q, reflecting the fact that the particle interaction is a constant for thin films, irrespective of the substrate. In this limit, our results are the same as those of the 2D electron system on helium films adsorbed on metallic substrate /4/. In RPA, $G(q)=0$. In Fig. 2 w present the Structure Factor in the self-consistent field approximation (SCF for $\Gamma=1$ and a sapphire substrate. For comparison, we plot the results within the framework of RPA. We have also calculated $S(\vec{q})$ for the other substrates. Sizable differences in $S(\vec{q})$ are found only between a nonmetallic substrate a a metallic substrate.

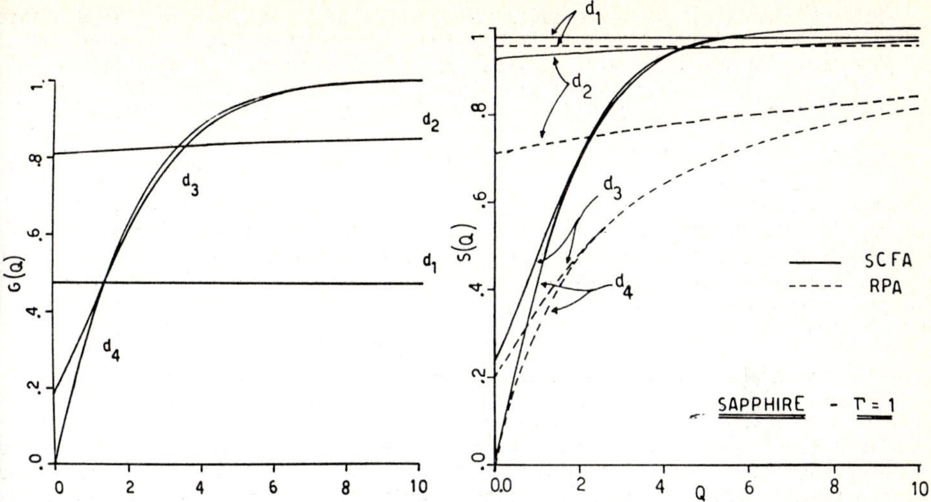

Fig.1-Local Field correction versus qa in the SCFA for several values of the film thickness (d_1=0.01, d_2=0.1, d_3=1.0, d_4=100), the plasma parameter Γ=1, and a sapphire substrate.

Fig.2-Structure Factor S(q) in both SCFA(——) and RPA(- - -) for a sapphire substrate, Γ=1. Values of d are the same as in Fig.1.

The Correlation Energy of the system is obtained from the Structure Factor as

$$E_C = \frac{n}{4\pi} \int_0^\infty d\vec{k} \left[S(\vec{k}) - 1 \right] \Phi(\vec{k}) \quad . \tag{5}$$

The results for E_c/nk_BT as a function of the distance between the electron layer and the ion layer is shown in Fig.3 for Γ=0.1 and several substrates. With decreasing d, the correlation energy decreases because of the enhanced screening of the particle potential by the substrate. We plot also the results from RPA. In the bulk helium film ($d \to \infty$), the correlation energy tends to different constant values depending on the substrate.

The collective excitation spectrum of the diplon plasma can be determined from the poles of the density-density response function. In the long wavelength limit, we obtain the dispersion relation in SCFA as

$$\omega_q^2 = \frac{n}{m} q^2 \psi(q) \left(1 + \frac{3k_BT}{n\psi(q)} \right) , \tag{6}$$

where $\psi(q)=\Phi(q)(1-G(q))$ is the effective diplon-diplon potential. Using the self-consistent $G(\vec{q})$, we have calculated the dispersion curves which are shown in Fig.4 for a neon substrate. Two distinct behaviors are clearly apparent in Fig.4: in the thin film limit, we found a sound-like mode $\omega_q = cq$ with $c = [3+4\Gamma(1-G(0)d)]^{1/2}/2\Gamma$ and in the opposite limit we obtain a typical 2D plasma mode $\omega_q=(2\pi n\, e^{*2}q/m)^{1/2}$.

In conclusion, we have studied the Struture Factor, Pair Correlation Function, the Correlation Energy and the Plasmon Spectrum of the liquid phase of a 2D system of dipolar complexes in helium film adsorbed on a solid substrate. The importance of the film thickness and the substrate on the screening properties of such a system have been shown.

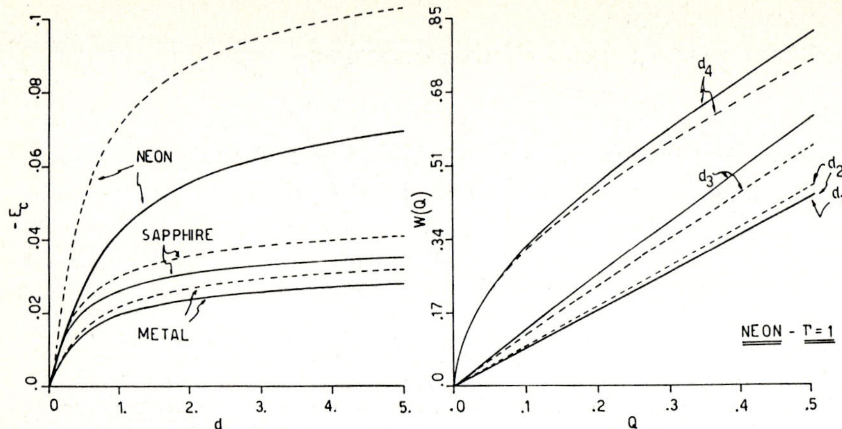

Fig.3-Correlation Energy as function of the film thickness d in units of the average particle-particle distance a for different substrates in both SCFA(——) and RPA(---).

Fig.4-Plasmon dispersion curves in units of $\omega=2\pi ne^2/(mk_BT)^{1/2}$ for a neon substrate. Dashed lines are the results from RPA. The parameters are the same as in Fig.2.

This work was supported in part by CNPq and CAPES. One of us (U.F.) acknowledges Universidade Federal da Paraība for leave of absence. We thank Dr. Gilmar Marques for a critical reading of the manuscript.

References

1. Yu.P.Monarkha, Fiz.Nizk.Temp.1526 (1975) [Sov. J. Low Temp. Phys. 1,258 (1975)].
2. V.I.Karamushko,Yu.Z.Kovdrya,F.F.Mende and V.A.Nikolaenko, Fiz. Nizk. Temp. 8,219 (1982) [Sov. J. Low Temp. Phys. 8,109 (1982)].
3. N.Studart and O.Hipolito,Rev.Bras.Fís., 16(2), (1986).
4. J.P.Rino, N.Studart and O.Hipolito,Phys. Rev. B29, 2584 (1984).
5. F.M.Peeters, Phys. Rev.B30,159(1984).
6. B.Maraviglia, Phys.Lett.25A, 99 (1967).
7. Yu.P.Monarkha and Yu.Z.Kovdrya,Fiz.Nizk. Temp. 8,215 (1982) [Sov. J. Low Temp. Phys. 8,107 (1982)].
8. Yu.P.Monarkha,Fiz. Nizk.Temp.,8,1133 (1982) [Sov. J. Low. Temp. 8, 571 (1982)].
9. K.S.Singwi, M.P.Tosi,R.H.Land and A.Sjolander, Phys. Rev. 176, 589 (1968).

Instrumental Aspects of Ultraviolet Inverse Photoemission

V. Dose

Max-Planck-Institut für Plasmaphysik, EURATOM Association,
D-8046 Garching/München, Fed. Rep. of Germany

Ultraviolet inverse photoemission has matured as the principal spectroscopic technique for the investigation of unoccupied electronic states in solids and at surfaces /1/. The basic phenomenon, namely emission of radiation from solids under electron bombardment has now been known for ninety years as x-ray emission. Its relation to the well-established photoelectron spectroscopy (PES) /2/ will be described by reference to fig. 1. The left hand panel shows an energy diagram for photoemission. In ultraviolet photoelectron spectroscopy a quantum $\hbar\omega$ of monochromatic radiation is absorbed by a valence electron in the solid at initial energy E_i.The electron is thereby raised to a final previously empty electronic state of energy

$$E_f = E_i + \hbar\omega . \quad (1)$$

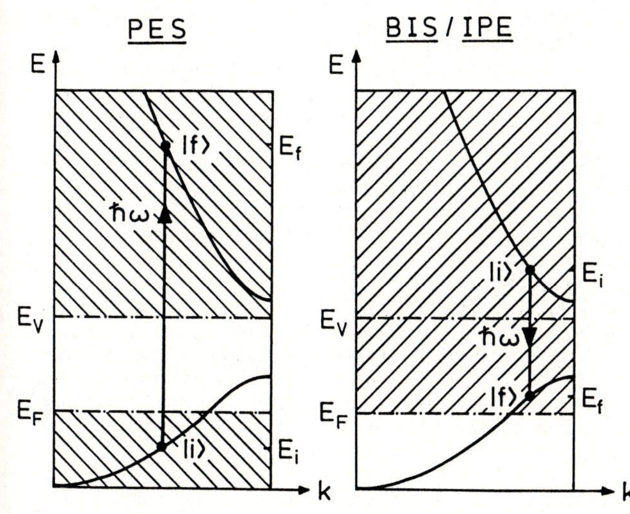

Fig.1: Schematic of the photoemission and the inverse photoemission processes.

If the final state energy E_f lies above the vacuum level of the solid, the electron may be emitted into the vacuum and can be subject to energy analysis with the help of some suitable electron spectrometer. From the measured final state energy E_f and the known quantum energy $\hbar\omega$ the energy of the initial state of the electron can then be deduced. In experiments which average over the electrons' momentum such as in PES from polycrystalline states, the intensity of electron emission at final state energy E_f is a measure of the initial state occupation, hence the density of occupied valence states. If on the other hand in PES from single crystals the electron energy analyzer accepts only electrons with a well-specified emission direction, the energy determination fixes the emitted electrons' momentum in the final state simultaneously. From the conservation of momentum

$$\underline{k}_f - \underline{k}_i = \underline{q} + \underline{G}, \tag{2}$$

where $\underline{k}_f$, $\underline{k}_i$ are the electron momenta in the initial and final electronic states, $\underline{q}$ the momentum of the photon and $\underline{G}$ a reciprocal lattice vector, the initial state momentum $\underline{k}_i$ can be deduced. This is particularly easy if the photon momentum can be neglected in the momentum balance.

This is always the case in the ultraviolet spectral range since the momentum of a photon of 1 atomic unit of energy (27.21 eV) is only 1/137 atomic units as compared to $|\underline{G}|$ which is always of the order of unity. Radiative transitions in UPS occur therefore vertically in a reduced zone scheme. The information provided by a PES experiment comprises the energy versus momentum dispersion of the initial occupied electronic band below the Fermi level and the final initially unoccupied electronic band above the vacuum level.

It is immediately obvious from the left hand panel of fig. 1 that electronic states with energy between the Fermi level E_F and the vacuum level E_V (unshaded area) are inaccessible to ordinary photoemission. These states can neither serve as initial states since they are unoccupied nor can electrons excited to this range leave the solid. The apparent gap of information is closed by inverse photoemission. A schematic representation of the process is given in the right hand panel of fig. 1. Electrons of initial energy E_i impinge on a solid and undergo radiative transitions emitting light of energy $\hbar\omega$. From a knowledge of E_i and the photon energy $\hbar\omega$ the energy of the final previously unoccupied electronic state can be deduced. In experiments which average over the initial state electron momentum such as in the case of polycrystalline samples the intensity of the emitted light provides a measure of the probability for the transition, which according to Fermi's golden rule is proportional to the density of final electronic states. This type of radiative transition which is to a certain degree the time reverse of the ordinary photoemission connects two previously empty electronic states of the solid with the only restriction that the initial state energy must exceed the vacuum energy E_V of the solid. No restriction (except of course for energy conservation) exists for the energy of the final state above E_F.

Exactly analogous to ordinary PES, IPE on well-ordered single crystals employing electron beams of well-defined energy and momentum discloses even further information. In such a case not only energy but also momentum of the final electronic state can be inferred from the electron beam data and the known transition energy. The restriction of the experiment to the ultraviolet spectral range is even more important in this case. It allows not only the simplified vertical transition interpretation, but also for photon collection over substantial solid angles without blurring the momentum conservation. This on the other hand is of crucial importance in setting up an IPE experiment since PES and IPE though governed by the same dipole matrix element differ in transition probabilites due to phase space factors by roughly α^2, where $\alpha = 1/137$ is Sommerfeld's fine structure constant. This is the well-known Milne relation connecting photoionization and radiative capture /3/.

Having established the full equivalency of PES and IPE, a final inspection of fig. 1 shows that the energetic range common to both spectroscopies lies above the vacuum level. Electronic states with energy below E_F are exclusively accessible by PES while those with energy between E_F and E_V are exclusively accessible to IPE. In view of the substantial activity persisting in PES now for roughly twenty years we can predict for IPE quite a prosperous future.

1. Experimental Requirements

The experimental requirements for the two spectroscopies, PES and IPE, result in a straightforward manner from the introductory discussion of the physical processes. PES requires a source of monochromatic radiation for excitation and an electron spectrometer for energy and momentum analysis of the emitted electron. IPE requires an electron beam of well-defined energy and momentum for excitation and an energy selective photon detector. The usual excitation sources in PES work are resonance lamps offering fixed energy highly monochromatic radiation /4/ or tunable monochromators coupled to "white light" synchrotron radiation sources /5/. The analogous instruments in IPE are photon detection at fixed energy usually accomplished with some sort of band-pass detector to be discussed below /6/, or photon detection employing a tunable energy photon detector based for example on either diffraction /7/ or refraction /8/. The simplest approach in PES/IPE is the resonance lamp/ band-pass detector excitation / detection scheme. The necessity for the considerably more complicated instrumentation for work at variable photon energy arises from the constraints imposed by conservation of energy in both processes: if we prefix the excitation/detection energy $\hbar\omega$ and the electron momentum in the final/initial state we will only accidentally find a pair of electronic bands at this particular reduced momentum which are separated by $\hbar\omega$. The system is overdetermined. If on the other hand the photon energy in the excitation/detection channel is at disposal, it can always be tuned to lock in a direct optical transition.

We shall abandon at this point the parallel discussion of PES and IPE which was pursued so far to point out the close mutual relationship. We shall rather concentrate in the following on photon detectors and electron sources so far employed in IPE.

2. Tunable Photon Detectors

IPE experiments date back to the early forties /9/. They were initially performed in the 1 keV region using crystal monochromators for energy selective photon detection. In particular K. Ulmer and coworkers have extended the range of detection energies down to 152 eV using a variety of crystals partly grown by highly sophisticated techniques /10/. All these early experiments were conducted in the so-called isochromat (constant colour) mode. The monochromator is operated at a fixed preset pass energy $\hbar\omega$ in this mode and the yield of photons of energy $\hbar\omega$ is measured as a function of the energy of the electrons impinging on the sample /11/. A modern version of such a keV isochromat spectrometer has been reported by Lang and Baer who employed the monochromator for an

Al K_α-source in an XPS spectrometer and thereby nicely combined valence band (XPS) and conduction band (BIS) density of states studies in the same experiment /12/.

The first experiment employing a vacuum ultraviolet reflection grating has been reported by Merz /13/. The mean detection energy was 16 eV and the resolution 2 eV. The experiment was quickly abandoned due to excessively low count rates. Chauvet and Baptist were the next to employ a monochromator in an ultraviolet inverse photoemission experiment /14/. The dispersive element in their experiment was a holographic grating recorded on a toroidal blank. The reported characteristics are 30 x 30 mm^2 area, a focal length of 320 mm and a groove density of 550/mm. The angle between the directions of light incidence and exit was 142°. This translates into an acceptance of f/33 using the above grating data. The energy range covered by this instrument is 20 to 100 eV. A sensitivity of 3 x 10^5 counts per Coulomb per eV in the d-band region of platinum was reported.

A suggestive further improvement of the monochromator approach is the parallel detection of different photon energies. Fauster et al. /7/ were the first who mounted a position sensitive detector consisting of two micro-channel plates for amplification and a resistive anode in the image focal plane of their Seya Namioka f/45 monochromator. The first channel plate was sensitized for the UV photon energy range by evaporation of CsI. The increase in sensitivity of the apparatus due to parallel detection was found to be two orders of magnitude. A performance figure of 1.8 x 10^4 counts per Coulomb per eV was reported for gold.

Instruments with higher sensitivity require gratings with better f-numbers. The latest generation of monochromators employ gratings with acceptance as large as f/3 in normal incidence /15/. The highest brightness is obtained with equal source and image distances. An example is shown in figure 2. The instrument is equipped with two holographic gratings of size 90 x 120 mm^2, focal length of 204 mm, and groove densities of 1200/mm and 2400/mm, respectively. Photons are detected with a position sensitive detector. The energy ranges are 8–18 eV and 16 - 40 eV, respectively. Several entrance slits between 0.1 mm and 0.8 mm are available for optimum compromise between sensitivity and resolution.

A certain disadvantage of monochromators with equal source and image distances at normal incidence is the immediate neighborhood of the electron gun exciting the sample radiation and the channel plate photon detector. Extreme precautions have to be taken in order to shield the detector from stray electrons from the gun. In the case that electric bias is employed one may encounter difficulties with ions released in getter ion pumped vacuum systems. Vacua well below 10^{-10} Torr are therefore mandatory. This in turn requires separate vacuum chambers for IPE and sample preparation and characterization associated with long and complicated transfer-mechanisms. A way out of these problems is a design with unequal source and image distances /16/. This sacrifice in brightness allows moreover a design of the monochromator as an add-on instrument to existing surface analytical systems.

Setting up a vacuum UV monochromator for IPE experiments requires substantial financial investments. It is therefore well worth mentioning that quite a different solution to a tunable photon detector has been proposed. Hulbert et al. /17/ and Childs et al. /8/ exploit the strong chromatic aberration of a LiF lens in the vicinity of the LiF transmission cutoff. Very briefly the principle of this instrument rests on the fact

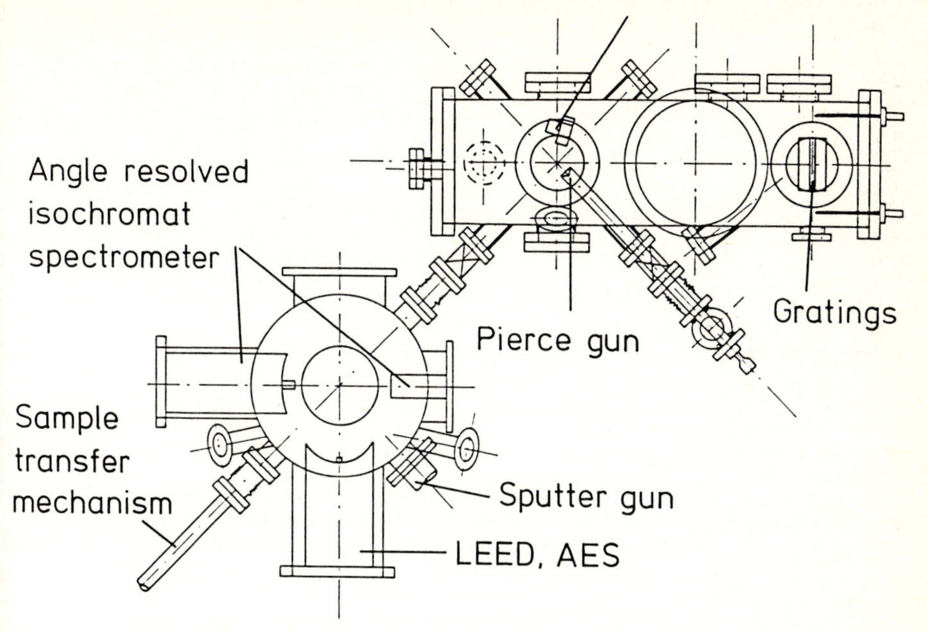

Fig.2: Spectrometer for inverse photoemission studies based on a VUV normal incidence monochromator.

that a point source of radiation at distance g will be imaged into a point a distance b from the lens where b, due to chromatic aberration, depends on the photon energy. In the 8 - 12 eV range, a linear dispersion of the image for fixed g and $g \simeq b$ at 10 eV of -0.6 f/eV is obtained where f is the focal length at 10 eV. An overall resolution of 0.7 eV was reported for a 90 mm diameter LiF lens with approximate acceptance of f/5. Though no direct sensitivity figures are given in /17/we expect a similar performance as for grating spectrographs of similar acceptance.

3. Band-Pass Detectors

In this section we shall consider quite a different approach to energy selective photon detection. Its application to IPE gave quite a substantial impetus to this field. Its obvious success encouraged a reconsideration of the classical monochromator experiment in spite of its low sensitivity and high costs.

Band-pass detectors for photon detection in the vacuum ultraviolet have been known for a long time in atomic collision and upper atmospheric physics /18/. The first application of such a device, to be specific, a He-I_2 Geiger counter with a CaF_2 entrance window, was reported in 1977 /6/. The counter was made from a 24 mm diameter stainless steel shell with central stainless steel electrode of 1.5 mm diameter. The counter filling consists of 400 Torr of helium and a few crystals of iodine. The entrance window is a 20 mm diameter 2 mm thick CaF_2 single crystal. The high energy limit of the counter is provided by the transmission cutoff of the

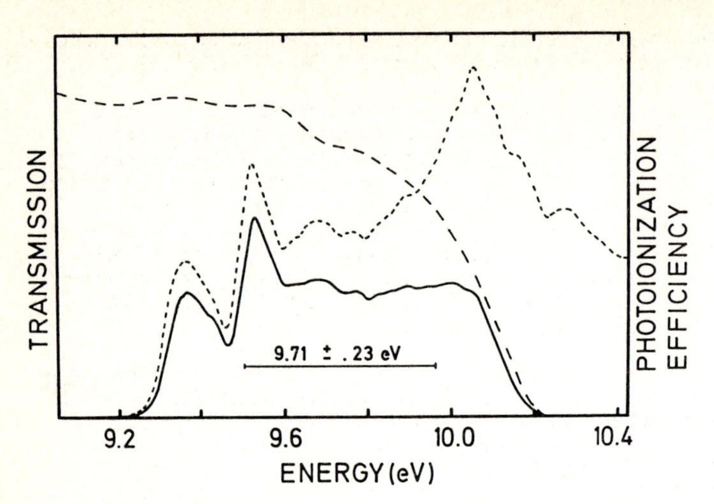

Fig.3: Band-pass characteristics of the He-I_2 Geiger counter.

CaF_2 entrance window at 10.2 eV /19/. The lower detection limit is given by the threshold for molecular photoionization of iodine /20/

$$I_2 + \hbar\omega \rightarrow I_2^+ + e^- \qquad (3)$$

at $\hbar\omega$ = 9.23 eV. The energy dependence of both the CaF_2 transmission and (3) and their combined action to form a band-pass with (9.7 $\pm$.23) eV mean energy are illustrated in fig. 3. Note that the rms deviation σ = 230 meV is given to characterize the resolution. Its relation to the widely used full width at half maximum is FWHM = 3.5 x σ for a rectangular distribution function and FWHM = 2.35 x σ for Gaussian.

The iodine filling also resolves in an elegant way the problem of photoelectrons released from the counter wall, also by radiation below the I_2 photoionization threshold. Such photoelectrons are a potential source for a deterioration of the counter resolution. Photoelectrons from the counter wall are thermalized by the helium filling and after having reached a sufficiently low energy captured in the dissociative attachment reaction

$$I_2 + e^- \rightarrow I + I^-. \qquad (4)$$

The cross section for this reaction has a threshold of (0.03 $\pm$ 0.03) eV and a maximum of the order of $10^{-13} cm^2$ at 340 meV /21/. The attachment process is therefore very efficient and can be further influenced by the iodine vapour pressure which is strongly temperature dependent. Slow negative iodine ions produced in the attachment process and also by dissociation photoionization

$$I_2 + \hbar\omega \rightarrow I^+ + I^-, \qquad (5)$$

which has a lower threshold than (3), do not lead to counter discharges due to the very low impact ionization cross sections at the velocities of the iodine ions. The counter central electrode has to be made sufficiently thick, however, in order to prevent electric field detachment of electrons from negative iodine ions in the vicinity of the central electrode.

Variations of the band-pass counter principle are possible by appropriate choice of window materials and photoionization materials, both gases and solids. Let us consider first other window materials. A suitable choice in combination with iodine is SrF_2 whose transmission cutoff lies substantially below that of CaF_2. In fact an iodine counter with a SrF_2 window has a mean pass energy of 9.43 eV with a rms resolution of σ = 113 meV. Further narrowing of the band-pass can be obtained if the SrF_2 window is kept at elevated temperature /22/. This moves the transmission cutoff of the window to even longer wavelengths. At T = 70°C a rms resolution of σ = 73 meV was observed.

Similar resolutions can, in principle, be obtained choosing other gases for photoionization. Allen et al. /23/ have used CS_2 in combination with CaF_2 and report a measured optical resolution of 0.1 eV FWHM. Funnemann and Merz employ acetone combined with CaF_2 /24/. The reported resolution is 0.4 eV FWHM. In both cases two comments apply: firstly such tubes cannot be operated in the Geiger discharge mode because the active gas CS_2 or acetone respectively would be irreversibly dissociated in the discharges. This restricts operation to the proportional region with gas multiplications of the order of 10^3. The disadvantage of low signal pulse processing is however compensated by the fact that no dead time problems arise. The second comment relates to wall photoelectrons. Since both filling gases lack electronegative properties, such electrons contribute as well to the measured signal. This effect is reflected in the measured spectra by a considerable tailing of the signal below the calculated threshold (Fermi level). A possible means of reducing the wall contributions by more than an order of magnitude has been suggested by Dose already in 1968 /25/.

In a variation of the above principle a fluoride window crystal is combined with a channeltron or multiplier. This design was first proposed by Kovacs et al. /26/. The high energy cutoff is as sharp as in the gas counter case. The low energy "threshold" is however provided by the strong variation of the spectral photoelectron yield of the electron multiplier. Babbe et al. /27/ report a 0.6 eV FWHM resolution at 9.8 eV for a CuBe multiplier combined with a CaF_2 window. The tailing of the sensitivity to lower energies is of course even more severe in this case.

The apparent advantage of band-pass counters over UV monochromators so far is the simplicity of design. A further point of considerable importance is that with such detectors spectrometers with very large solid angles for photon collection may be constructed. A design taking advantage of this possibility is shown in fig. 4 /28/. The counter is mounted in a standard UHV system equipped with the usual facilities for sample preparation and analysis. The sample can be rotated through 360° in the figure plane. It is shown in the IPE position opposite to a small electron gun mounted coaxially to a vacuum ultraviolet mirror. The mirror is made from a spherical glass blank by evaporation of aluminium and subsequent coating with magnesium fluoride. Such mirrors reach reflectivities of the order of 70 % at 10 eV photon energy. Radiation produced by electrons hitting the sample at polar angle Θ is then imaged into the band-pass counter opposite to the mirror.

This design principle was recently optimized by employing an ellipsoidal mirror offering 2π solid angle to the sample /29/. Such an arrangement offers of course the ultimate limit of sensitivity. A figure of "several

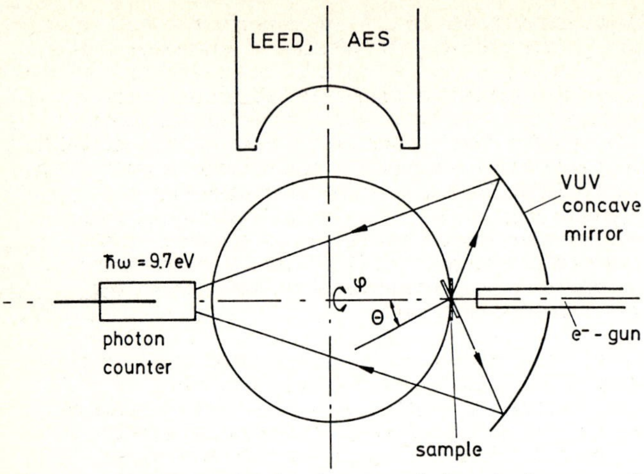

Fig.4: Spectrometer for inverse photoemission experiments based on a VUV band-pass photon detector.

times" 10^8 per Coulomb per eV was reported, which exceeds the performance of state of the art spectrographs by two orders of magnitude. This is particularly welcome if not crucial in IPE work on adsorbate systems.
In such arrangements, the angle of incidence of the electrons impinging on the sample can be varied with little if any effect on the photon collection angle. The setup is practically insensitive to radiation polarization.
The change of structural features in the observed spectra as a function of Θ can therefore uniquely be attributed to the variation of momentum of the primary electrons with respect to the sample normal.

On the other hand, the polarization of the emitted radiation allows one to deduce symmetry information of the electronic bands participating in the radiative transition. In the absence of effective polarizers in the ultraviolet spectral range information on the polarization of the emitted radiation can be deduced from its angular distribution. An experimental setup employing Geiger counters offering this possibility has been described by Donath et al. /30/. They use two photon counters co-planar with the electron gun and enclosing fixed angles of $\alpha_1 = 37^o$ and $\alpha_2 = 90^o$ with the electron beam direction. Obviously for an angle of incidence of $\Theta = 37^o$ of the primary electrons with respect to the sample normal (z-direction) the first counter mounted under $\alpha_1 = 37^o$ is blind to z-polarized light. The data presented by Donath et al. show an additional benefit of polarization sensitive arrangements. The apparent instrumental resolution increases considerably in certain cases.

4. Electron Sources

The overall resolution of an IPE experiment comprises contributions from the resolution of the photon detector σ_{ph} and from the width of the energy distribution of the electrons $\sigma_{electr.}$ impinging on the sample. Pushing the overall resolution of an IPE experiment

$$\sigma_{tot} = (\sigma_{ph}^2 + \sigma_{electr.}^2)^{1/2} \qquad (6)$$

requires simultaneous improvement of both σ_{ph} and $\sigma_{electr.}$ while both are preferably kept at the same value. Furthermore in work on single crystal samples where momentum resolution is required not only the energy but also the angular spread of electron beams becomes important.

The mean source for the energy spread of electron beams is the cathode temperature. Improvements are therefore expected when moving from high temperature equipotential tungsten cathodes to low temperature BaO, osmium coated BaO or LaB_6 cathodes. This is not always the case. Except for a very few setups the electron source has mostly been operated in a space charge mode which may result in a narrowing of the energy distribution expected from the cathode temperature only. In fact, experience in my own laboratory with a four electrode electron gun showed no detectable difference in energy resolution between a LaB_6 and a W filament cathode. In such cases W is to be preferred as a cathode material due to its insensitivity to background gas composition and pressure.

Still another type of electron emitter with quite different properties is the GaAs negative electron affinity (NEA) photoemitter. The energy distribution of electrons emitted from such devices can be well below 100 meV. More important, the beams can be produced with electron spin-polarization without compromising on the total emission intensity /31/. This is of extreme importance for IPE on magnetically ordered samples. This is believed to be the most fruitful future field of application for IPE since PES suffers from the substantial sensitivity reduction of the order of 10^4 involved in spin analysis. This factor compensates nearly excactly the α^2 phase space difference between PES and IPE and puts the two techniques as far as sensitivity is concerned in competitive positions.

One usual approach for preparing intense electron beams especially required for monochromator based spectrometers is the Pierce diode gun. The emission from a relatively large planar or preferably concave cathode is radially compressed and accelerated by the inhomogeneous field generated by the special geometric structure of the diode. The Pierce diode delivers a parallel beam with a momentum resolution of $|\Delta K| = 0.1\ Å^{-1}$ /7/ in the anode plane. The disadvantage of the Pierce configuration is that it cannot produce well-collimated beams at larger distances from the cathode. The majority of measurements with instruments using a Pierce gun has therefore been taken at normal incidence, where the sample can be kept in or close to the anode plane. Furthermore, the total emission of a Pierce gun is space charge limited thus varying as $U^{3/2}$. These limitations can be largely overcome with a gun design based on the acceleration/deceleration principle. Accelerating in the cathode region provides for a high cathode brightness. A properly designed deceleration and refocusing section can even provide a reduction of the thermally induced momentum spread of the electrons leaving the cathode./32/. There is only one drawback inherent in the acceleration/deceleration principle. Photons with energies in the range of the sensitivity of the photon detector are produced in the gun already at final energies well below the threshold for corresponding radiative transitions in the sample.

Acknowledgements

I want to acknowledge my coworkers for their contributions to this article. Thanks are also due to the Deutsche Forschungsgemeinschaft for continuous support of our inverse photoemission work.

Abbreviations

PES	photoelectron spectroscopy
UPS	ultraviolet photoelectron spectroscopy
IPE	inverse photoemission
VUV	vacuum ultraviolet
XPS	X-ray photoelectron spectroscopy
BIS	Bremsstrahlung isochromat spectroscopy
FWHM	full width at half maximum

References

1. V. Dose: Progr. Surf. Sci. 13, 225 (1983); J. Phys. Chem. 88, 1681 (1984); Surf. Sci. Rep. 5, 337 (1985). N.V. Smith : Vacuum 33, 803 (1983). F.J. Himpsel and Th. Fauster;J. Vac. Sci. Technol. A2, 815 (1984) .
2. F.J. Himpsel: Adv. Phys. 32, 1 (1983)
3. E.A. Milne; Philos. Mag. 47, 209 (1924) . J.B. Pendry: Phys. Rev. Lett. 45, 1356 (1980)
4. M. Cardona and L. Ley:in "Photoemission in Solids I" M. Cardona and L. Ley eds. (Springer Verlag, Berlin, Heidelberg, New York 1978) p. 52.
5. C. Kunz:in "Photoemission in Solids II", L. Ley and M. Cardona eds. (Springer-Verlag,Berlin, Heidelberg, New York 1979) p.299
6. V. Dose : Appl. Phys. 14, 117 (1977)
7. Th. Fauster, F.J. Himpsel, J.J. Donelon, and A. Marx: Rev. Sci. Instrum. 54, 68 (1983)
8. T.T. Childs, W.A. Royer and N.V. Smith: Rev. Sci. Instrum. 55,1613 (1984)
9. P. Ohlin: Ark. Mat. Astr. Fys., A29, 3 (1942)
10. G. Böhm and K. Ulmer: Z. Phys. 228, 473 (1969)
11. H. Scheidt: Fortschr. Phys. 31, 357 (1983)
12. J.K. Lang and Y. Baer: Rev. Sci. Instrum. 50 221 (1979)
13. H. Merz, private communication 1977
14. G. Chauvet and R. Baptist: J. Electron Spectros. Relat. Phenom. 24, 255 (1981)
15. F.J. Himpsel and Th. Fauster: J. Vac. Sci. Technol. A2, 815 (1984)
16. P.D. Johnson, S.L. Hulbert, R.F. Garrett, and M.R. Howells: Rev. Sci. Instrum. 57, 1324 (1986)
17. S.L. Hulbert, P.D. Johnson, N.G. Stoffel, W.A. Royer, and N.V. Smith: Phys. Rev. B 31, 6815 (1985)
18. T.A. Chubb and H. Friedman: Rev. Sci. Instrum. 26, 493 (1955)
19. W.R. Hunter and S.A. Malo : J. Phys. Chem. Solids 30, 2739 (1969)
20. V.H. Dibeler, J.A.Walker, K.E. McCulloh and H.M. Rosenstock: Int. J. Mass Spectrom. Ion Phys. 7 ,209 (1971)
21. H. Neuert, private communication
22. V. Dose, Th. Fauster, and R. Schneider: Appl. Phys. A40, 203 (1986)
23. P.M.G. Allen, P.J. Dobson, and R.G. Egdell: Solid State Commun. 55, 701 (1985)
24. D. Funnemann and H. Merz: J. Phys. E: Sci. Instrum. 19 (1986) (in press)
25. V. Dose: Nucl. Instrum. Meth. 59, 322 (1968)

26. A. Kovacs, P.O. Nilsson, and J. Kanski: Physica Scripta 25, 791 (1982)
27. N. Babbe, W. Drube, I. Schäfer, and M. Skibowski: J. Phys. E.: Sci. Instrum. 18, 158 (1985)
28. V. Dose, M. Glöbl, and H. Scheidt: Phys. Rev. B 30, 1045 (1984); K. Desinger, V. Dose, M. Glöbl, and H. Scheidt: Solid State Commun. 49, 479 (1984)
29. R.A. Bartynski and T. Gustafsson: Phys. Rev. B 33, 6588 (1986)
30. M. Donath, M. Glöbl, B. Senftinger, and V. Dose: Solid State Commun. (in press 1986)
31. D.T. Pierce, R.J. Celotta, G.-C. Wang, W.N. Unertl, A. Galeijs, C.E. Kuyatt and S.R. Mielczarek: Rev. Sci. Instrum. 51, 478 (1980)
32. N.G. Stoffel and P.D. Johnson: Nucl. Instrum. Meth. Phys. Res. A234, 230 (1985)

Hydrogen Adsorption Studies on Transition Metals: The Role of Surface Resonances in Vibrational Cross Section Enhancements

H. Conrad, M.E. Kordesch, and W. Stenzel

Fritz-Haber-Institut der Max-Planck-Gesellschaft, Faradayweg 4–6, D-1000 Berlin 33

1. Introduction

High resolution electron energy loss spectroscopy (HREELS) has been developed into a well established technique for the investigation of the vibrational properties of adsorbed molecules [1,2]. Most studies are focussed on the identification of the adsorbed species (fingerprinting) and the modification of their bond character due to chemisorption. These studies used mainly the dipole scattering mechanism. In this process the long range electric field of the electron interacts with the dynamic dipole moment of the adsorbed particle whereby its normal modes can be excited.

Theoretically this interaction is well understood [3,1] and it leads to the same surface selection rules as are utilized in infrared reflection absorption spectroscopy (IRAS) [2]. Experimentally, the electrons which are inelastically scattered via the dipole mechanism are easily identified by their angular distribution which is strongly peaked near the specular direction.

The much less well understood impact scattering process is caused by short range interactions and leads generally to a more complicated angular dependence distributed mostly over the off-specular directions and becoming more important at higher electron energies.

Recently, another inelastic channel for vibrational excitations by electrons has been identified for weakly physisorbed molecules [4]. The incoming electron is temporarily trapped on the molecule and forms a negative ion. This excited state decomposes into a free electron and the neutral adsorbate which is left with high probability in an excited state where rather high vibration states are excited. The existence of these negative ion resonances is well established and understood in electron-free- molecule scattering experiments [5]. At surfaces, they are only observable for molecules which are weakly perturbed by the adsorption, i. e. physisorption.

In this article we want to demonstrate that the two-dimensional analogue of negative ion resonances [6] leads to a strong enhancement of the vibrational cross sections of chemisorbed hydrogen, thereby emphasizing an additional channel for the interaction of electrons with adsorbed particles which is specific to surfaces. It will be shown that the observed effects are not only due to a prolonged interaction time with the electron trapped at the surface, but that the inelastic processes involved are more complicated and need further experimental and theoretical investigation.

2. Experimental Results and Discussion

The measurements were carried out with a hemispherical spectrometer described elsewhere [7] in a standard UHV system provided with LEED and TDS to monitor the substrate conditions and the hydrogen coverages which were adsorbed at about 110 to 120 K. It was first determined that changes in the electron primary energies only weakly influenced the transmission function of the instrument. This constancy was provided by increasing the pass energy of both hemispheres which thereby decreased the resolution to about 12 - 13 meV.

An immediate impression of the resonant behavior of the inelastic cross section as a function of the primary energy can be obtained from fig. 1, which shows HREEL spectra of hydrogen at full coverage adsorbed at 110 K on a Pd(111) surface. The angle of incidence was 57° to the surface normal, and measured in the specular direction.

There are several differences between the spectra in fig. 1 and standard HREEL spectra worthy of notice. First, the range of primary energies used is rather high compared with the usually applied energies. Second, the normalized loss intensities (I_{loss}/I_0) show extreme variations as functions of primary energy as is evident from the magnification factor which changes more than an order of magnitude. The curve at 5.7 eV with the dominant loss at 96 meV is replotted with a factor

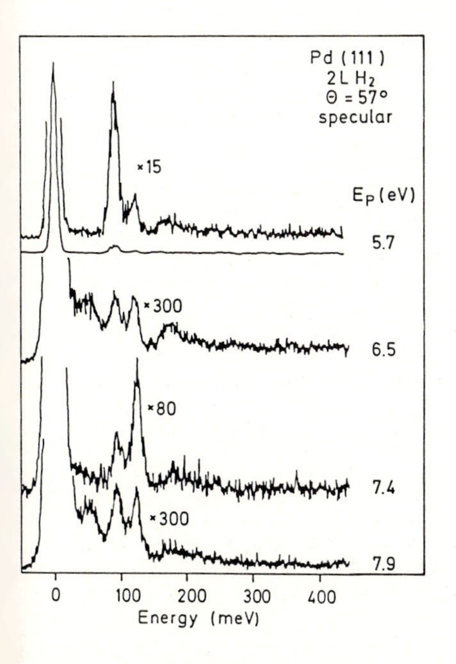

Figure 1.
Energy loss spectra of the hydrogen-covered Pd(111) surface at different primary energy, E_P.

of 1 just below the first spectrum and the loss is still visible. At 7.4 eV the loss at 124 meV is maximum, whereby in the spectra at 6.5 and 7.9 eV both losses are equally intense. At the latter energies a peak at about 50 meV is discernible which is due to a negligible amount of water adsorbed from the residual gas which shows up at this high magnification because of its high dynamic dipole moment. The broad humps at higher loss energies will be discussed later.

The third distinctive feature in the spectra is the fact that two losses are exhibited. In view of the known threefold H adsorption site [8,9] the symmetry group of the adsorption complex is C_{3v}. The normal modes are correspondingly separated into A_1 modes (perpendicular to the surface) which, according to the surface selection rules for dipole scattering, are dipole active [1], and into degenerate E modes (parallel to the surface) which are dipole inactive and only excitable via the impact mechanism. Obviously, the surface dipole selection rules do not apply in this case. The last point to mention is that the relative intensity of the two modes changes, with the intensity of the two peaks maximized at different energies.

An important effect not apparent from the normalized spectra of fig. 1 is the concomitant variation of the (0,0) beam or, equivalently, the specularly reflected elastic intensity. It exhibits strong fluctuation with energy and reaches minima at exactly those energies where the loss peaks are maximum. For a more complete overview in the energy range used fig. 2 shows

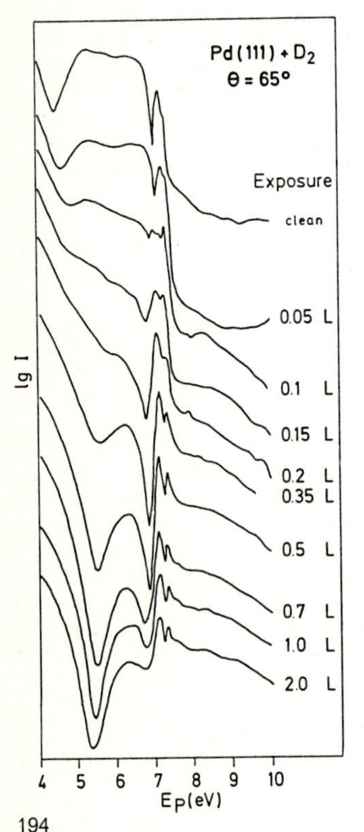

Figure 2.
Intensity of the (0,0) beam as a function of primary energy and exposure.

the (0,0) beam intensity (logarithmically scaled) as a function of primary electron energy. The series presented starts from the clean surface (topmost curve) and displays the modifications induced by deuterium adsorption up to full coverage. The use of deuterium is irrelevant for the results, since hydrogen gives identical curves.

In order to compare the elastic intensity variation with the spectra of fig. 1, the last curve of the sequence will be discussed. The intensity of the specularly reflected electrons decreases by more than two orders of magnitude from the lowest energy shown and exhibits a broad minimum at about 5.4 eV. Towards higher energy this minimum is followed by two additional narrower ones (see 0.5 L curve) and ends rather smoothly above these structures. Correlation with the data of fig.1 shows that the peaklike enhancement of the loss intensities takes place at energies corresponding to the reflectivity minima. The slight shift of the energy in both figures is connected to the dependence on the incidence angle which is 57° in fig.1 and 65° in fig. 2 (see below).

The structures in the reflectivity curve of clean Pd(111) are qualitatively very similar to the fluctuations of the deuterium covered surface, but the minima are located at different energies and the characteristic intensity drop at the high energy edge of the structured region of the clean surface gets washed out with increasing deuterium coverage. The existence of these (0,0) beam variations, at least for clean single crystal surfaces, has been known since the beginning of LEED measurements and has been demonstrated at the W(001) surface by Adnot and Carrette [10]. The interpretation given by McRae [6] and other authors [11] ascribes these effects to interference between the (0,0) and the first order diffracted LEED beams. The intensity drop at high energy occurs at the emergence threshold of the (1,0) LEED beam, which can be uniquely identified in the experimental geometry of the results presented. At this threshold the surface lattice diffracts an incoming electron into a state where its kinetic energy perpendicular to the surface is zero so that it moves only parallel to the surface.

When the electron wavefunction is separated into a parallel and perpendicular part, as is natural for a surface, it is described as two-dimensionally free and its energy perpendicular to the surface (only one-dimensional) is at the vacuum level. Below the threshold, corresponding diffraction processes can only occur if unoccupied surface states exist into which the electron can be scattered. The possibility of pure multiple scattering effects [12] can be neglected in the present case (see discussion below). Such surface states can be either intrinsic [13] or image potential induced [6,14] provided a band gap exists which confines the electrons to the surface. Even if there is no band gap an increased density of states can be located in the surface region which can be described as a surface resonance.

The image potential induced states are energetically located below the vacuum level and have an Coulombic energy spectrum,

$\sim 1/n^2$, which converges at the vacuum level. Correspondingly the structures near the threshold in the reflectivity curve of the clean surface in fig. 2 are identified as image potential states. Experimentally, this assignment is proved by measuring the changes of the threshold energy with the angle of incidence. The variations are easily calculated from the diffraction conditions. The image potential states are pinned to the vacuum level,and follow these shifts. This dependence has been verified [15] and converted into a dispersion curve, E vs k_{par}, which is a free electron parabola: the electrons in image potential states are essentially two-dimensionally free.

The limited energy range of the image potential states [6] prohibits a similar interpretation for the low energy minimum of the clean surface reflectivity curve. It is therefore assigned to an intrinsic surface state which has also been observed with inverse photoemission spectroscopy [16,17].

With increasing coverage the variations of the clean surface reflectivity are first suppressed (fig. 2, small exposures), but new minima develop at different energies. The near threshold features shifted to lower primary energies are still assigned to image potential states with higher binding energies (relative to the vacuum level) with free-electron -like dispersion [15]. The broad minimum dominating at full coverage exhibits in contrast nearly no dispersion [15] and is interpreted as due to a hydrogen/deuterium induced state localized near the hydrogen cores.

In the light of the interpretation presented above, the specific features of fig. 1 become clear. The resonant enhancements of the cross sections occur under conditions where the electrons are diffracted into surface states. The lower energy cross section increase corresponds to the hydrogen induced, the higher energy maximum to an image potential state, respectively. The primary energies are sufficient to satisfy the diffraction conditions in the experimental geometry. The two spectra measured between the resonances correspond to the reflectivity maxima following the minima of fig.2. In fig. 1 the normalized spectra are shown, but because the elastically reflected intensity exhibits such strong fluctuations the strength of the loss features should be measured with the primary current kept constant. The resulting curves are shown in fig. 3. The dashed curve corresponds to the 124 meV loss, the full curve to the 96 meV loss, respectively. The range of the intensity variation (note again the logarithmic scale) and the opposite energy dependence of loss and (0,0) beam change are evident.

In a first order interpretation this behavior can be correlated with the fact that the trapping of the electron at the surface must lead to an extended interaction time between the adsorbed molecule and the electron as compared to direct reflection. This prolonged interaction time is reflected in an enhancement of the probability for vibrational excitation. From figs. 1 and 3 it is evident that this interpretation does not explain why, in the hydrogen induced state, the lower energy mode is mostly affected, while in the image state the higher

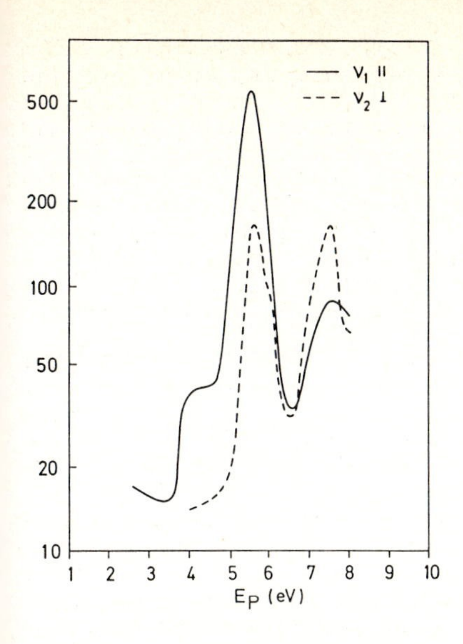

Figure 3.
Intensity of the loss peaks of the two fundamentals as a function of primary energy. Normalized to constant primary current.

energy loss dominates. Only an overall enhancement is compatible with the lifetime argument.

Another important question is still open, namely, why the dipole selection rules do not apply. It is evident for the surface resonances in which a one-dimensionally localized electron excites the vibration. Between the minima both modes are apparently equally intense. It must be concluded that the dynamic dipole moment of the adsorbed hydrogen is so small that even in specular direction the short range (impact) scattering dominates. It is therefore not possible to directly assign the losses to the normal modes of the adsorbed hydrogen which have been described above. In cases where small dipole contributions to the inelastic cross sections of one mode are detectable by means of a maximum loss intensity in the specular direction, this must be the perpendicular vibration. For the Pt(111)/H system [18] it was found that the lower energy loss is due to the dipole active perpendicular mode of hydrogen in the threefold hollow site. In a simple spring model of bonding it can be shown [1] that for a reasonable H bond length it should always be the symmetric (dipole active) mode which shows the lower frequency, while the asymmetric (dipole inactive) mode is at higher loss energy.

A similar interpretation for the Pd(111)/H system would be suggestive but a more direct approach to the assignment can be based on the observation of the higher harmonics. In fig.1 the broad maxima above the peaks discussed so far are due to overtone excitations of the fundamentals. As the frequencies of the hydrogen fundamentals on Pd(111) are very close together, it is difficult to disentangle the frequencies of the higher harmonics. As an example of the use of the overtones in determining the H coordination an essentially identical system is chosen with experimentally more suitable frequencies. In

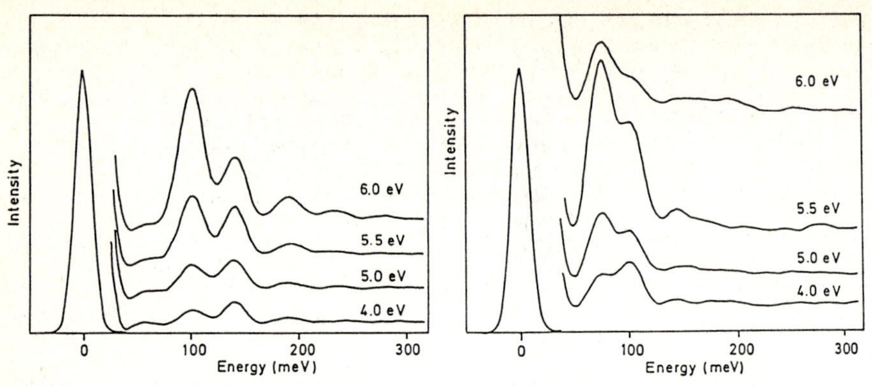

Figure 4. Energy loss spectra at different primary energy of 4a (left): hydrogen and 4b (right): deuterium. Measured at full coverage, 10° off-specular.

fig. 4 the HREEL spectra of hydrogen (4a) and deuterium (4b) adsorbed on Ru(001) are shown. The (001) surface is the hcp basal plane of the Ru lattice and directly comparable with an fcc (111) surface. For the Ru(001)/H system it was found [19] that the cross sections for vibrational excitation of adsorbed hydrogen also exhibit a resonant enhancement as a function of primary electron energy. This is demonstrated in fig. 4 for hydrogen and deuterium; and is directly comparable to fig 1. The energies of the normal modes are at 102 meV and 141 meV with the corresponding deuterium frequencies isotopically shifted. In the topmost curve of fig. 4a the overtones are much better separated and more intense than those in fig. 1. All energies are listed in table 1.

In the harmonic approximation the overtones are integer multiples of the fundamentals, the numbers of which are also given in table 1. From the assignment, it is seen that the

Table 1. Vibration frequencies of hydrogen and deuterium on Ru(001). The left column of the hydrogen frequencies lists the overtone and combination values in harmonic approximation.

Vibrational mode	Hydrogen		Deuterium
ν_1		102 meV	74 meV
ν_2		141 meV	101 meV
ν_3 ($2\nu_1$)	204 meV	192 meV	142 meV
ν_4 ($\nu_1+\nu_2$)	243 meV	235 meV	
ν_5 ($2\nu_2$)	282 meV	281 meV	

three additional losses correspond to the overtones of ν_1 and ν_2 plus the cross excitation of both. The comparison of the experimental values and those obtained from the harmonic approximation shows that only ν_2 can be described as harmonic. Both other overtones exhibit considerable anharmonicity. From an analysis of the anharmonic deviations the potential surface of the adsorbed hydrogen can be constructed with first order perturbation theory around the equilibrium H position [19]. The qualitative features are similar to the potentials of (111) surfaces obtained in theoretical calculations [20]. A crucial result is the low potential barrier for translations parallel to the surface whereas the potential well perpendicular is on the order of eV. Both agree very well with the experimental results. The binding energy for hydrogen on transition metals is typically above 2 eV [21] and the shallowness of the parallel potential is reflected in the experimentally observed high mobility of hydrogen on transition metals [22] which leads to order-disorder transitions in the hydrogen adlayer. This low diffusion barrier is just the potential determining the parallel vibration frequency. The parallel mode correspondingly exhibits strong anharmonic deviations from the pure harmonic behavior.

The assignment in terms of the above argumentation is obvious. The parallel mode is identified with ν_1, i.e. the energetically lower mode, and the perpendicular vibration with the ν_2 loss. The analysis of the Pd(111)/H modes yield the identical assignment which is already indicated in fig.3. These results are different from the assignments of the Pt(111)/H system discussed above, for which no corresponding analysis of the overtones was carried out. In the spectra presented [18], no loss features due to higher harmonics are visible, which implies a mainly harmonic shape of the hydrogen binding potential. The spring model seems to be the appropriate approach for this system.

An example of a dipole active mode of adsorbed hydrogen on Pd is found on the (100) surface. In this system only one loss at 63 meV has been observed in the specular direction. It exhibits an angular dependence which agrees very well with a model calculation for a dipole mode [23]. The adsorption site of hydrogen is the fourfold hollow site [24]. The normal modes of the C_{4v} symmetry group are separated into one dipole active A_1 (perpendicular) and two degenerate dipole inactive E (parallel) modes. The parallel modes can be observed via the impact scattering channel in the off-specular direction. This is demonstrated in fig.5 which shows for an ordered adlayer (c2x2 structure) HREEL spectra measured specularly and at 25° from the specular direction. The parallel modes at 76 meV are even more intense than the dipole active mode.

To investigate the influence of the surface states on H vibrations on this surface the reflectivity curves (fig.2) have been measured in a sequence of increasing hydrogen coverage as for the Pd(111) surface . The resulting curves (fig.6) also show the sharp intensity fluctuations characteristic of the image potential states. The emergence threshold energy, identified as the high energy limit of the narrow structures,

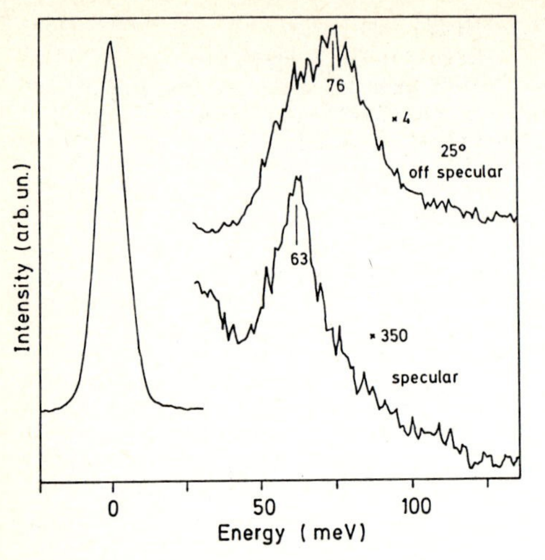

Figure 5. Energy loss spectra of H on Pd(100) for the c(2x2) LEED structure. Primary energy 2.2 eV.

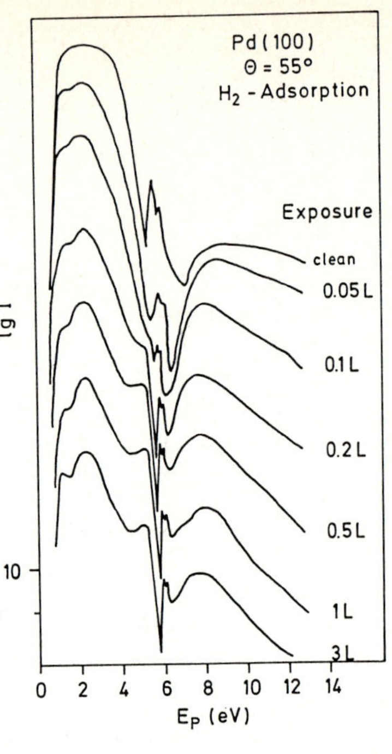

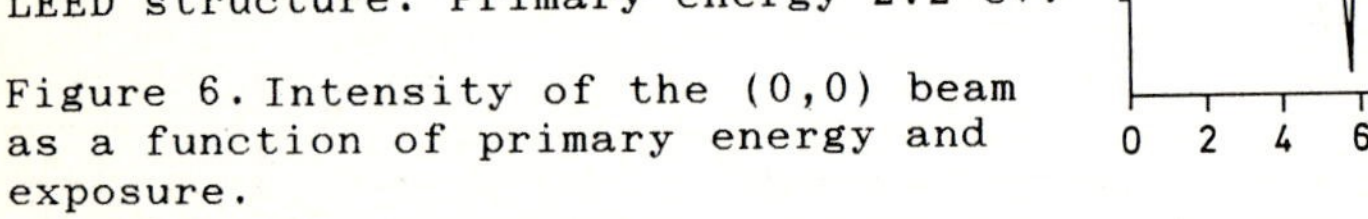

Figure 6. Intensity of the (0,0) beam as a function of primary energy and exposure.

is lower in energy than on the (111) surface in agreement with the smaller surface reciprocal lattice vector of the (100) plane. The dispersion curves, derived by measuring the reflectivity at different incidence angles, again exhibit free electron behavior [25]. An intrinsic surface state is not observed. With increasing hydrogen coverage the clean surface structures are suppressed and new features at new, shifted energies emerge, indicating a modification of the image potential but keeping their two-dimensional free electron character [25]. At about 2 eV an additional structure appears which, in analogy with the Pd(111) surface, is interpreted as a hydrogen induced surface state. The low energy cut-off is due to the transmission function of the spectrometer and has no physical significance.

Measurements of HREEL spectra at that energies where the variations of the reflectivity occur reveal that although the dipole loss always dominates in the specular direction, cross section resonances can be observed. In the off-specular directions, this effect is more pronounced and an example is presented in fig.7. The spectra shown here are measured at full hydrogen coverage. The vibration frequency of the parallel modes is shifted to 83 meV which indicates an interaction in the adlayer. The spectra of fig.7 show only the narrow energy range where the parallel loss exhibits a maximum. Like the Pd(111)/H system this enhancement is connected with the hydrogen induced surface state. The maximum of the cross section is located at 2.25 eV primary energy which, however, does not coincide with the minimum of the hydrogen induced

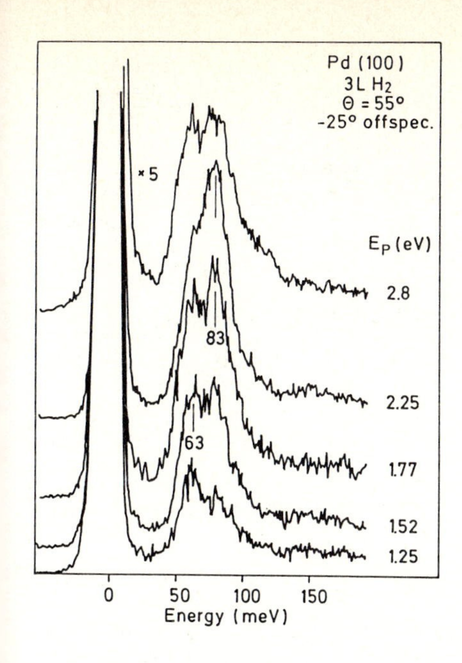

Figure 7.
Energy loss spectra at different primary energy. Normalized to constant primary current.

structure but lies on the increasing flank nearer to the following maximum as can be seen from the 3 L curve in fig.6. This result demonstrates that the energetic location of a surface state is not necessarily at the minimum of reflectivity as was implicitly assumed in the discussion of the results for the Pd(111) surface. The fluctuations of the (0,0) beam are caused by interference effects between the directly reflected and the temporarily trapped electrons. They can essentially be described as Feshbach resonances [6] and the shape of the fluctuations can continuously vary between maxima and minima. The inelastic channel offers an appropriate experimental probe to determine the surface state energies as was shown in the studies of the Pd(111) and Pd(100) surfaces presented here.

3. Summary

It has been shown that surface resonances where the scattering electron is trapped in surface states are additional channels for vibrational excitations of adsorbed hydrogen. The measurement of the (0,0) beam intensity of the clean and adsorbate - covered surface allows the identification of unoccupied surface states which can be intrinsic, image potential derived, or hydrogen induced. It has also been demonstrated that the enhanced scattering probability can be used to detect all vibrational modes of hydrogen, and additionally, by analysis of the overtone frequencies, the mode assignment is possible in cases where no dipole active mode can be identified. A still open question is the detailed description of the interaction of an electron captured in a surface state with adsorbed hydrogen, which is found to be mode selective.

4. References.

1. H. Ibach and D.L. Mills, Electron Energy Loss Spectroscopy and Surface Vibrations (Academic, New York, 1982).
2. R.F.Willis, A.A.Lucas and G.D.Mahan, in The Chemical Physics of Solid Surfaces and Heterogeneous Catalysis, edited by D.A.King and P.D.Woodruff (Elsevier, New York, 1983), vol.2.
3. A.A.Lucas and M.Sunjic, Progr.Surf.Sci. 2 (1972),75.
4. D.Schmeisser, J.E.Demuth and Ph.Avouris, Phys.Rev. B 26 (1982),4857.
5. G.J.Schulz, Rev.Mod.Phys. 45 (1973),378.
6. E.G.McRae, Rev.Mod.Phys. 51 (1979),541.
7. R.Unwin, W.Stenzel, A.Garbout and H.Conrad, Rev.Sci.Instr. 55 (1984),1809
8. S.M.Foiles and M.S.Daw, J.Vac.Sci.Techn. A3 (1985),1565
9. T.E.Felter and R.H.Stulen, J.Vac.Sci.Techn. A3 (9185),1566
10. A.Adnot and J.D.Carrette, Phys.Rev.Lett. 39 (1977),209
11. P.M.Echenique and J.B.Pendry, J.Phys. C11 (1978),2065
12. J.C.LeBossé, J.Lopez, C.Gaubert, Y.Gauthier and R.Baudoing J.Phys. C15 (1982),6087
13. W.Shockley, Phys.Rev. 56 (1939),317
14. J.B.Pendry,C.G.Larsson and P.M.Echenique, Surf.Sci. 166 (1986),57
15. H.Conrad, M.E.Kordesch, R.Scala and W.Stenzel, J.Electr.Spectr. 38 (1986),289
16. S.L.Hulbert, P.D.Johnson and M.Weinert, Phys.Rev. B34 (1986),3670
17. K.H.Frank and J.Wilder, unpublished
18. A.M.Baro, H.Ibach and H.D.Bruckmann, Surf.Sci. 88 (9179),384
19. H.Conrad, R.Scala, W.Stenzel and R.Unwin, J.Chem.Phys. 81 (1984),6371
20. P.Nordlander, S.Holloway and J.K.Norskov Surf.Sci. 136 (1984),59
21. See ref. 20, fig.2 and references cited
22. K.Christmann, R.J.Behm, G.Ertl, M.A.Van Hove and W.H.Weinberg, J.Chem.Phys 70 (1979),4168
23. C.Nyberg and C.G.Tengstal, Sol.St.Comm., 44 (1982),251
24. R.J.Behm, K.Christmann and G.Ertl, Surf.Sci., 99 (1980),320
25. H.Conrad, M.E.Kordesch,W.Stenzel,M.Sunjic and B. Trninic-Radja, Surf.Sci., in press

Molecular Bonding and Decomposition at Metal Surfaces

J. Küppers

Institut für Physikalische Chemie, Universität München, D-8000 München 2, Fed. Rep. of Germany

1. Introduction

A better understanding of the elementary steps in heterogeneous catalytic reactions is an important aspect of chemistry-devoted surface science. As modern spectroscopies which are capable to characterize solid surfaces and adsorbed layers generally are not compatible with those surfaces and environments which are used in practical catalysis, real systems are modeled by single crystals and an ultra high vacuum (UHV) environment.

At first sight the "gap" between a heterogeneous surface at high pressure and a single crystal surface contained in an UHV system seems too big to allow reliable conclusions to be drawn for real systems from studies at model systems. However, this is not the case, as has been shown for example by the methanation reaction at Ni surfaces [1]. This does not exclude that in more complex surface reactions at heterogeneous surfaces under high-pressure reaction steps occur which cannot be seen at model systems.

A further complication arises from the fact that only a few catalytic reactions can be actually performed under low-pressure conditions, $p < 10^{-6}$ torr. Examples are

$$CO + \frac{1}{2} O_2 \longrightarrow CO_2$$

and

$$H_2 + \frac{1}{2} O_2 \longrightarrow H_2O$$

at noble transition metal surfaces, eg. Pd, Pt and Ru. In these cases the reaction probability per incoming particle is so big that the reaction product, CO_2 or H_2O, can be easily detected. With the important ammonia synthesis reaction,

$$\frac{3}{2} H_2 + \frac{1}{2} N_2 \longrightarrow NH_3$$

this is not the case, as well as with many other hydrogenation or dehydrogenation reactions of practical importance. In these cases the reaction probability is so small, $\sim 10^{-6}$, that reaction products or even reaction intermediates cannot possibly be detected at low pressure. However, a study of the reverse reaction, i.e.

$$NH_3 \longrightarrow \frac{1}{2} N_2 + \frac{3}{2} H_2$$

can be used to get insight into the microscopic processes of relevance for the ammonia synthesis, especially with respect to the reaction intermediates. The principle of microscopic reversibility assures that these intermediates will also occur in the synthesis reaction.

Due to this, the study of molecular decomposition at surfaces becomes a powerful technique to elucidate catalytic reactions at surfaces.

2. Theoretical Considerations

The energetics which controls the decomposition of a molecule at a surface can be understood from a diagram as shown in fig. 1 [2].

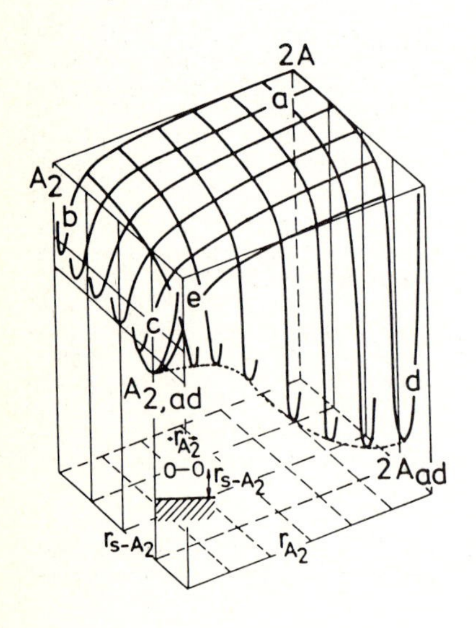

Fig. 1 Potential energy surface for a diatomic molecule interacting with a surface.

Here, as the most simple case, the dissociation of A_2 into 2 A is assumed. The four corners a, b, c and d of the cube which contains the energy surface as a function of the interatomic spacing, r(A-A), and the surface-molecule separation, $r(S-A_2)$, correspond to (a) two A atoms far away from the surface, (b) an isolated A_2 molecule, (c) A_2 adsorbed at the surface and (d) two A atoms adsorbed at the surface. In an experiment, (b) is the starting point: gas phase molecules and a bare surface.

The interaction of A_2 with the surface may be controlled by chemical or dispersion-type forces, in any case upon adsorption of A_2 the energy is lowered by up to some 10 kcal /mole. Excluding noble atoms, the energy gained by adsorption of atoms is much higher, up to 100 kcal/mole.

What happens to A_2 approaching the surface is essentially controlled by the shape of the energy surface around (e), see fig. 1. If the molecule can pass region (e) without facing an energy barrier, it will circumvent point (c) and proceed to point (d). This spontaneous dissociation process will also be observed if there is an energy barrier E* at (e), but the ratio E*/kT is small. An example for that latter case is the system O_2/Ni(111), for which only recently the barrier height has been determined to E* ≈ 13 meV [3]. Accordingly, beyond cryogenic temperatures oxygen will always dissociatively adsorb at Ni.

If the barrier height is bigger, i.e. E*/kT ≫ 1, the incoming molecule will be trapped in the well at (c) to form A_2(ad). That situation is most commonly observed for simple molecules, eg. CO, N_2, NO, NH_3, C_nH_m, at not too high temperatures, T ≈ 100 K.

Consider a molecule bonded to a surface which should undergo a decomposition reaction. Experimentally this can be performed by a temperature increase which accordingly decreases E*/kT. Assuming that the decomposition kinetics is correctly expressed by a first order rate law, one has

$$-dn/dt = k_x\, n \;, \quad k_x = \nu_x \exp -(E^*/kT) \quad ; \qquad (1)$$

dn/dt is the decrease of the number of adsorbed molecules, n, per unit time, k_x the rate constant which depends on a preexponential ν_x and the Boltzmann factor exp-(E*/kT).

Consulting fig. 1 it is seen that competitive to the decomposition reaction (c)-(d), desorption of the molecule (c)-(b) can occur which also may be described by a first order rate law:

$$-dn/dt = k_d\, n \;, \; k_d = \nu_d \exp-(E_d/kT). \qquad (2)$$

Obviously, the relation between k_x and k_d determines which fraction of the adsorbed amount decomposes at a given temperature and which fraction desorbs.

The experimental verification of desorption/dissociation usually is performed by increasing the temperature of an adsorbate covered surface in a controlled manner, $T = T_o + \beta t$. In order to isolate the decomposition products, the temperature must be ramped to above the desorption temperature. The result to be expected for a particular choice of the parameters occurring in equ. 1 and 2 is shown in fig. 2. The curves displayed correspond to: amount adsorbed, desorbed and decomposed, and the partial pressure increase as observed by a mass spectrometer tuned to the mass of the adsorbed molecules.

From fig. 2b, which corresponds to the situation where the adsorption energy decreases with coverage, $E_d = E_d^o\,(1-f\cdot\theta)$, it is seen that decomposition only occurs when the coverage has decreased considerably. In this case desorption is favoured over decomposition for high coverage, whereas at constant adsorption energy decomposition parallels desorption throughout the whole temperature range, see fig. 2a.

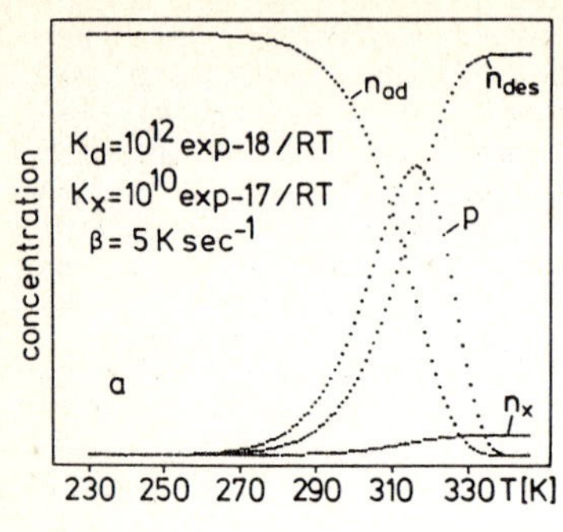

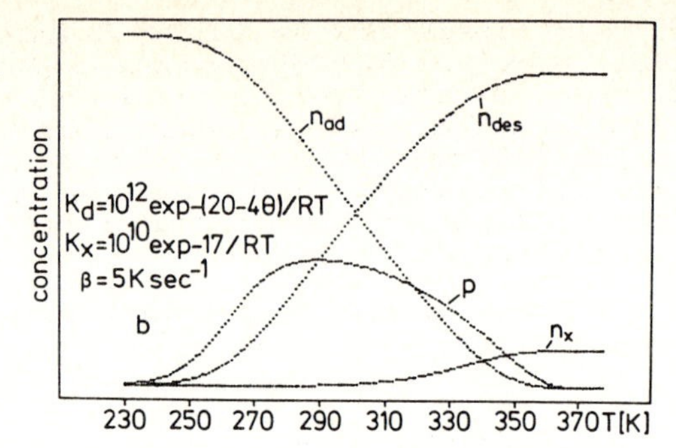

Fig. 2 Concentration of adparticles, n_{ad}, desorbed particles, n_{des} and decomposed particles, n_x, when the temperature of the substrate is increased linearly with time. Energies in kcal/mol. a) constant adsorption energy; b) adsorption energy decreases with coverage.

In reality commonly a decrease of the adsorption energy with coverage is observed which suggests an alternative technique to prepare decomposition products at a surface. Instead of covering the surface at low temperature and subsequently increasing the temperature, the surface is exposed to an ambient pressure of the reacting gas at elevated temperatures. By this procedure one has

$$n_{gas} \underset{k_d}{\overset{k_a}{\rightleftharpoons}} n_{ad} \xrightarrow{k_x} n_x \quad . \tag{3}$$

Adsorption-desorption processes establish a pre-equilibrium, controlled by respective rate constants. The decomposition rate constant, k_x, determines what fraction of the adsorbed molecules decomposes.

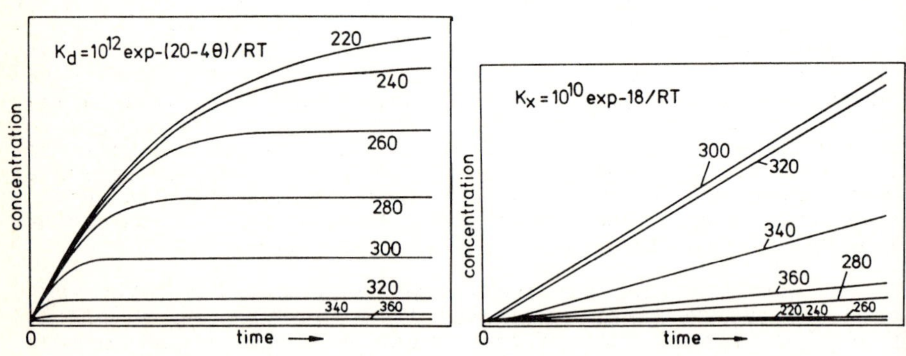

Fig. 3 Concentration of adparticles, n_{ad}, (left) and decomposed particles, n_x, (right), when a surface is exposed to a constant pressure of 10^{-8} torr at t = 0. Langmuir type adsorption is assumed. Energies in kcal/mol, T in K.

The result to be expected from a measurement performed in this way is displayed in fig. 3 for various temperatures. At t = 0 a pressure of p = 10^{-8} torr is installed above the surface. Langmuir-type adsorption is assumed for simplicity. At low temperature only the formation of an adsorption layer is observed, decomposition is negligible. Increasing temperature decreases the steady-state coverage but favours decomposition. With the rate constants used, product formation has its optimum around 300 K, above that temperature it decreases again as the equilibrium coverage drops considerably.

3. Experimental Considerations

Characterization of molecular decomposition products can be performed in principle with any technique which is capable to identify adsorbate particles. However, it is essential that the technique can discriminate between molecular and decomposed species. Accordingly, the spectroscopy involved should exhibit both, high resolution and bandwidth. Auger and XP spectroscopy offer a sufficient bandwidth, but the resolution of these methods is either limited by the physical process underlying the technique (AES) or by the spectral width of the incoming radiation (XPS) [4]. However, in selected cases both techniques have been successfully applied to chemical reactions at surfaces.

Two spectroscopies are available which probe matter at low excitation energies, namely TDS [5] and HREELS [6]. The examples discussed below demonstrate that these techniques are ideally suited to serve as monitors of reactions at surfaces.

4. Case Studies

The results reported in this section are all obtained at well defined surfaces. Surface preparation, characterisation and handling was performed using standard techniques.

4.1 C_2H_2/Fe(111)

Adsorption of acetylene at Fe(111) surfaces at low temperature is molecular [7] as also observed at other metal surfaces [8]. All vibrational features in the HREELS spectra displayed in fig. 4 can be correlated with vibrations of C_2H_2. Most prominent are the ν(CH) mode at 2850 cm^{-1}, the ν(CC) mode at 1145 cm^{-1} and δ(935 cm^{-1}) and ς(660 cm^{-1}) modes. The loss at 470 cm^{-1} corresponds to the (CFe) vibration.

Thermal desorption shows that molecular desorption occurs around 200 K. However, mass 2 desorption spectra as displayed in fig. 5 suggest that competitive to desorption decomposition had occurred. As molecular hydrogen desorption is completed at 380 K, the peaks at 490 K and 580 K are caused by decomposition of a dissociation product, note that molecular C_2H_2 cannot exist at the surface above 250 K.

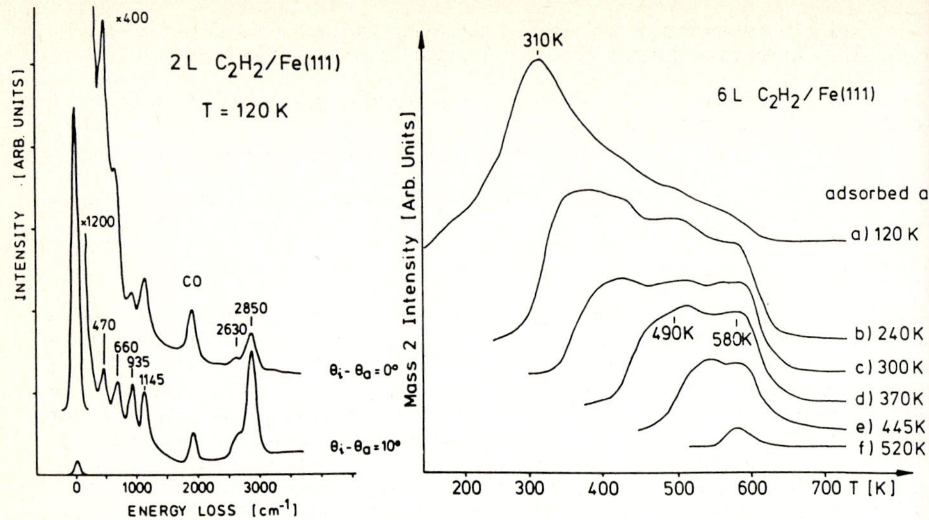

Fig. 4 (left) HREEL spectra measured specular and off specular at a C_2H_2 adlayer on Fe(111) at 120K

Fig. 5 (right) H_2-desorption spectra obtained after exposing a clean Fe(111) surface to 6L C_2H_2 at various temperatures.

Accumulation of the decomposition products, by exposing C_2H_2 to Fe(111) at elevated temperatures (this procedure has been considered above) leads to an identification of these intermediates. The most prominent features in fig. 6, losses at 2930 cm^{-1} and 3015 cm^{-1} are due to the ν(CH) vibration in CH_2 and CH. The CH_2 wagging mode at 1130 and the CH bending mode at 795 cm^{-1} are also seen upon dissociation of C_2H_2 at various metal surfaces [9]. The prominent loss at 1290 cm^{-1} which remains even at high temperature may be interpreted as due to a C-metal vibration of carbon bonded at a low coordination site.

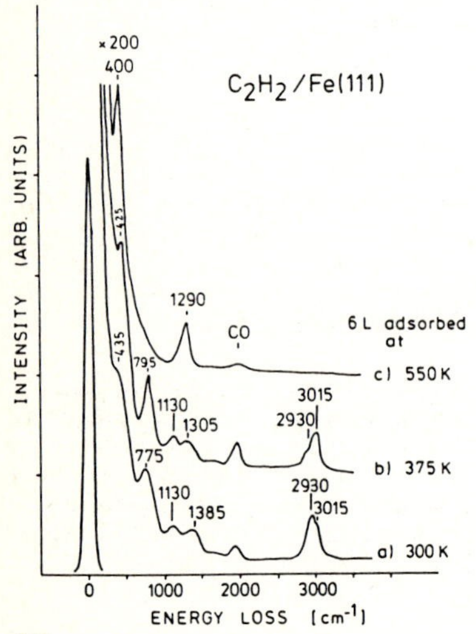

Fig. 6 HREEL spectra which characterize the decomposition of C_2H_2 on Fe(111) surfaces at elevated temperature.

As the spectra in fig. 6 suggest that CH_2 and CH coexist, the decomposition pathway of C_2H_2 can be specified as

$$C_2H_2(gas) \leftarrow 200\ K - C_2H_2(ad) - 230\ K \rightarrow \begin{pmatrix} CCH \\ CCH_2 \end{pmatrix} \begin{matrix} - CH - 580\ K \rightarrow C \\ - CH_2 - 400\ K \rightarrow C \end{matrix}$$

The intermediates CCH and CCH_2 may exist, but are not isolated at Fe(111).

4.2 NH_3/Ni(110)

Ammonia adsorbs molecularly at low temperature [10]. The HREEL spectrum, see fig. 7, is dominated by an intense loss around 1100 cm^{-1}, originating from the umbrella mode (δ_s). Less intense are the δ_d mode at ~1700 cm^{-1} and the ν(N-H) mode around 3300 cm^{-1}. Collecting data in vacuo at higher temperatures causes only the molecular derived vibrations to lose intensity, as the molecules desorb. Desorption is completed around 330 K as judged from TDS, leaving a bare surface behind: desorption is faster than decomposition under the given conditions.

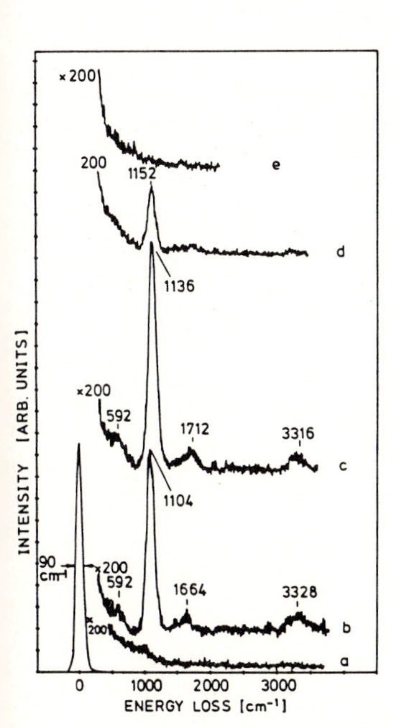

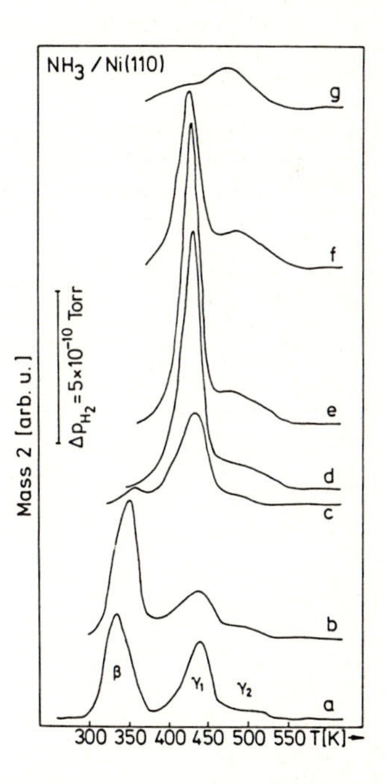

Fig. 7 (left) Loss spectra of clean Ni(110) (a), a saturated ammonia layer at 186 K (b), and subsequently raising the temperature to 245 K (c), 300 K (d) and 493 K (e).

Fig. 8 (middle) H_2-desorption traces measured after exposing clean Ni(110) to 30 L of ammonia at 217 K (a), 269 K, 328 K (c), 345 K (d), 357 K (e), 370 K (f) and 393 K (g).

As ammonia exhibits repulsive particle-particle interaction in the adsorbed phase, the adsorption energy increases with decreasing coverage. This suggests the application of the procedure outlined in section 2, fragment accumulation at elevated temperatures and low stationary coverage. As these fragments may be NH_2, NH, N and H, mass 2 (H_2) and mass 28 (N_2) desorption can be used to monitor indirectly the existence of these fragements. The family of desorption spectra shown in fig. 8 indeed signal molecular decomposition.

If ammonia is exposed to clean Ni(110) surfaces at T 300 K clearly two H_2 desorption peaks develop, γ_1 and γ_2 which cannot be correlated with "normal" hydrogen desorption from H(ad), as this shows up in the ß-peak. The trend observed in fig. 8 gives evidence for two reaction products, one favoured around 345 K, the other around 393 K.

The spectroscopic identification of these intermediates, as will be seen NH_2 and NH, is performed using HREELS. In fig. 9 spectra b, c and d are measured at 344 K and ambient NH_3 pressure, p = 2 10^{-8} torr, after backfilling the system with ammonia. It is clear from the loss intensities in the ν(N-H) region that NH_3 cannot possibly be the related species, compare with the δ_s/ ν(N-H) intensities of molecular NH_3 in fig. 7. Identification of the loss features is as follows: NH_2 : ν(Ni-N) 504 cm^{-1}, δ(HNH) 1520 cm^{-1}, ν(NH) 3290 cm^{-1}; NH_3: δ_s 1184 cm^{-1}; and CO: ν(C-O) 1870 cm^{-1}. The CO related vibration is shifted as compared with CO/Fe(111) by the presence of coadsorbed NH_3, just as observed with elec-

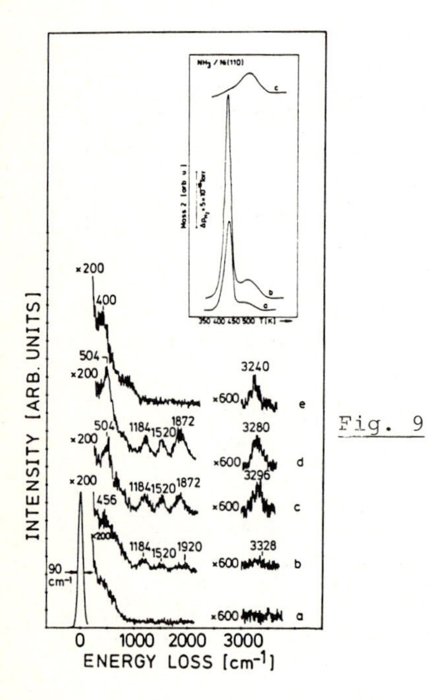

Fig. 9 (right) HREEL spectra measured at 344 K, (a) at clean Ni(110), 15 min (b), 30 min (c) and 45 min (d) after installation of 2·10^{-8} torr NH_3, at 393K (e) after completion of spectrum d. The inset shows mass 2 TD spectra corresponding to surface conditions as characterized by HREELS: b-TDS a, d-TDS b, e-TDS c

tropositive promoting elements, e.g. K [11]. The topmost spectrum only exhibits a ν(N-H) vibration at 3240 cm^{-1} and a weak ν(Ni-N) vibration at 400 cm^{-1} which are due to adsorbed NH. The inset in fig. 9 displays supporting TDS data which are collected at specific stages of the ammonia decomposition reaction as followed by HREELS. These TDS spectra justify the assignment of the γ_1, γ_2 peaks as noted above and allow to specify the reaction pathways as: $NH_3 \longrightarrow NH_2 \longrightarrow NH \longrightarrow N$. A detailed analysis [10] suggests that $NH_2 \longrightarrow N$ is an important parallel reaction to NH_2 NH (the fate of hydrogen is omitted in this notation).

4.3 H_2O/O/Ni(110)

As seen with the examples described above, decomposition of molecules often proceeds via intermediates towards the atomic constituents. As known from practical catalysis, coadsorbates referred to as promotors can be used to manipulate surface reaction schemes, either by changing structural or electronic conditions these intermediates met at a surface. In the present section we will discuss this aspect.

As seen from the vibrational spectrum in fig. 10, at low temperature water adsorbs molecularly at Ni(110) surfaces. Most prominent are the libration mode at 730 cm^{-1}, the scis-

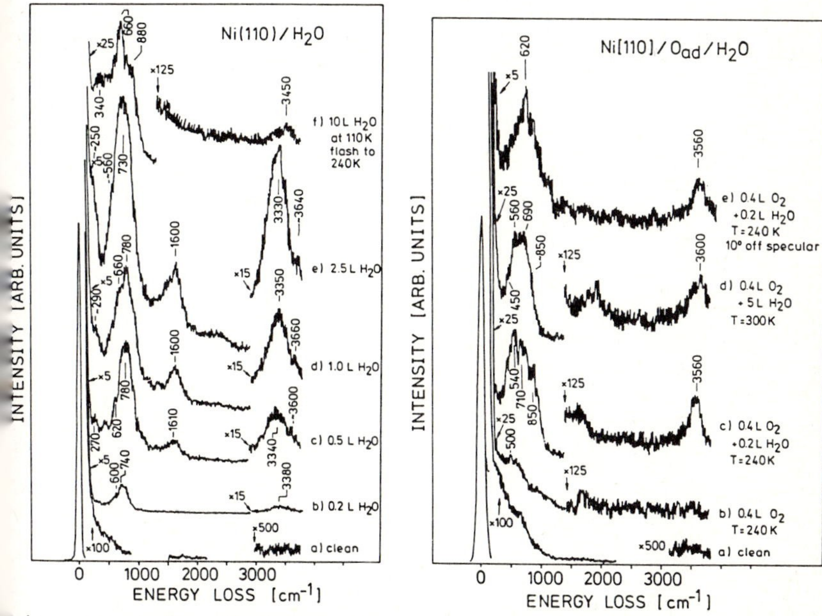

Fig. 10 Left: HREEL spectra measured after exposing clean Ni(110) to increasing amounts of H_2O at 110 K. The top spectrum is obtained after flashing to 240 K. Right: HREEL spectra measured after exposing oxygen-precovered Ni(110) surfaces to water at elevated temperature.

soring mode at 1600 cm^{-1} and the OH stretch at 3330 cm^{-1}. The position of this last mode is considerably shifted from its gas phase value (3700 cm^{-1}) and due to water clustering which is commonly observed at water adsorption layers [12]. Upon heating to 240 K the molecular features disappear as desorption has occurred. However, this is accompanied by water dissociation, as seen from the mode at 3450 cm^{-1} which is due to the OH vibration of OH(ad).

In a recent ESDIAD study [13] it was observed that exposure of water to oxygen-precovered Ni(110) surfaces leads to a well-ordered adsorbate layer, which was attributed to a fully developed OH(ad) layer. The vibrational spectrum in fig. 10 measured at an adsorbed layer prepared in the proposed manner [14] indeed reveals only OH(ad) related features. Moreover, the OH bending mode at 620 cm^{-1} exhibits an intensity variation with respect to the scattering angle which is in accordance with an inclined OH adsorbed at Ni(110), as already proposed in the ESDIAD study [13].

4.4 CO/H/Ni(110)

In this section the question is addressed by which mechanism a molecule fragments at a surface. The coupling of the molecular orbitals to the substrate electronic states opens decomposition pathways which are not present with a gas phase molecule. An illustrative example is CO, for which the 5σ-metal donation, metal-2π backdonation scheme [15] generally is accepted. As the antibonding 2π is unoccupied in free CO, the degree of its occupation by metal electrons will affect the stability of adsorbed CO.

Inverse photoemission (IP) as developed by DOSE and coworkers [16] has proven to allow a spectroscopic investigation of unoccupied electronic states at surfaces. Applied to adsorbate covered surfaces, states above the Fermi level, E_F, which are introduced by the adsorbate can be identified.

As an example, fig. 11 shows IP spectra of a clean and CO covered Ni(100) surface. The clean surface spectrum is dominated by unoccupied d-states right above E_F. Upon CO adsorption an intense broad peak centered at 4.5 eV develops, which is easily identified as due to the 2 electron orbital of CO(ad). When CO is adsorbed at a hydrogen-saturated Ni(100) surface, the 2π derived peak is observed at 4.2 eV, i.e. shifted by 0.3 eV towards the Fermi level. This favours d-2π backdonation and accordingly destabilizes the C-O bond. By this the tendency for CO dissociation is expected to increase, which is an important prerequisite in the CO hydrogenation reaction.

5. Conclusion

Decomposition of molecular species at surfaces generally proceeds via intermediates. Characterization of molecular and fragmented adparticles can be performed with spectroscopic techniques, most favourably HREELS and TDS, which allow con-

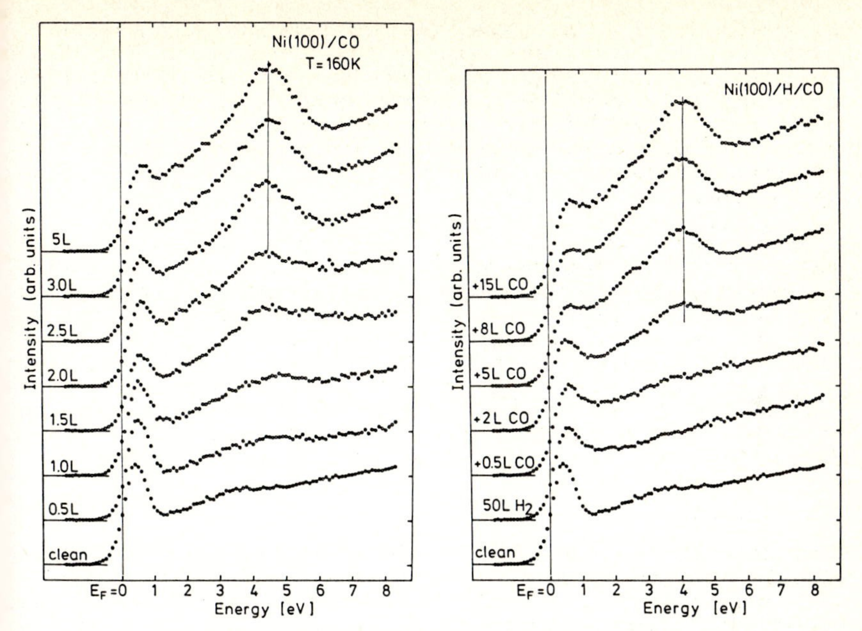

Fig. 11 Inverse photoemission spectra of (left) CO- and (right) H/CO-covered Ni(100) surfaces.

clusions to be drawn with respect to elementary steps in heterogeneous catalytic reactions.

Support of the Deutsche Forschungsgemeinschaft (SFB 128) and Fonds der Chemischen Industrie is gratefully acknowledged.

References

[1] R. D. Kelley and D. W. Goodman: In The Chemical Physics of Solid Surfaces and Heterogeneous Catalysis, ed. by D. A. King and D. P. Woodruff: Vol. 4 (Elsevier, Amsterdam 1982) p. 425
[2] S. Holloway and J. W. Gadzuk: J. Chem. Phys. 82 (1985) 5203
[3] M. Shayegan, J. M. Cavallo, R. E. Glover and R. L. Park: Phys. Rev. Lett. 53 (1984) 1578
[4] G. Ertl and J. Küppers: Low-Energy Electrons and Surface Chemistry, 2nd ed. (VCH Verlagsgesellschaft, Weinheim (1985)
[5] D. Menzel: In Interactions on Metal Surfaces, ed. by R. Gomer, Topics in Applied Physics, Vol. 4 (Springer, Berlin, Heidelberg 1975)
[6] H. Ibach and D. L. Mills: Electron Energy-Loss Spectroscopy and Surface Vibrations (Academic Press, New York (1982)
[7] U. Seip, M.-C. Tsai, J. Küppers and G. Ertl: Surface Sci. 147 (1984) 65
[8] H. Ibach and S. Lehwald: J. Vacuum Sci. Technol. 15 (1978) 407

J. A. Gates and L. L. Kesmodel: J. Chem. Phys. 76 (1982) 4281
[9] J. E. Demuth and H. Ibach: Surface Sci. 78 (1978) L238
C. Backx, B. Feuerbacher, B. Fitton and R. F. Willis, Surf. Sci. 63 (1977) 193
[10] J. C. Bassignana, K. Wagemann, J. Küppers and G. Ertl: Surface Sci. in press
[11] H. P. Bonzel: J. Vacuum Sci. Technol. A2 (1984) 866
[12] H. Ibach and S. Lehwald: Surface Sci. 91 (1980) 187
[13] C. Benndorf, C. Nöbl and T. E. Madey: Surface Sci. 138 (1984) 292
[14] M. Hock, U. Seip, K. Wagemann and J. Küppers: Surface Sci. submitted
[15] G. Blyholder: J. Phys. Chem. 68 (1964) 2772
[16] V. Dose: Progr. Surface Sci. 13 (1983) 225
[17] W. Reimer, Th. Fink and J. Küppers: to be published

Thermal Desorption of Alkali A
from Metal Surfaces

E.V. Albano

Instituto de Investigaciones Fisicoquímicas Teóricas y Apl
(INIFTA), Facultad de Ciencias Exactas, Universidad Nacion
C.C. 16, Sucursal 4, 1900 La Plata, Argentina

1. Introduction

The study of the adsorption of alkali metals on transition metal surfaces has attracted growing attention because alkali atoms and (or) alkali compounds are used in order to enhance the electron emission of surfaces as well as to promote heterogeneously catalysed reactions /1-6/. The aim of the present work is to discuss thermal desorption mass spectra (TDS) of alkali atoms from transition metal surfaces based on the calculation of the coverage dependence of the activation energy of desorption for a planar array of electrical dipoles.

2. The Model

The rate of desorption R(N,T) of alkali atoms from a surface may be written as /1 and references therein/:

$$R(N,T) = -dN/dt = \nu_o N \exp -[(E_{d0} - E_R(N))/RT] \ , \qquad (1)$$

where N is the number of adalkalis per unit surface area, ν_o is the pre-exponential factor of the desorption rate, T is the absolute temperature, R is the gas constant, E_{d0} is the activation energy of desorption in the limit N = 0 and $E_R(N)$ is the coverage-dependent repulsion energy on the adlayer.

In order to evaluate $E_R(N)$ it has been assumed that the adsorbed atom can be treated as a polarizable entity with a constant polarizability (α) and a dipole moment of magnitude P(N) pointing away from the surface. Due to the repulsive forces acting on the adlayer the dipoles are uniformly spaced. Under these assumptions and for an hexagonal array, the depolarization field $F_d(N)$ in the position of each dipole is given by /1/:

$(n^2 + m^2 - nm)^{-3/2}$, (2)

s (i.e. the coordinates of each
not be equal to zero simulta-
ummation of (2), it follows that

(3)

depolarized by $F_d(N)$, its dipole

$^{2})$, (4)

where P(N ... e moment of an isolated adalkali. Due to its position ... ield created by the surrounding adparticles, each dipole has a potential energy $E_R(N) = -P(N)F_d(N)$ which can be evaluated using (3) and (4). Hence,

$$E_R(N) = 9P(N=0)^2 N^{3/2}/(1 + 9\alpha N^{3/2})^2 \quad . \tag{5}$$

The insertion of (5) in (1) gives explicitly the dependence of the activation energy of desorption on coverage, and can be used in order to evaluate TDS traces.

3. Comparison with Experimental Results

All spectra were computed assuming $\nu_0 = 10^{13}\ s^{-1}$ and increasing uniformly the temperature of the sample. For a given alkali/transition metal system, TDS traces were evaluated for different initial coverages and the best set of parameters (P(N=0), α and E_{d0}) was selected when the difference between experiments and theory lay within the error range for all the desorption peaks. As an example of the above-mentioned procedure Figs. 1a and 1b show TDS traces of K from Fe(110) and Fe(100), respectively.

The results obtained for K/Fe(poly) and K/Pt(111) have been listed in the following Table I.

After the determination of the best set of parameters (P(N=0), α and E_{d0}) capable to describe TDS traces of a given system, one can evaluate the dependence of the desorption energy $E_d(N)$ on the coverage. In order to easily compare the results obtained for

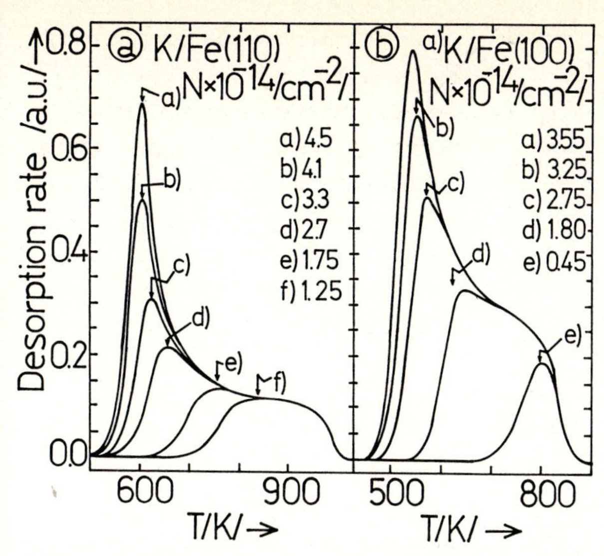

Figure 1. TDS traces of K adsorbed on Fe at various coverages. a) K/Fe(110), α = 11.8 Å^3, P(N=0) = 8.5 D, and E_{d0} = 61.0 kcal/mol. b) K/Fe(100), α = 11 Å^3, P(N=0) = 7.5 D and E_{d0} = 52 kcal/mol. The arrows indicate the maximum desorption rate experimentally determined /2/ for the same initial surface coverage as that listed in the figures.

Table I. The desorption peaks (T_p) of K/Fe(poly) with E_{d0} = 54 kcal/mol, P(N=0) = 8.1 D and α = 11.3 Å^3; and K/Pt(111) with E_{d0} = 63.7 kcal/mol, P(N=0) = 11.5 D and α = 17.2 Å^3. Experimental data from Ref. /3/# and Ref. /4/*.

Fe(polycrystalline)			Pt(111)		
$Nx10^{-14}$ /cm^{-2}/	T_p /K/ Exp.#	T_p /K/ Theory	$Nx10^{-14}$ /cm^{-2}/	T_p /K/ Exp.*	T_p /K/ Theory
0.42	839	835	0.32	1033	1035
0.75	814	805	0.76	959	960
0.95	779	785	1.30	801	800
1.30	728	730	1.62	735	735
1.75	664	660	3.78	595	590

different systems, the ratio $E_d(N)/E_{d0}$ has been plotted against N in Fig. 2. It is worth mentioning that the calculated activation energies of desorption are in good agreement with experimental data obtained from TDS traces and P(N=0) and the polarizability also agree with data obtained fitting alkali-induced work function changes on transition metals /1/.

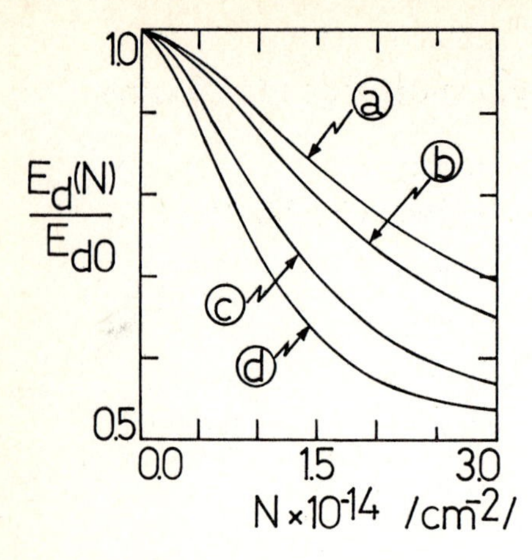

Figure 2. The coverage dependence of the desorption energy relative to its value at N=0 for various systems. a) Na/Ni(111) and Ni(100), E_{d0} = 58 kcal/mol, P(N=0) = 7.3 D and α = 9 $Å^3$ (Ref. /5/); b) K/Fe(poly) see Table I; c) K/Pt(111) see Table I and d) Cs/W(110) E_{d0} = 73 kcal/mol, P(N=0) = 15 D and α = 24.5 $Å^3$ (Ref. /6/)

4. Conclusions

The successful comparison between the computed spectra and available experimental data for K/Fe(110), K/Fe(poly), K/Pt(111), Na/Ni(111), Na/Ni(100) and Cs/W(110) allows us to demonstrate that the mutual dipolar repulsion within the adlayer is responsible for the coverage dependence of the thermal desorption spectra characteristic of the above-mentioned systems, suggesting that indirect through-bond interactions should contribute to a much lesser extent. Shifts of the desorption peaks up to about 450 K due to the lowering of about 50% in the energy of desorption are well described for the proposed model.

Acknowledgments

This work was financially supported by the CONICET, R. Argentina.

References

1. E.V. Albano: J.Chem.Phys. In press.
2. S.B. Lee, M. Weiss and G. Ertl: Surface Sci. 108, 357 (1981)
3. Z. Paál, G. Ertl and S.B. Lee: Appl.Surface Sci. 8, 231 (1981)
4. E.L. Garfunkel and G.A. Somorjai: Surface Sci. 115, 441 (1982)
5. R.L. Gerlach and T.N. Rhodin: Surface Sci. 19, 403 (1970)
6. J.L. Desplat and C.A. Papageorgopoulos: Surface Sci. 92, 97 (1980) and 92, 119 (1980)

The Work Function Change During the Adsorption of Oxygen on Pd(111) Films

L.M. de la Garza

Instituto de Física, UNAM, Laboratorio de Ensenada, P.O. Box 2681, 22800, Ensenada, B.C. México

The work function change ($\Delta\phi$) has been studied by contact potential difference (CPD) measurements using a Kelvin Probe during the adsorption of oxygen on a Pd(111) film of 2000 Å deposited on moscovite.

The Pd(111) film was deposited at 1 X 10^{-9} torr of residual gas pressure and with a substrate temperature of 150°C. The rate of deposition was 0.2Å/sec. This film is almost free of chemical impurities and defects like steps and grain boundaries.

RESULTS

After the deposition of the Pd(111), its surface is cleaned of the residual gases by heating it at 300°C in ultra-high vacuum (UHV) the main impurities observed were carbon and chlorine, heating the sample at 150°C in 1 X 10^{-8} torr of oxygen and then increasing the temperature at 300°C in UHV was the best method to get rid of carbon from the surface. We determined the cleanness of the surface by Auger Electron Spectroscopy (AES). Figure 1 shows the Auger Spectrum of the clean Pd(111) surface.

The results obtained from the CPD measurements during the adsorption of oxygen show that $\Delta\phi$ is lower ($\sim$ 500 mV) than the value observed for

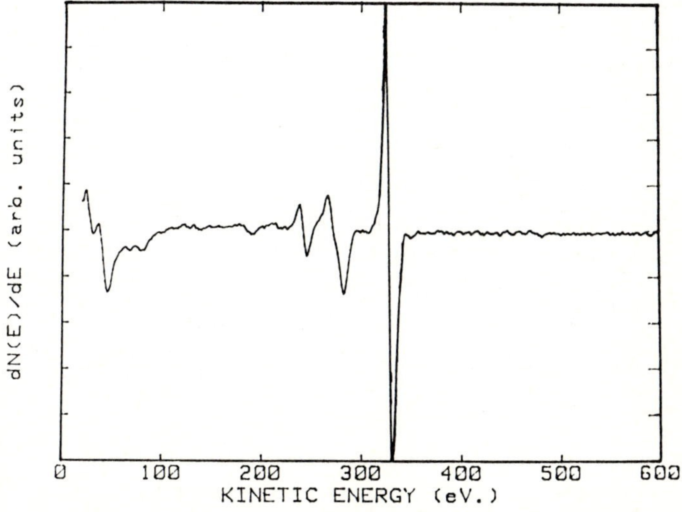

Fig. 1 Auger Spectrum of the clean Pd(111) film surface.

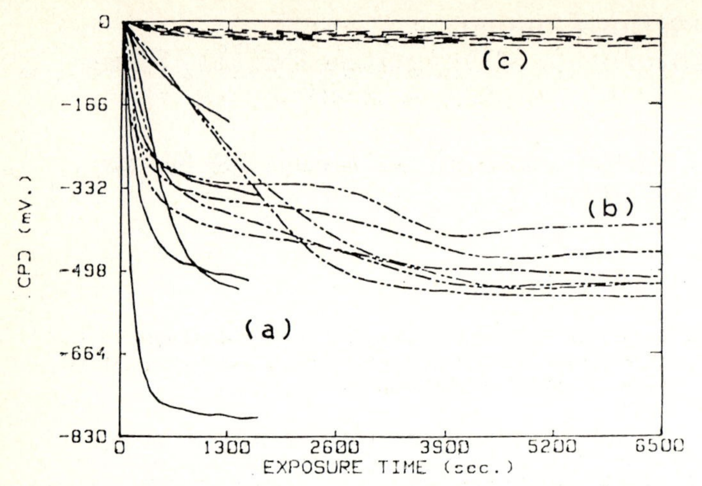

Fig. 2(a) adsorption of O_2 on freshly deposited Pd(111) film, (b) film is saturated with O_2 and CPD is independent of pressure and (c) oxide Pd(111) film. Pressures of O_2 were 1×10^{-8}, 2×10^{-8}, 5×10^{-8}, 1×10^{-7}, 5×10^{-7} and 7×10^{-7} torr.

Pd(111) on a bulk crystal by Surnev et. al. [1] (∿ 800 mV). It is also noted that the Pd(111) film had three different stages during the O_2 adsorption experiments. The first stage is of the freshly deposited Pd(111) film (Fig. 2(a)); the second, the oxygen saturated film (Fig.2(b)), and third the oxide Pd(111) film (Fig. 2(c)).

A dependence of $\Delta\phi$ with the pressure of O_2 is observed in the first stage of the Pd(111) film (Fig. 2(a)). In general, the higher the pressure, the higher the work function change.

In the second stage, $\Delta\phi$ is independent of O_2 pressure (Fig. 2(b)) and has an average value of ∿ 500 mV

In the third stage, (Fig. 2(c)) the surface is inert to oxygen adsorption which indicate that the Pd(111) surface is oxidized.

CONCLUSION

Due to the method used to grow the film, it was found that the Pd(111) surface in the film is very smooth compared with a Pd(111) from a bulk crystal. The step and grain boundary density on the Pd(111) film is much lower than in bulk crystal. As a consequence, the contribution to $\Delta\phi$ due to these defects of higher energy is lower, giving a lower value of $\Delta\phi$.

The different CPD values observed during the first stage of the Pd(111) film may indicate that the oxygen is diffusing into the Pd matrix. Other groups [1,2] have observed this in bulk Pd(111) but the difference in values is not so high because the bulk Pd(111) is higher in impurities than the Pd(111) film.

Once the film has been exposed to oxygen for a long period in the first stage, $\Delta\phi$ becomes invariant to O_2 pressure and the CPD obtained is about 500 mV.

From the last stage of the sample it can be concluded that the film was subjected to an oxidation process during the first and second stages and as a consequence the Pd(111) surface became inert to O_2 adsorption and the observed change in $\Delta\phi$ of 50 mV is mainly due to residual gases in the chamber.

As a general conclusion, it can be said that the morphology of the surface influences greatly the observed value of $\Delta\phi$ of about 500 mV on Pd(111) film compare with the higher value on bulk Pd(111) of 800 mV.

ACKNOWLEDGEMENTS

The author wishes to acknowledge Prof. R.W. Vook and Prof. James A. Schwarz of Syracuse University, N. Y. for facilities and financial support for this work during his stay in Syracuse.

REFERENCES

1. Surnev L., Bliznakov G. and Kiskinova M.
   Surface Science 140 (1984) 294-260.
2. Engel T. and Ertl G.
   Advances in Catalysis 28 (1979) 2-73.

Electrical Conductivity and Photoelectric Work Function During the Growth and Annealing of Polycrystalline Copper Films

M.C. Asensio, J.M. Heras, and L. Viscido

Instituto de Investigaciones Fisicoquímicas Teóricas y Aplicadas (INIFTA), Facultad de Ciencias Exactas, Universidad Nacional de La Plata, C.C. 16, Sucursal 4, 1900 La Plata, Argentina

1. Introduction

Thin evaporated metal films have a microcrystalline structure showing some special optical and electronic properties widely used in microelectronics and as model catalysts. These films can be seen as an intermediate stage between isolated atoms -or small clusters- and bulk material. There are many papers dealing with the growth of Cu layers over metals -mostly above room temperature- specially on W /1/, but we know of only one paper in which the growth on a Cu substrate has been studied /2/, employing a (100) single crysta Hence, as a first step in gaining knowledge of the electrical and adsorptive behaviour of polycrystalline thin Cu films, we have studied their growth at low temperatures onto two types of substrates: Pyrex glass and polycrystalline thick Cu films with a strong (111) fiber texture with the axis normal to the surface. Complementary structure studies through X-Ray diffraction (XRD) and scanning electron microscopy (SEM) were also performed.

2. Experimental Details

The experiments were carried out in a conventional glass-metal UHV-apparatus having an Omegatron-type mass spectrometer to control residual gas composition. The films were obtained by condensation of the Cu vapour coming from a W filament resistively heated, onto a cold substrate with a type K thermocouple and four Pt contacts to measure the electrical resistance (R). Film growth at 77 and 196 K was monitored through R and the photoelectric work function (WF). Mean deposition rate, about 1.09 x 10^{17} $m^{-2}s^{-1}$, was estimated after the experiment from the mass deposited (determined by atomic absorption spectroscopy), and the deposition time. Afterwards, SEM and XRD studies were performed on films deposited simultaneously on separate substrate disks in the same chamber. The coverage θ was determined from the mean deposition rate, assuming a roughness factor of 2.5 in the case of the Cu base film /3/.

3. Results and Discussion

3.1 Pyrex glass substrate

Cu films were grown on the clean substrate at 77 or 196 K up to thicknesses d_1 between 20-80 nm, monitoring R. In the first stages ($\sim$ 8 monolayers) R oscillates between 10^{13} and 10^5 Ohms. Afterwards, R decreases monotonically up to the final values. The oscillations in R point to a high Cu-adatom diffusion even at 77 K accounted for by the low diffusion energy of Cu over glass (3.4 kJ/mol), which favours nucleation at low critical coverages. The films obtained were highly porous, as the adsorptive behaviour for water molecules demonstrated later. The absence of electrical conductivity in the first growth stages prevented photoelectric WF measurements up to $\sim$10 monolayers. Then, in spite of thickness increase, the WF took a constant value characteristic for each deposition temperature: at 77 K, 4.40 ± 0.02 eV and at 196 K, 4.53 eV. After stabilization at deposition temperature, the films were annealed by increasing temperature at a rate of 4 K/min up to 473 K. During annealing the WF increased, reaching a steady value of 4.94 eV. This increase is not reversible by lowering the measuring temperature. A WF increase is always related to film surface rearrangements exposing close atom-packed planes, as we confirmed through XRD. The WF values found for films annealed at 77, 196 and 473 K are in very good agreement with those reported for the planes (110), (100) and (111) for Cu single crystals, respectively /4/. The fact that films expose these crystallographic planes following the increase in annealing temperature was also neatly proved later in similar films when the changes in WF observed during oxygen adsorption were compared with those reported for the above-mentioned single crystals /5/.

After one hour at 473 K we consider the films to be well annealed since they showed no change in WF, though R still decreased at a very low rate; one tenth of a percent per minute. As a consequence of the annealing process R diminished ∿ 88 %, indicating a bulk rearrangement, possibly a recrystallization, as XRD and WF suggest. However, from a kinetic study monitoring R changes during isothermal annealing at low temperatures (77 and 196 K), the activation energy associated with the ordering process was determined to be 20-40 kJ/mol. This indicates mainly the migration of point defects as we have previously stated for Fe, Co, and Ni films /6/. The values obtained for Cu agree with those found for Cu wires bombarded with 12 MeV deuterons /7/, which it is said creates point defects.

3.2 Copper Substrate

Over the well annealed Cu layers of thickness d_1 just described, other layers were deposited also at 77 and 196 K, monitoring R and WF as a function of their increasing thickness d_2 (see Fig. 1). At 77 K, R increased initially showing a maximum at $\theta = 2.5$ ($d_2 = 1.5$ nm), which was greater the thinner the base layer. Thereafter, R diminished monotonically. This behaviour is similar to the parabolic dependence of thin metal film resistivity on surface contamination, sometimes referred to as the Nordheim effect /8/. Conversely, the WF did not undergo any change for a while, and then diminished steadily from 4.94 to 4.40 eV as θ increased, thus indicating a continuous decrease in surface ordering. Films condensed at 196 K behave similarly. However, the initial increase in R was very small with the maximum at $\theta = 0.5$, and the final WF was 4.70 eV. This WF value is higher than that observed in films condensed at 196 K on glass and actually corresponds to the value of such films annealed at 373 K, suggesting a restricted epitaxy. A maximum WF decrease of 0.8 eV has been reported recently during deposition of Cu on W(110) /1/ at 300 K when $\theta = 2$. Taking the WF of W(110) as 5.47 eV /9/, the WF of the Cu film should be 4.67 eV, a value in excellent agreement with our results in both, the deposition on Cu at 196 K -including the value of θ (see Fig. 1b, curve 2)- and the deposition on glass at 77 K after annealing at 300 K. Figure 1 shows our results.

Transport properties of thin metal films can be described in the size-effect model through the fraction p of the conduction electrons specularly reflected by the film boundary surfaces. Hence, the increase in R during the growth of the second layer can be explained by a decrease in p because the arriving atoms stick to the Cu surface without diffusion and remain isolated or build widely spaced clusters, acting in this way as scattering centers for the conduction electrons. The maximum R value can be ascribed to $p = 0$. From this point on, R diminishes as deposition progresses because of thickness increase. Following these assumptions /10/ a value $p = 0.75$ was calculated for the clean and well-annealed base Cu layer. This value compa-

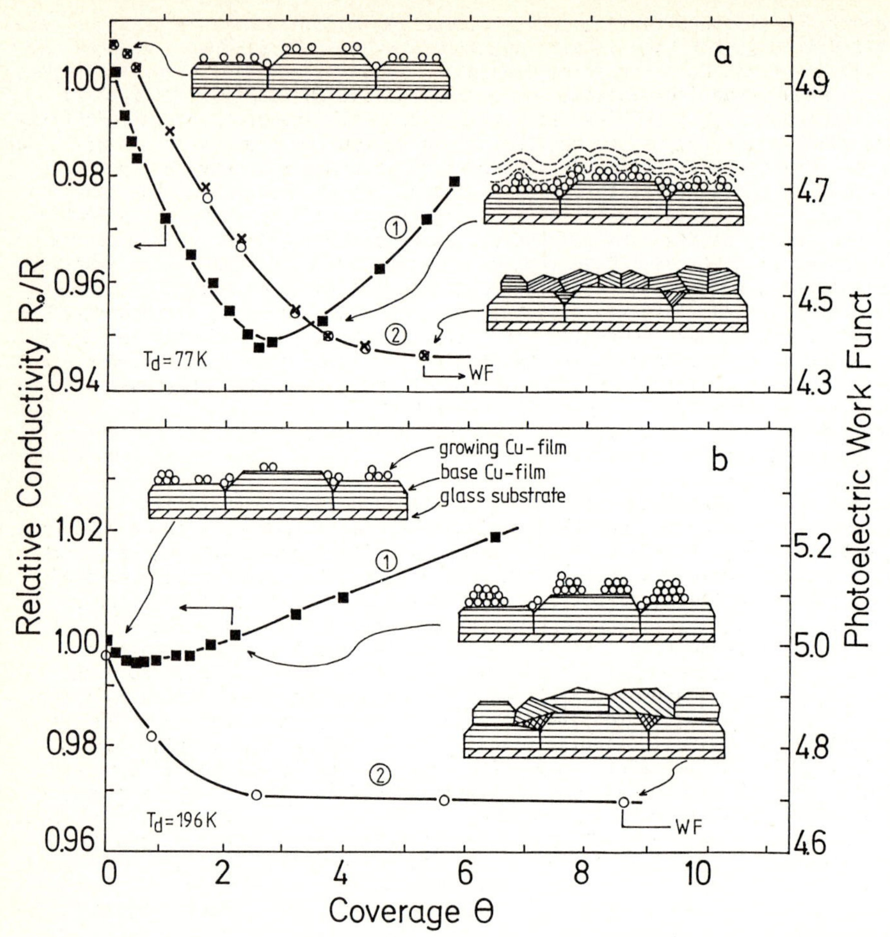

Figure 1. Changes in relative conductivity (1) and WF (2) during growth of a Cu overlayer onto a well-annealed Cu base layer at a) 77 K and b) 196 K. The relative conductivity equals the initial resistance of the base layer, R_0, divided by the overall resistance R during film growth. The crosses in curve 2a result from model fitting. Insets are schematic representations of three growth stages: Notice the difference between temperatures.

res well with that of 0.8 calculated for Cu films on Si(111) /11/. High p values can be associated with the extended (111) terraces we found with XRD and SEM, (Fig. 1, inset schemes for the base layer).

The WF decrease during the growth of the second layer can be interpreted also assuming localized adsorption of isolated atoms followed by clustering due to capture of impinging atoms. The contribution of the adatoms to the WF can be expressed as usual by a Helmholtz-type equation; whereas clusters modify substrate WF proportionally to θ and to the difference between their WF and that of the clean surface. The explicit dependence on θ of the contribution of isolated atoms and clusters, was statistically derived taking into account an hexagonal lattice /12/. Fitting the model to our experimental data at 77 K (crosses in Fig. 1a) the dipole moment of an isolated Cu adatom can be calculated to be μ = 0.27 D. Due to the S-shape of the curve, with increasing θ, this value increases to μ = 0.92 D. As the Cu substrate, though highly textured, is polycrystalline, at very low θ the impinging Cu atoms stick preferentially at valley sites due to the high trapping "cross-section", causing a small change in WF (increase in ordering). A similar observation has been reported in the system Cu/W(110) in spite of the single crystal surface employed /1/. In this paper, from the linear part of the curves WF-change vs. θ at different temperatures, μ-values have been derived which we extrapolate to μ = 0.64 D at 77 K. This value agrees with ours, the difference is ascribable to a different substrate and θ scale. In other words, μ is not very sensitive to substrate type and crystallinity. Perhaps, as suggested in /1/, the μ values are better correlated to the density of those electrons which can be displaced easily. The fitting of the WF changes observed during the growth at 196 K does not give meaningful values, possibly because of the scarcity of data.

4. Conclusions

Cu films grow over glass at 77 or 196 K with a mechanism involving a high nucleation rate and the building of 3-dimensional islands (Volmer-Weber mechanism), which results in a highly porous deposit. Conversely, over a Cu substrate this mechanism is found at 196 K, while at 77 K, as a consequence of the large diffusion energy Cu/Cu, (190 kJ/mol), a localized adsorption of atoms takes place, which reduces nucleation rate and prevents the building up of large islands. These films, though having rough surfaces and a microcrystalline structure, are not porous as we have proved since through the adsorption of water molecules. A critical temperature at which the film growth mechanism changes has also been observed in the systems Ni/Ni(100) /13/ and Cu/Cu(100) /2/. In this last paper -an He-scattering study- the Cu adatom diffusion was so restricted below 235 K that during deposition of several monolayers an accumulation of surface disorder took place, until a steady state of surface roughness was reached, consisting of a large number of small islands with a characteristic dimension of 6 nm. These results support the WF changes we observed during deposition over a Cu film substrate.

5. References

1. J. Kolaczkiewicz, E. Bauer: Surf.Sci. 160, 1 (1985)
2. L.J. Gómez, S. Bourgeal, J. Ibáñez, M. Salmerón: Phys.Rev. B31, 2551 (1985)
3. J.A. Allen, C.C. Evan, J.W. Mitchell: in Structure and Properties of Thin Films, ed. by C.A. Neugebauer, J.B. Newkirk, D.A. Vermilyea, (J. Wiley and Sons, Inc., New York, 1959) p. 46
4. G.A. Haas, R.E. Thomas: J.Appl.Phys. 48, 86 (1977)
5. T.A. Delchar: Surf.Sci. 27, 11 (1971); J.E. Boggio: J.Chem.Phys. 57, 4738 (1972); H. Niehus: Surf.Sci. 130, 41 (1983)
6. J.M. Heras, E.V. Albano: Thin Solid Films 106, 275 (1983)
7. A.W. Overhauser: Phys.Rev. 90, 393 (1953)
8. D.L. Lessie, E.R. Crosson: J.Appl.Phys. 59, 504 (1986)
9. R.S. Polizzotti, G. Ehrlich: Surf.Sci. 91, 24 (1980)
10. B. Fischer, G.V. Minnigerode: Z.Phys. B42, 349 (1981)
11. J. Feder, P. Rudolf, P. Wissmann: Thin Solid Films 36, 183 (1976)
12. H. Schmidt: Thesis (Hannover University, West Germany, 1973)
13. P. Schrammen, J. Hölzl: Surf.Sci. 130, 203 (1983)

NMR for the Study of Physisorbed and Chemisorbed Matter on the Surface of Solids

B. Boddenberg

Lehrstuhl für Physikalische Chemie II, Universität Dortmund,
Otto-Hahn-Str., D-4600 Dortmund 50, Fed. Rep. of Germany

1. Introduction

Nuclear magnetic resonance (NMR) has been applied in the field of surface science and heterogeneous catalysis for more than thirty years. Until about 1970, however, for sensitivity reasons the proton and to some extent the fluorine nucleus were practically the only nuclei detected in such studies for which, moreover, high surface area adsorbents had to be used. Since then the enormous progress in NMR spectrometer instrumentation (superconducting magnets, Fourier transform technique, high speed and high capacity computers) as well as the development of new NMR techniques especially in the field of solid state NMR in the last decade have made it possible to observe most nuclei of importance on a reasonable sensitivity level even with microcrystalline solid supports of specific surface areas of several m^2/g.

The most important feature of NMR which renders it extremely useful as well as complementary rather than competitive to other spectroscopies is to provide a tremendous dynamic range from picoseconds to seconds or even longer. It is in this range of time scale that such dynamic processes as molecular translational and rotational motions as well as atomic and molecular exchanges proceed. This dynamic view provided by NMR of an interphase system seems to have been paid too little attention in comparison with the static structural view, and is presumably essential for the understanding of e.g. the kinetics of phase changes in two-dimensional (2D) surface films and the mechanism of catalytic reactions.

In studies of hydrogen-containing species on solid surfaces, either intrinsic or intentionally introduced, the proton is still a very valuable NMR probe nucleus in many respects. However, the great variety of magnetic couplings of comparable strengths the proton is subject to, e.g. dipolar with homo- and/or heteronuclear spins and with electronic spins of impurities present in the surface layer of the solid substrate, almost unavoidably cause the interpretation of the proton resonance data obtained to run into severe difficulties.

Replacing hydrogen H by deuterium (D) which because of the low natural abundance of D can be done even selectively, and looking for the deuteron (^2H) instead of the proton (^1H) resonance almost entirely removes the above-mentioned difficulties in most cases. This is due to the almost completely dominating interaction energy of the electric quadrupole moment of the deuteron with the electric field gradient tensor at the nuclear site in comparison with the internal magnetic coupling energies. This advantage over the proton has, of course, to be paid for by leaving such types of motions undetected which do not change the quadrupole coupling energy, e.g. pure translational motions.

The present contribution is intended mainly to demonstrate the use of the deuteron as probe nucleus to reveal the dynamics of species adsorbed on a variety of solid surfaces which are different with respect to structure, morphology, capability for chemisorption and catalysis, and providing support for 2D phases in the thermodynamic sense. Emphasis is further laid on the application of the modern techniques of solid state deuteron magnetic resonance spectroscopy which are most suited for the study of chemisorbed species and the properties of physisorbed layers in the 2D solid state. Surprisingly little work using these techniques has appeared in the literature. Besides our own contributions /1-5/ only three publications being concerned with the study of molecules on high area aluminas and alumosilicates have come to our attention /6-8/.

2. Theoretical Basis for the Interpretation of ^2H Spectra

The deuteron (^2H) has spin $I = 1$, and therefore on account of its magnetic dipole moment (μ) and its electric quadrupole moment (Q) is subject to interactions with the magnetic field and the electric field gradient at the nuclear site, respectively. In the present context the deuteron is considered to be contained in a hydrogen-carbon bond of a molecule adsorbed on the surface of a diamagnetic solid.

For the examples to be presented in this contribution the electric field gradient (EFG) tensor at the deuteron site can be assumed to be practically axially symmetric, and to be almost completely determined by the intramolecular charge distribution. This entails that the distinct principal axis EFG tensor component (q) is in line with the C-D bond axis and is the only tensor component affecting the spectral features.

In bulk phases the internal magnetic couplings such as dipolar and chemical shielding are negligible in comparison to the electric quadrupole interaction unless the latter is averaged to low values by rapid rotational types of motion, or unless special NMR techniques, e.g. double quantum coherence excitation, are applied. In the case of adsorbed species on solid surfaces, however, magnetic interactions of considerable strength due to shielding fields may occur which are set up by the action of the strong Zeeman field on the diamagnetic susceptibility tensor of the substrate, This is notably the case with graphite owing to its comparatively large values of the isotropically averaged susceptibility $\bar{\chi}$ and the susceptibility anisotropy $\Delta\chi = \chi_{\parallel} - \chi_{\perp}$. For graphite, which to date is the only substrate where such a magnetic interaction has been detected /5/, the shielding tensor is axially symmetric with its distinct principle axis oriented normal to the basal planes exposed.

On the basis of the foregoing discussion the deuteron spin Hamiltonian from which the spectrum is to be calculated may be set up as

$$H = H_Z + H_Q + H_S \ , \tag{1}$$

where H_Z represents the Zeeman interaction with the applied static magnetic field B_0 (8.3 T in the present study), and H_Q and H_S are the quadrupole and shielding Hamiltonians, respectively. In any case, $|H_Z| \gg |H_Q|, |H_S|$ which allows the internal interactions to be treated as first-order perturbations to H_Z. In section 2.1 the basic case of negligible shielding interaction will be treated which pertains to the majority of adsorption systems that have been studied. This is followed by section 2.2 where H_S is taken full account of, and which serves as the basis for the interpretation of the spectra of systems with graphite as solid substrate. In section 2.3 the most important situation is covered where the quadrupole Hamiltonian becomes time dependent as consequence of rotational motions of the adsorbed molecules or parts of them.

2.1 Quadrupole Interaction

In Fig. 1 a and b, is depicted schematically the energy level diagram and the NMR spectrum of a deuteron in a C-D bond with fixed orientation in space being acted on by the Zeeman field B_0 and the electric field gradient. The splitting of the

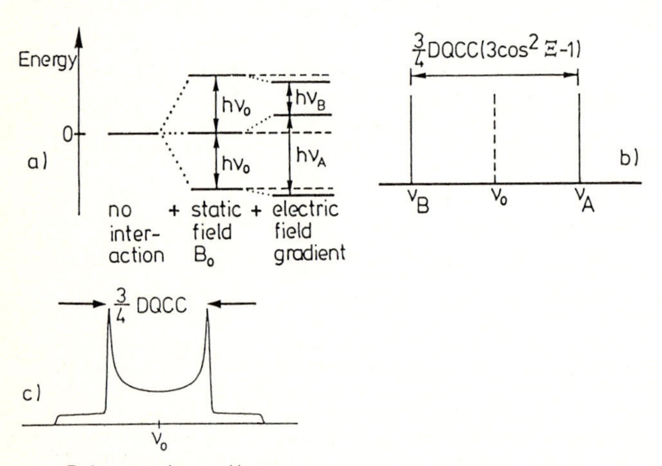

Fig. 1a-c. Schematic energy level diagram (a) and NMR spectrum (b) of a deuteron acted on by a strong static magnetic field and an axially symmetric EFG tensor. (c) Pake powder pattern

doublet lines ν_A and ν_B which are symmetrically displaced about the central frequency is seen to depend on the quadrupole coupling constant DQCC = e^2qQ/h as well as on the angle Ξ between the C-D bond direction and the magnetic field $\vec{B}_0$. This angular dependence entails that the spectrum of an ensemble of randomly oriented C-D bonds such as is realized in the case of deuterated hydrocarbon molecules adsorbed on an adsorbent powder with random distribution of surface orientation is the sum of doublets of frequency splittings between zero and the quadrupole frequency $\nu_Q = (3/2)(e^2qQ/h)$. This Pake powder pattern is shown in Fig. 1c. The prominent edges have splitting $\Delta\nu = (1/2)\ \nu_Q$ from which relation the quadrupole coupling constant can most reliably be determined. With typical values of e^2qQ/h in the range 160 to 200 kHz /9/ $\Delta\nu$ amounts to 120 to 150 kHz.

2.2 Combined Quadrupole and Shielding Interactions

In order to calculate the NMR resonance frequencies for this case from the Hamiltonian (1) the EFG and shielding tensors have to be expressed in the laboratory coordinate frame of reference (L) the Z axis of which is chosen to be parallel to the static field $\vec{B}_0$. The principal axis system of the EFG tensor (P) and of the shielding tensor (C) are fixed with respect to the molecular framework of the adsorbed species and the substrate crystalline lattice, respectively. Choosing the Z axes of these latter frames parallel to the C-D bond axis and the surface normal, respectively, the P, C, and L frames are interrelated by sets of Eulerian angles according to

$$\text{P-frame} \xrightarrow{(0,\theta,\phi)} \text{C-frame} \xrightarrow{(\alpha,\beta,\gamma)} \text{L-frame.} \qquad (2)$$

Using these rotation transformations the resonance frequencies yielding the doublet are calculated /5/ as

$$\nu_A = \nu_0(1-s_{iso}) - 3a_Q - a_S \qquad (3)$$

$$\nu_B = \nu_0(1-s_{iso}) + 3a_Q - a_S \quad \text{with}$$

$$a_Q = \frac{1}{16}\,\text{DQCC}\left[(3\cos^2\theta-1)\,(3\cos^2\beta-1) + 3\sin^2\theta\,\sin^2\beta\,\cos2(\alpha+\phi) - 3\sin2\theta\,\sin2\beta\,\cos(\alpha+\phi)\right] \qquad (4)$$

$$a_S = \frac{1}{3}\,\nu_0\,\Delta s\,(3\cos^2\beta-1). \qquad (5)$$

In (3) to (5) ν_0 is the Zeeman frequency (cf. Fig. 1), s_{iso} the isotropic shielding constant, $\Delta s = s_{\parallel} - s_{\perp}$ the shielding anisotropy and DQCC = e^2qQ/h the deuterium quadrupole coupling constant.

Using known procedures /10, 11/ the powder pattern is readily calculated from (3) to (5) by integrating over the domain of definition of the Eulerian angles, α, β, and γ.

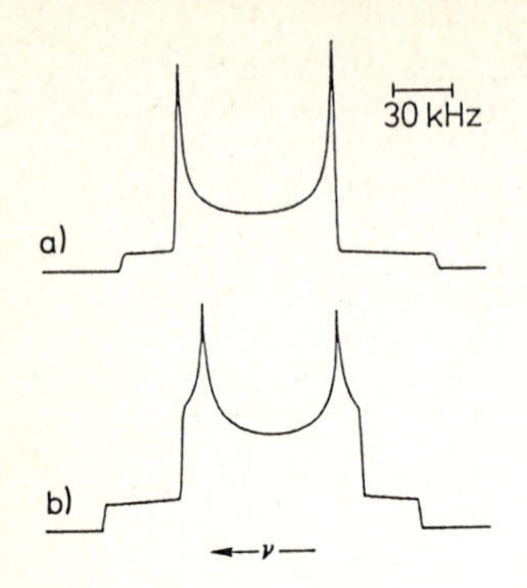

Fig. 2a and b. Calculated ^2H powder patterns for combined action of an axially symmetric EFG tensor, DQCC = 100 kHz, and an axially symmetric shielding tensor, $\Delta s = + 200$ ppm. The tensors are assumed to be parallel (a) or perpendicular (b) to each other

Figure 2 shows the result of the calculation of such powder spectra with parameters DQCC = 100 kHz, $\Delta s = + 200$ ppm, and $\theta = 0^o$ (Fig. 2a), $\theta = 90^o$ (Fig. 2b). The two cases correspond to situations where the tensors are parallel and perpendicular to each other, i.e. the C-D bonds are normal and parallel to the surface plane. It is clearly seen that the patterns are unsymmetric as compared with the Pake pattern displayed in Fig. 1 and depend on the relative orientation of the EFG and shielding tensors.

2.3 ^2H Spectra in the Presence of Rotational Motions

The treatments in Sections 2.1 and 2.2 have assumed that the C-D bonds do not undergo any directional change in space. For molecules adsorbed on particulate adsorbents two types of motion may be envisaged which change the C-D bond orientation: rotation of the molecule as a whole or of part of it, and translational motion which carries the molecules around the particles by surface diffusion. Both types of motion lead to fluctuations of the quadrupole interaction energy with correlation times τ_r and τ_t, respectively, whereas the shielding interaction energy is modulated by the translations only. Since for the examples to be discussed the shielding interaction is either negligible or of lower strength as compared with the quadrupole interaction the latter only will be taken into account in the further treatment.

The time scale of NMR for the presently discussed detection of ^2H spectra is given by $1/\nu_Q$, where ν_Q is the previously introduced quadrupole frequency coincident with the powder spectrum width. With the data given before, the time scale is in the range of several microseconds. If $1/\tau_r, 1/\tau_t \ll \nu_Q$ the results obtained in the previous sections remain valid. For $1/\tau_t \gg \nu_Q$ in which case the adsorbed molecules sample rapidly on the NMR time scale the faces of a microcrystallite crystal the powder pattern collapses into a singlet. The same is true if some rotational type of motion moves the C-D bond considered isotropically in space.

An interesting case very often encountered in adsorption systems arises if $1/\tau_t \ll \nu_Q \ll 1/\tau_r$ where τ_r is the correlation time of the quadrupole energy fluctuation brought about by some anisotropic molecular rotation about a space fixed axis. Such situations occur with chemisorbed species and physisorbed layers in the 2D solid state. In this case H_Q in (1) can be replaced by its mean over the motion, $\overline{H}_Q$ /11/, with the consequence that an axially symmetric averaged EFG tensor instead of the (rigid) EFG tensor itself appears to be operative. This average tensor is aligned along the rotation axis and has its distinct principle axis component reduced by the factor $(1/2)(3\cos^2\Delta-1)$ where Δ is the angle between the C-D bond and the rotation axis. This situation is depicted in Fig. 3 for a deuteron contained in a methyl group. The explicit calculation of the powder spectrum may be carried out with the aid of (3), (4) and (5) interpreting DQCC = $(1/2)(3\cos^2\Delta-1)(e^2qQ/h)$ and the set of Eulerian angles $(0, \theta, \phi)$ as accomplishing the rotational transformation from the principle axis system of the averaged tensor to the C-frame /5/. It is seen from the example in Fig. 3 that with methyl group rotation being fast on the NMR time scale the Pake pattern is narrowed to just one third ($\Delta = 109.5^0$) of its width in the slow rotation case.

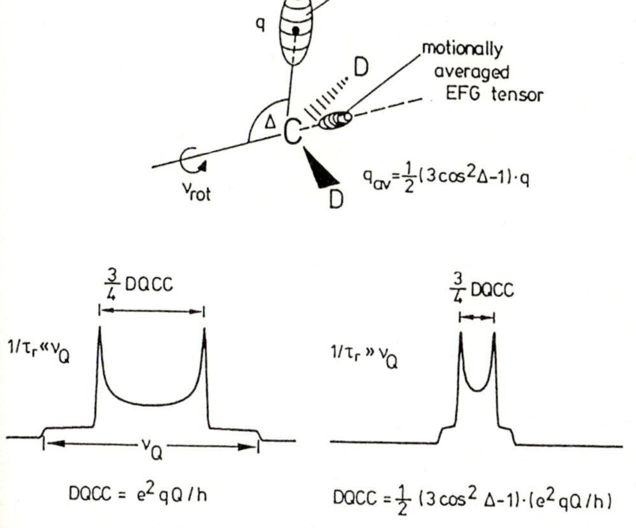

Fig. 3. Motional averaging of EFG tensor by rapid single axis rotation and resulting powder pattern, exemplified for a deuteron in a methyl group

The case of multiple single axis rotations can be treated in a similar way provided that the relevant rotations proceed either fast or slow on the NMR time scale (Fig. 4).

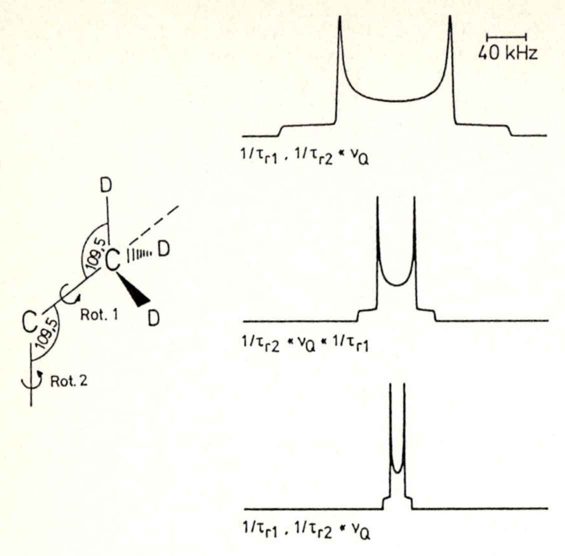

Fig. 4. Effect of multiple single axis rotation on ^{2}H spectra, exemplified for a deuteron in a methyl group of a carbon skeleton fragment

3. Illustrative Examples for the Applications of ^{2}H FT NMR in Surface Science

The deuteron spectra which are shown in chapters 3 and 4 were measured at resonance frequency 52.7 MHz in a magnetic field B_0 = 8.27 T. They are the Fourier Transform (FT) of free induction decay and quadrupole echo signals in cases where isotropically averaged singlets and solid state patterns were observed, respectively. Quadrupole echo sequence /12/ excitation and quadrature detection of the echoes was applied. Following common parlance the characterization of a motion to proceed slowly or fast is always understood as being referred to the NMR time scale.

3.1 Trimethylsilyl Groups on Silica

In Fig. 5a are shown, at several selected temperatures between 273 and 75 K, the ^{2}H spectra of trimethylsilyl(TMS)-d_9 groups anchored on a silica surface /4/. Although substantially rounded off the spectra are of the Pake pattern type with edge separation $\Delta\nu$ = 14±1 kHz from which DQCC = 18.5±1.5 kHz is obtained. Since the value of e^2qQ/h can be assumed to be about 176 kHz /9/ this result is readily explained /4/ by the notion of rapid rotations about both the O-Si and Si-C bonds (Fig. 5b) each of which rendering a reduction factor of 3 if the relevant bond angles are taken to be tetrahedral.

3.2. Benzene on η-Al_2O_3 and Pt/η-Al_2O_3

Figures 6 and 7 display the ^{2}H spectra of a monolayer of benzene-d_6 adsorbed on η-alumina and on a platinum/η-alumina catalyst, respectively /13/. In both cases nearly the same sequence of spectral shapes as a function of temperature

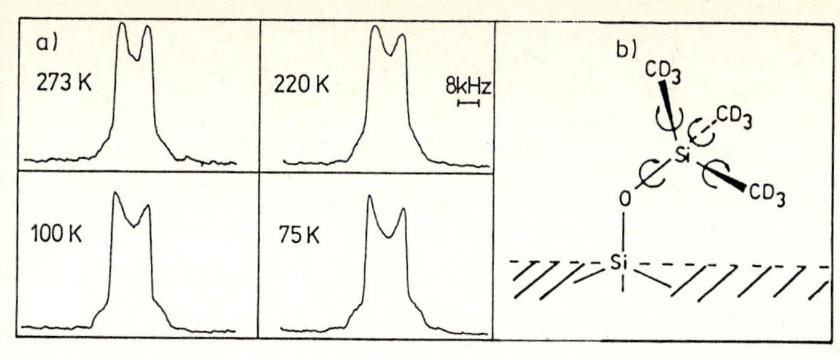

Fig. 5a and b. ^{2}H spectra of trimethylsilyl-d_9 groups anchored on a silica surface (a) and proposed model of molecular rotations (b)

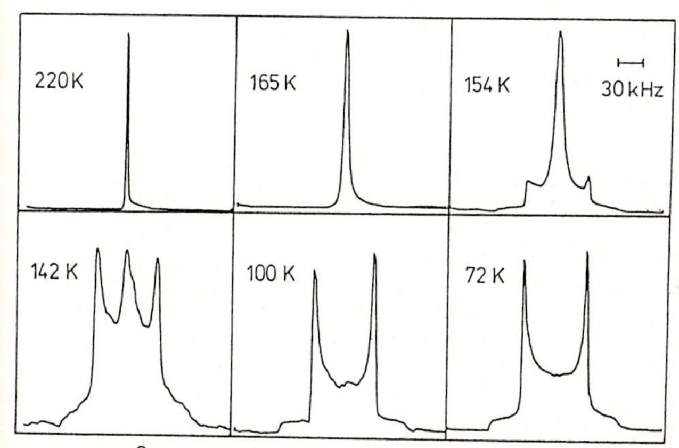

Fig. 6. ^{2}H spectra of a monolayer of benzene-d_6 adsorbed on η-alumina

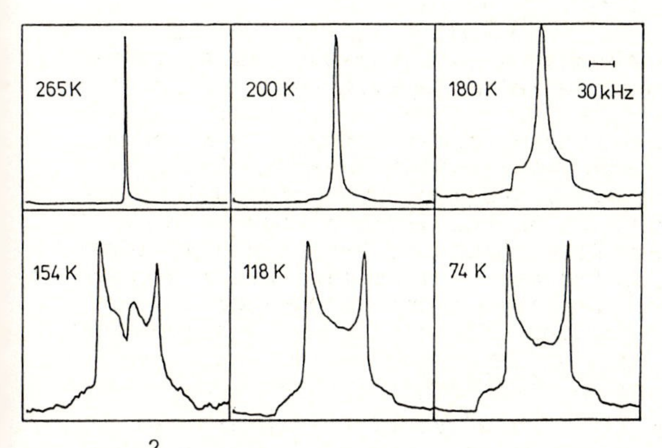

Fig. 7. ^{2}H spectra of a monolayer of benzene-d_6 adsorbed on a 5% Pt/η-alumina catalyst

is observed going from singlets via superpositions of a singlet with a Pake pattern to, finally, clearly developed Pake patterns of edge slittings $\Delta\nu$ = 70 kHz. However, for the catalyst the sequence appears to be shifted by about 30 K to higher temperatures as compared to the pure alumina support. A similar behaviour has recently been detected with the aid of ^{1}H NMR spectroscopy for benzene on η-alumina and the corresponding platinum catalyst /14/.

With DQCC = 93 kHz from the edge splittings of the low temperature powder patterns and e^2qQ/h = 186 kHz /15/ the angle Δ is readily evaluated to be 90° indicating rapid hexad rotation even at temperatures as low as 72 K. In the high temperature regime the singlet line shapes reveal isotropic averaging of the quadrupole interaction energy as a consequence of rapid rotational motion and/or rapid diffusion within the voids of the mesoporous alumina. The forementioned high temperature shift of the line shape developments when platinum is deposited on the support is an indication of lower mobility of the benzene molecules when adsorbed on platinum rather than on the oxide.

3.3 Isopropanol on η-Al_2O_3

Figure 8 shows the ^{2}H NMR spectra of various selectively deuterated isopropanols chemisorbed on η-alumina at several selected temperatures in the range 292 to 73 K. The samples were prepared by contacting the oxide outgassed at 450°C under high vacuum with isopropanol vapour at ambient temperature followed by 24 hrs evacuation at 300 K /16/. By this procedure only chemisorbed species at a very low concentration level are left on the surface.

Assuming that the isopropanol molecule skeleton is essentially conserved on chemisorption the low temperature Pake patterns of Fig. 8a immediately reveal that at 100 K and below the molecules are rigidly held on the surface as may be concluded from the 130 kHz edge splittings corresponding to 160 to 180 kHz rigid quadrupole coupling constants typical for deuterons in sp^3 hybrized C-D bonds /17/. At higher temperatures the typical Pake pattern shape is lost and the line width somewhat decreased, which is probably due to some types of torsional motions and heterogeneity effects.

Similar conclusions concerning the rotational freedom of the isopropanol skeleton have to be drawn from the spectra shown in Fig. 8b. In this case, however, the spectral shapes are determined predominantly by the rotation of the methyl groups, the rate of which goes from fast to slow on the NMR time scale with decreasing temperature. This conclusion is suggested by the enlargement of the Pake pattern edge separation from about 40 to 130 kHz between 100 and 73 K. The appearance of the central portions of the 100 K spectrum is not fully understood as yet, presumably heterogeneity effects are responsible.

In contrast to the ^{2}H spectra shown in Fig. 8a and b the spectra of the O-D group deuteron (Fig. 8c) neither exhibit

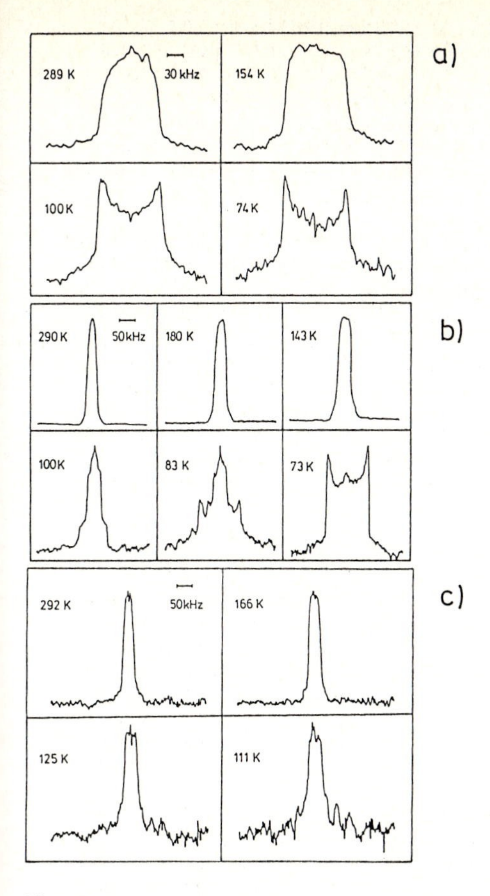

Fig. 8a-c. ^2H spectra of chemisorbed isopropanol on η-alumina.
(a) $(CH_3)_2CDOH$,
(b) $(CD_3)_2CHOH$,
(c) $(CH_3)_2CHOD$

the Pake pattern shapes at any temperature investigated nor the large widths observed with the other deuterated species although the electric field gradient at the deuteron site in an O-D bond is generally larger than in a C-D bond /18/. This behaviour could be explained either by a very strong reduction and non-axial symmetry of the EFG tensor, or by strong motional narrowing as a consequence of some rapid deuteron movement other than rotation of the molecule as a whole, or by both effects. In any case, as expected for chemical reasons the alcoholic group deuteron deserves special attention, its partial or complete detachment from the molecule seems probable.

3.4 Acetone and Isopropanol on Titania (Rutile)

Acetone-d_6 and isopropanol-d_6 of which ^2H spectra are shown in Figs. 9 and 10, respectively, were chemisorbed on microcrystalline rutile with two differently conditioned surfaces: bare (standard pretreated /19/) and preloaded with irreversibly adsorbed water H_2O /19/. The loadings with acetone and isopropanol were performed at ambient temperature by contacting the evacuated adsorbents with the respective adsorp-

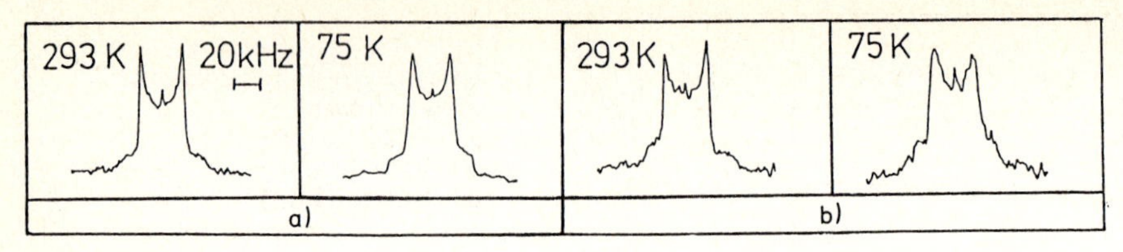

Fig. 9a and b. ^2H spectra of acetone-d_6 chemisorbed on microcrystalline rutile. (a) Bare surface, (b) surface preloaded with irreversibly adsorbed water (H_2O)

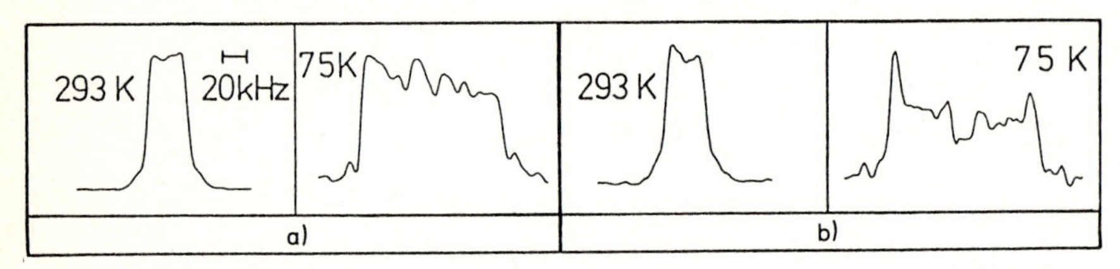

Fig. 10a and b. ^2H spectra of isopropanol $(CD_3)_2CHOH$ chemisorbed on microcrystalline rutile. (a) Bare surface, (b) surface preloaded with irreversibly adsorbed water (H_2O)

tive vapours and subsequent 24 hrs evacuation under high vacuum /20/.

For acetone well developed Pake patterns are obtained, the edge splittings (40 kHz) of which are almost independent of both temperature and surface pretreatment. The spectra immediately reveal that the molecules are orientationally fixed in space with allowance given for rapid methyl group rotation. These results suggest that onto the bare surface each acetone molecule is coordinatively bonded via its oxygen atom to a Ti^{4+} ion site /21/. It is further suggested that the same type of bonding occurs on the preloaded surface by displacement of water molecules occupying these sites /22/ by acetone since otherwise, e.g. with the notion of hydrogen bonding, the spectra could hardly be explained. Actually, this conclusion is at variance with the view of MUNUERA and STONE /23/ based on adsorption isotherm and TPD results who favour the notion of displacement of acetone by water.

In comparison to acetone the ^2H spectra of isopropanol on rutile are more involved (Fig. 10). The development of the edge splittings of the powder patterns from about 40 kHz at 293 K to about 130 kHz at 75 K indicates that the rate of methyl group rotation runs through the NMR time scale from fast to slow, which behaviour is similar to that found for isopropanol chemisorbed on η-alumina (Fig. 8b). The practically same spectral shapes of isopropanol on the bare and water preloaded rutile samples suggest displacement of water by isopropanol to occur. This is, indeed, the case, as has been shown /19/ by monitoring the change of the ^1H resonance signal intensity if a rutile sample preloaded with water (H_2O) is brought in contact with isopropanol $(CD_3)_2CDOD$ and subsequently evacuated.

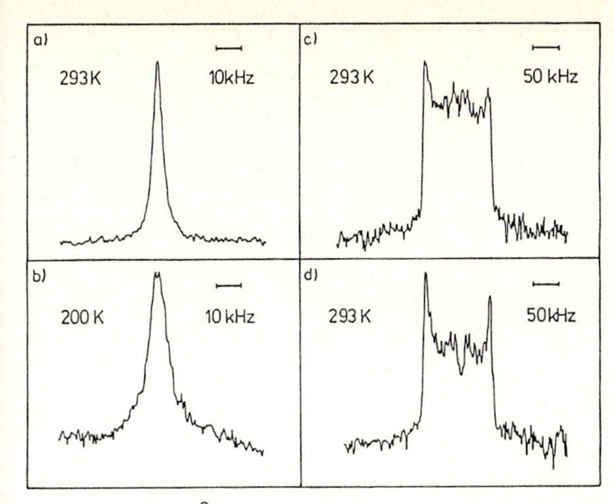

Fig. 11a-d. ^2H spectra of (a and b) water-d_2 and (c) pyridine-d_5 chemisorbed on bare anatase surface, and (d) pyridine-d_5 chemisorbed on anatase preloaded with H_2O

3.5 Water and Pyridine on Titania (Anatase)

Figure 11 shows the ^2H spectra of water-d_2 and of pyridine-d_5 chemisorbed on microcrystalline anatase. The samples were prepared by exposing standard pretreated anatase /24/ to the vapours of D_2O and C_5D_5N and pumping down the equilibrated samples subsequently for 24 hrs under high vacuum. The pyridine sample with water preloading was prepared by carrying out this procedure twice, first with H_2O, then with C_6D_5N.

The surprisingly narrow lines of adsorbed D_2O strongly indicate rapid deuteron movement, presumably by exchange between various oxygen ion sites and chemisorbed water molecules, either dissociated or undissociated. A similar conclusion has been drawn for water chemisorption on rutile /22/ from ^1H resonance measurements in which case the lines obtained are also substantially narrowed in the relevant temperature range.

The edge splittings $\Delta\nu$ = 125 kHz of the pyridine ^2H spectra corresponding to DQCC = 170 kHz in agreement with literature data /17/ immediately reveal firm bonding of the molecules to the surface with no rotational motion being feasible on the NMR time scale. Most probably, the bonding occurs via the electron lone pair at the nitrogen atom to the Ti^{4+} Lewis acidic centers of the surface. Consequently, displacement of water by pyridine must have occurred. It is hardly conceivable that hydrogen bonding of the pyridine molecules to the hydrated surface could explain the complete restriction of rotational mobility.

3.6 Ethene on Silver Exchanged NaX Zeolite

Figure 12 shows a selection of ^2H NMR spectra at several temperatures and degrees of silver versus sodium cation exchange

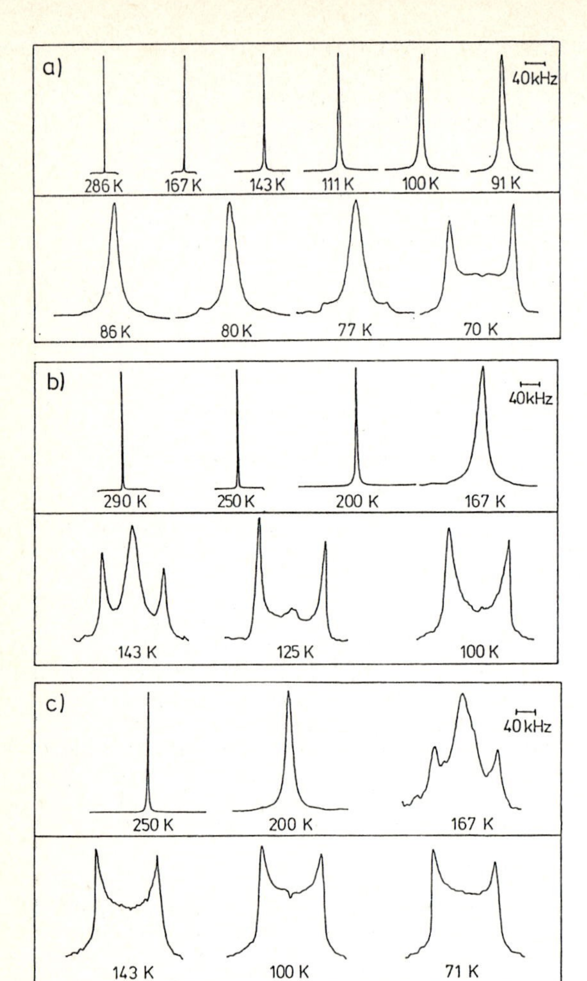

Fig. 12a-c. ^{2}H spectra ethene-d_4 in silver-exchanged NaX zeolite, degree of pore filling $\theta = 0.8$. Degrees of ion exchange: (a) 0%, (b) 60%, and (c) 100%

between 0 and 100% of ethene-d_4 adsorbed in the cavities (supercages) of synthetic Faujasite type X zeolite of pore filling degree $\theta = 0.8$ /25/. Independent if cation exchange the spectral shapes develop from singlets of increasing widths into Pake patterns of edge splitting $\Delta\nu = 130$ kHz after crossing an intermediate range of temperature with more complicated appearance. This transition range is continuously shifted with silver content between temperatures centered at about 80 to 230 K for the pure sodium and silver forms, respectively.

Since the bonding of ethene in zeolites can be assumed to be largely determined by the interaction of the π-electrons with the cations /26/, preferred rotation about the two-fold symmetry axis normal to the molecular plane was expected to show up in a Pake pattern of edge splittings $\Delta\nu$ = 65 kHz in the motional averaging case based on e^2qQ/h = 175 kHz for ethene deuterons /27/. Actually, such a behaviour was never observed but the spectra are rather indicative of vastly isotropic rotational motion of the molecules within the zeolite cages. Sticking to the view that the cations act as adsorption sites the spectra could be explained by considering the molecular exchange among these sites to be the only cause of C-D bond reorientation in space. If this notion holds true then the correlation time τ_r of the reorientational motion can roughly be identified with the residence time on a site which takes on the value $1/\nu_Q \simeq 4$ µs at some temperature in the range where the spectra are transformed from singlets into solid state patterns. The shift of this temperature to higher values with the silver content then simply reflects the increase of the mean residence time when sodium is replaced by silver. This conclusion is in full accordance with the well-known stronger bonding of olefins to silver than to alkali ions.

4. Benzene on Graphite and Boron Nitride

Graphitized carbon blacks and hexagonal boron nitride are known to possess energetically very homogeneous surfaces on account of exposing almost entirely the (0001) crystallographic planes. This property renders these substances especially well suited for the study of physisorbed layers with respect to molecular dynamics, orientation of molecules, two dimensional phase changes, etc.

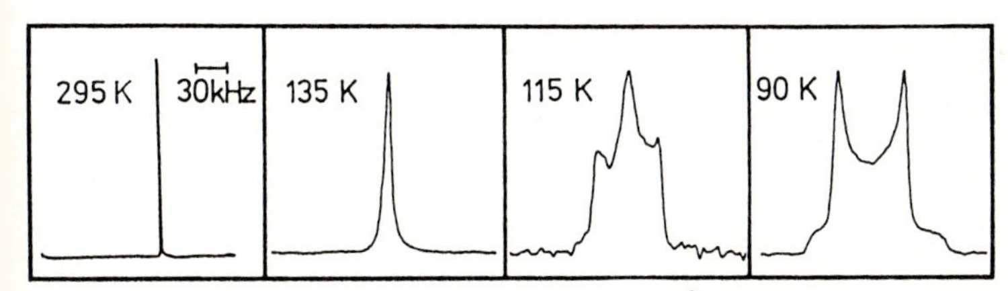

Fig. 13. Temperature dependence of ^{2}H spectra of one monolayer of benzene-d_6 adsorbed on Graphon

As an example for such studies Fig. 13 shows the temperature dependence of the ^{2}H spectra of a benzene-d_6 monolayer on the graphitized carbon black Graphon /28/. The singlet lines of increasing line widths (295 to 130 K) are typical for the 2D fluid state. After passing through a narrow range of temperature of rather involved line shapes, in which according to neutron diffraction studies /29, 30/ a 2D fluid/solid phase transition takes place, Pake type patterns of edge separation $\Delta\nu$ = 70 kHz are obtained below 110 K. Most interestingly, these patterns exhibit distinct asymmetries which become apparent in the outer spectral portions. Such asymmetries are also observed with other adsorbates on gra-

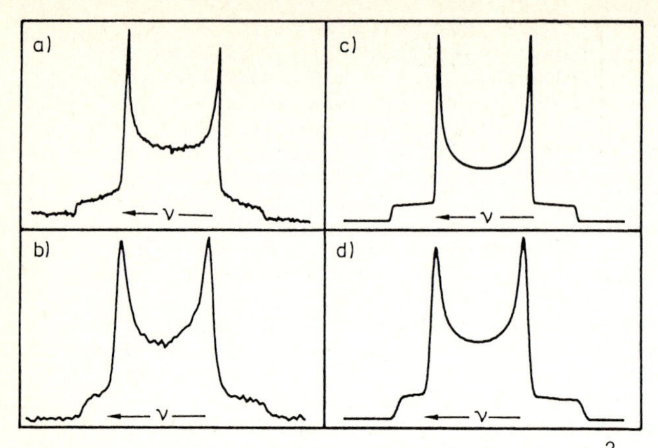

Fig. 14a-d. Experimental and theoretical ^{2}H spectra of one monolayer of benzene-d_6 adsorbed on boron nitride and Graphon. (a) C_6D_6 on BN, 75 K; (b) C_6D_6 on Graphon, 90 K; (c) Pake powder pattern, DQCC = 93 kHz, $\Delta s = 0$; (d) Powder pattern, DQCC = 93 kHz, $\Delta s = +150$ ppm, $\theta = 0^\circ$

phite, e.g. toluene /31/, but do not show up with other adsorbents studied so far, e.g. η-alumina (Fig. 6), Pt-η-alumina (Fig. 7) and boron nitride. This peculiarity of the graphitized carbons may be explained /5/ by the large magnetic shielding anisotropy at the adsorbate nuclear sites provided by graphite in the strong static field B_0.

In order to demonstrate this effect quantitatively, in Fig. 14 low temperature ^{2}H spectra of one-monolayer samples of benzene-d_6 on boron nitride and Graphon are shown and compared with theoretical spectra calculated on the basis of (3) to (5) with DQCC = (1/2) e^2qQ/h, e^2qQ/h = 186.6 kHz /15/ as well as $\Delta s = 0$ and Δs = 150 ppm, $\theta = 0^\circ$ for boron nitride and Graphon, respectively. The excellent agreement between theory and experiment reveals that the adsorbed benzene molecules perform rapid hexad axis rotation in both cases. For graphite the further conclusion can be drawn that the benzene molecules are adsorbed flat on the surface in agreement with neutron diffraction /29, 30/ and LEED /32/ studies because the successful fit with $\theta = 0^\circ$ stands for parallel orientation of the motionally averaged tensor, i.e. the hexad axis direction, with respect to the surface normal. A conclusion about the benzene molecule orientation on boron nitride is not possible. However, the flat configuration is most likely given as well /3/.

Very interesting ^{2}H spectra are observed if a Graphon sample carrying ten layers of adsorbed benzene-d_6 is cooled down to low temperatures /33/. At 90 K (Fig. 15a) a spectrum appears which, obviously, is the superposition of two Pake type patterns of edge separations 70 to 140 kHz. This spectrum was recorded with a cycle time of 10 seconds. If the cycle time is set to 0.2 seconds the 140 kHz becomes satura-

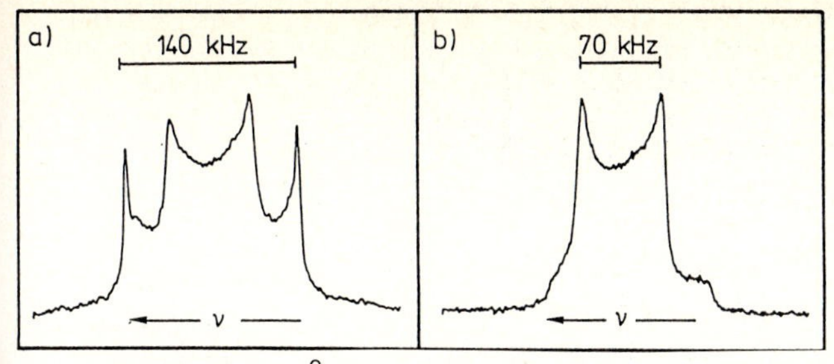

Fig. 15a and b. ^{2}H spectra of ten layers of benzene-d_6 on Graphon at 90 K. (a) Cycle time 10s, (b) Cycle time 0.2s

ted leaving the 70 kHz pattern (Fig. 15b) which exhibits the typical asymmetry found for the one-monolayer sample. The interpretation is straightforward: at 90 K one monolayer of benzene in the 2D solid state exists in equilibrium with all the rest of benzene in the 3D crystalline state. This conclusion is in full accordance with previous findings by a variety of techniques such as thermodynamic /34/, NMR /35, 36/, and neutron diffraction /30/ that below about 240 K the adsorption capacity of graphite for benzene is limited to a one-monolayer equivalent and further amounts of benzene being present or introduced yield 3D benzene crystallites. In fact, using the correlation time τ_r of hexad axis rotation as determined from ^{1}H measurements on solid benzene /37/ the estimates $1/\tau_r << \nu_Q$ and $T_1(^2H) \approx 20s$ for the 3D solid phase can readily be carried out /33/. This result substantiates the view that the 140 kHz Pake pattern is due to 3D solid benzene. In addition it points out that this pattern is only partly relaxed with the 10s cycle time applied explaining the non 1:9 ratio of the pattern areas.

5. Conclusions

Deuteron solid state NMR spectroscopy, which in recent years has proved a very valuable tool in the study of molecular solids, crystal hydrates, polymers, macromolecules, biological membranes, and other related systems, has only very recently been applied to problems in surface science. It was the aim of the present contribution to explore the field of application of this technique and to point out the type of information obtainable rather than to go into details.

For studies in the field of surface science the deuteron is especially useful as a NMR probe because the electric quadrupole interaction energy in most cases studied so far is almost completely dominant over the versatile internal magnetic couplings experienced by this nucleus on behalf of its magnetic moment. Moreover, the single spin type of the Hamiltonian representing the quadrupole interaction as well as the axial symmetry of the EFG tensor at the deuteron site of C-D bonds considerably simplifies the analysis and interpretation of the experimental results obtained. This statement also applies to the ^{2}H spin lattice relaxation behaviour of adsorbed species /38/ by which the molecular dynamics on the nanosecond time scale is accessible.

Acknowledgements

Financial support of this work by Deutsche Forschungsgemeinschaft, Land Nordrhein-Westfalen, and Fonds der Chemischen Industrie is gratefully acknowledged.

References

1. B. Boddenberg, G. Neue, R. Grosse: J. Chem. Phys. 79, 6418 (1983)
2. G. Neue, B. Boddenberg: Surf. Sci. 129, L256 (1983)
3. B. Boddenberg, R. Grosse, W. Horstmann, G. Neue: Colloids Surf. 11, 265 (1984)
4. B. Boddenberg, R. Grosse, U. Breuninger: Surf. Sci. 173, L655 (1986)
5. B. Boddenberg, R. Grosse: Z. Naturf., in press
6. H.E. Gottlieb, Z. Luz: J. Magn. Res. 54, 257 (1983)
7. R. Eckmann, A.J. Vega: J. Am. Chem. Soc. 105, 4841 (1983)
8. D.L. Hasha, V.W. Miner, J.M. Garcês, S.C. Rocke: In Catalyst Characterization Science, ed. by M.L. Deviney and J.L. Gland, ACS Symposium Series No. 288 (American Chemical Society, Washington, D.C. 1985) p. 485
9. H.H. Mantsch, H. Saito, I.C.P. Smith: Progr. NMR Spectrosc. 11, 211 (1977)
10. U. Haeberlen: In Advances in Magnetic Resonance, ed. by J.S. Waugh, Suppl. 1 (Academic, New York 1976)
11. H.W. Spiess: In NMR Basic Principles and Progress, ed. by P. Diehl, E. Fluck, R. Kosfeld, Vol. 15 (Springer, Berlin, Heidelberg 1978) p. 155
12. J.H. Davis, K.R. Jeffrey, M. Bloom, M.I. Valic, T.P. Higgs: Chem. Phys. Letters 42, 390 (1976)
13. B. Beerwerth: Dissertation, Universität Dortmund 1986
14. C.F. Tirendi, G.A. Mills, C. Dybowsky: J. Phys. Chem. 88, 5765 (1984)
15. F.S. Millett, B.P. Dailey: J. Chem. Phys. 56, 3249 (1972)
16. H.-J. Wertgen: Dissertation, Universität Dortmund, in preparation
17. M. Rinnê, J. Depireux: In Advances in Nuclear Quadrupole Resonance, ed. by J.A.S. Smith, Vol. 1 (Heyden, London, Philadelphia, Rheine 1974) p. 357
18. A. Weiss, N. Weiden: In Advances in Nuclear Quadrupole Resonance, ed. by J.A.S. Smith, Vol. 4 (Heyden, London, Philadelphia, Rheine 1980) p. 149
19. W. Horstmann: Dissertation, Universität Dortmund 1982
20. W. Horstmann, G. Auer, B. Boddenberg: Z. Phys. Chem. NF (Frankfurt), submitted for publication
21. P. Jones, J.A. Hockey: Trans. Far. Soc. 67, 2679 (1971)
22. B. Boddenberg, W. Horstmann: To be published

23. G. Munuera, F.S. Stone: Disc. Far. Soc. 52, 205 (1971)
24. K. Eltzner: Dissertation, Universität Dortmund, in preparation
25. B. Boddenberg, R. Burmeister: Proc. XXIII Congr. Ampere, Roma, September 15-19, 1986; R. Burmeister: Staatsexamensarbeit, Universität Dortmund 1986
26. J.L. Carter, D.J.C. Yates, P.J. Lucchesi, J.J. Elliot, V. Kevorkian: J. Phys. Chem. 70, 1126 (1966)
27. J. Kowalewski, T. Lindblom, R. Vestin, T. Drakenberg: Mol. Phys. 31, 1669 (1976)
28. R. Grosse, G. Auer, B. Boddenberg: Z. Phys. Chem. NF (Frankfurt), submitted for publication
29. M. Monkenbusch, R. Stockmeyer: Ber. Bunsenges. Phys. Chem. 84, 808 (1980)
30. P. Meehan, T. Rayment, R.K. Thomas, G. Bomchil, J.W. White: J. Chem. Soc., Faraday Trans. I, 76, 2011 (1980)
31. B. Boddenberg, R. Grosse: Unpublished
32. U. Bardi, S. Magnanelli, G. Rovida: Surf. Sci. 165, L7 (1986)
33. B. Boddenberg, R. Grosse: In preparation
34. Y. Khatir, M. Coulon, L. Bonnetain: J. Chim. Phys. 75, 789 (1978)
35. B. Boddenberg, J.A. Moreno: Z. Naturf. 31a, 854 (1976)
36. B. Boddenberg, J.A. Moreno: Ber. Bunsenges. Phys. Chem. 87, 83 (1983)
37. E.R. Andrew, R.G. Eades: Proc. Roy. Soc. A218, 537 (1953)
38. B. Boddenberg, G. Neue: In preparation

Formation of Paramagnetic Centres on Surfaces of Zeolite Systems

M. Hunger, A. Martinez, A. Diaz, and D. Moronta

Departamento de Física, Facultad de Ciencias, Universidad Central de Venezuela, Caracas 1020A, A.P. 21201, Venezuela

1. Introduction

The faujasitic zeolites are crystalline aluminosilicates of high porosity and are of wide use in petrochemistry. Generally, these synthesized zeolites possess organic compounds which under certain temperature and pressure conditions can build paramagnetic centres on the lattice surface. These centres can be of great importance in chemical reactions.

2. Sample Preparation

The samples were prepared from NaX zeolites synthesized by the University of Leipzig in Germany with a Si/Al ratio of 1.37, a Fe_2O_3 content of less than 3 ppm and a carbon concentration of 0.1%, gasometrically measured. Other paramagnetic impurities higher than 10 ppm could not be detected either in the NaX zeolite or in the other reagents used. The preparation of the zeolite samples was always done under high vacuum if not otherwise mentioned. The two prepared sample sets differ basically in the thermal pretreatment: one group was heated for two days at 260°C and the other group at 400°C. The ammonium exchange of the NaX-$(DMSO)_d$-NH_4-zeolite system was 15%.

3. Experimental Results

The first sample set showed the following results:
a) No EPR signal was observed in the original NaX zeolite that was thermally treated and sealed at room temperature at atmospheric air pressure (fig. 1, spectrum 2) or heated up to 260°C and sealed in high vacuum (fig. 1, spectrum 3). Similary, both the NaX-DMSO- and the NaX-$(DMSO)_d$-NH_4-zeolite systems exhibited no detectable paramagnetic signals if they were prepared and also sealed in high vacuum (fig.1, spectrum 6 and fig. 2, spectrum 5). For the sample mentioned first, this result was verified by NMR pulse spectrometry /1/.
b) On the contrary, the NaX-DMSO-zeolite system showed a structureless wide EPR line (ΔH=1000G±25G; fig. 1, spectrum 4), if it was filtered and dried at atmospheric pressure and not sealed hermetically. The concentration of these paramagnetic centres was approximately $N_p=10^{16}$ cm^{-3} /2/. Two and a half months later, it was impossible to detect this signal again (fig. 1, spectrum 5). Nevertheless, eight and a half months before this measurement, the same air-sealed system had an impurity concentration of about $N_p=10^{20}$ cm^{-3}.
c) Likewise, the NaX-$(DMSO)_d$-NH_4-zeolite system showed a broad structureless EPR line (ΔH=1000G±25G) with a $N_p=10^{16}$ cm^{-3} overlapped by a sharp free radical line (g=2,01; fig. 2, spectrum 1). This sample was filtered and dried under atmospheric air pressure and not sealed hermetically. The broad line decreased continuously with time, leading finally after six months to a nearly constant value. Meanwhile, the free radical line had disappeared

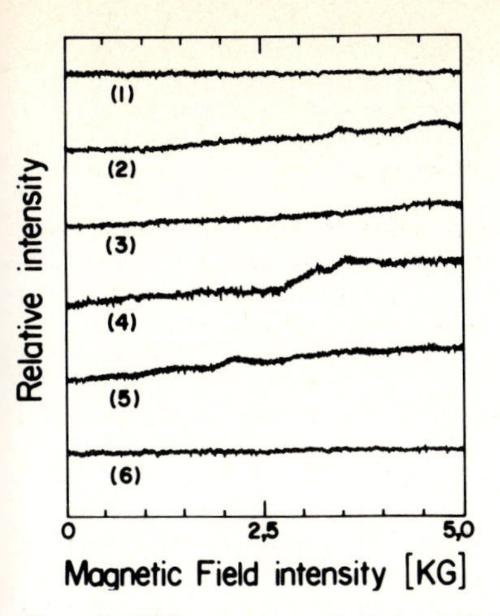

Fig. 1: EPR spectra of NaX- and NaX-DMSO-zeolite. Spectrum 1 is the background of the cavity.

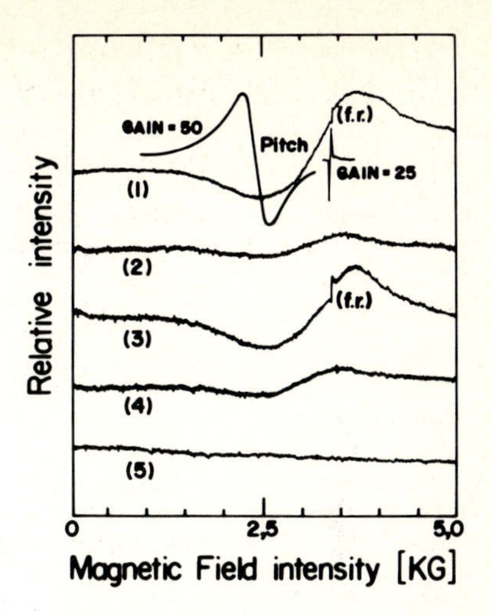

Fig. 2: EPR spectra of the NaX-$(DMSO)_d$-NH_4-zeolite system.

completely (fig. 2, spectrum 2). If the same type of sample is sealed hermetically at atmospheric air pressure after its preparation the EPR signal maintains its intensity (fig. 2, spectrum 3). However, if the seal is removed and the system is not closed hermetically the broad line drops off after two months to nearly the value mentioned above. The sharp signal could not be detected after that time (fig. 2, spectrum 4).

The second sample set showed the following experimental data:
a) The pure NaX zeolite displays no paramagnetic signal when it was heated

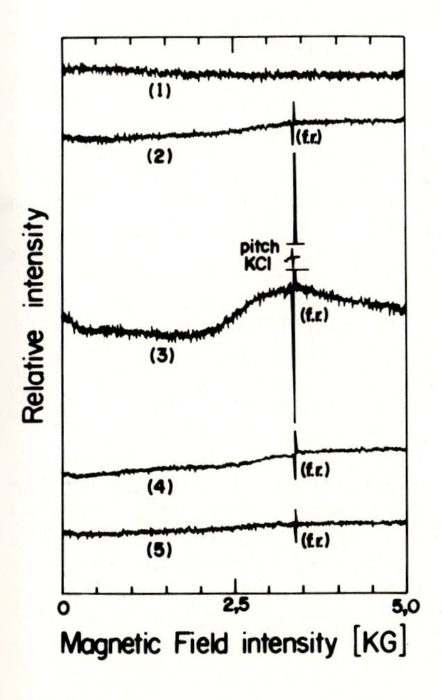

Fig. 3: EPR spectra of NaX-zeolites under different conditions of pretreatment.

in air up to 400°C at a rate of 2°C/min and sealed at atmospheric air pressure and room temperature (fig. 3, spectrum 1). To the contrary, if this sample was warmed up again to 400°C with the same heating rate and in high vacuum a sharp free radical line and a brad line were observed. The signal-to-noise ratio of the broad line was too low to be analysed (fig. 3, spectrum 2).

b) Both above-mentioned EPR signals could be detected again in a NaX zeolite (fig. 3, spectrum 3) when the sample was warmed up in high vacuum with a heating rate of 15°C/min and hermetically sealed in high vacuum or atmospheric pressure. These two signals lost their intensities considerably when the heating rate was reduced to 2°C/min independent of the sealing procedure of the sample (fig. 3, spectra 4 and 5).

c) By means of mass spectroscopy the carbon monoxide desorption was studied, i.e. the reaction product of the organic molecules contained in the organic zeolite which desorbed with increasing intensity from 30°C to 200°C and drops off to zero at 250°C (fig. 4a).

d) With the same technique it was found that the water of an original NaX zeolite desorbs continuously between room temperature and 200°C with a maximum at 150°C (fig. 4b).

e) Using a calorimetric test /3/ it was possible to prove the presence of carbon monoxide in NaX zeolite samples heated up to 400°C in high vacuum.

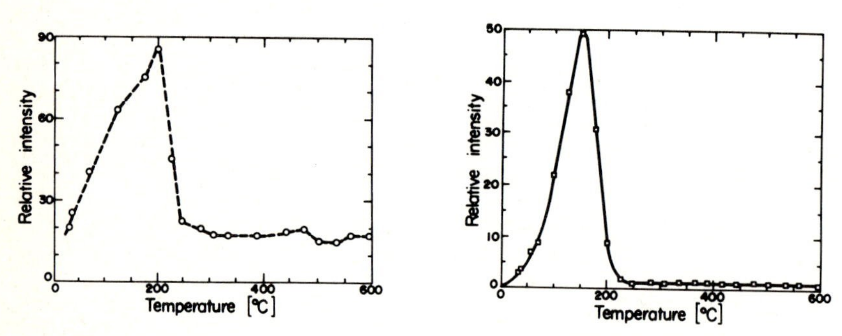

Fig. 4a(---) and 4b(——): Desorption curves of carbon and water of a NaX-zeolite.

4. Discussion

Samples heated up to 260°C:

Since the samples prepared and sealed under high vacuum showed an extremely low EPR signal, one has to conclude that the air presence generates and destroys the paramagnetic centres. This means that the formation of the broad paramagnetic signal in the NaX-DMSO- and in the NaX-$(DMSO)_d$-NH_4-zeolite systems has to be attributed to oxygen molecules being in contact with the inner zeolite surface. In these systems the broad EPR line conserves its magnitude when the samples are closed hermetically, and reduces its intensity with the time to a constant value when the samples are air filled. From these facts it has to be concluded that there are two opposing mechanisms in a NaX-DMSO-system: a fast one which creates paramagnetic centres and a slow one which destroys them. The reversibility of these processes suggests to postulate a physical adsorption of the oxygen molecules. The last assumption agrees with published results /4/. The reversibility is complete for the free radical centre and partial for the broad EPR signal which arrives finally at a dynamic equilibrium between the molecular oxygen complex and a molecular configuration without paramagnetic effect. In the same way it is possible to explain the behaviour of the free radical signal, with the additional condition that these centres require a well-defined electronic structure and the presence of the NH_4^+ ions that favours its generation and stability.

Samples heated up to 400°C:
For the interpretation of the experimental results of this set of samples the presence of carbon atoms in the zeolitic systems is decisive. If the NaX-zeolite is heated up to 400°C at atmospheric air pressure the carbon atoms oxidize completely to a non-paramagnetic carbon oxide, whereas this heating process in high vacuum forms carbon monoxide. Simultaneously paramagnetic ions are produced, generating the broad EPR line as well as the sharp free radical peak, too. The concurrent absorption of the water and the carbon monoxide makes necessary to assumed an incomplete oxidation of the zeolite included carbon according, for instance, to the following reaction equation:

$$C + H_2O \xrightarrow[\text{zeolite}]{\text{energy}} H_2 + CO \, .$$

Taking into account the high electric affinity of faujasitic zeolites for polar molecules, the presence of the carbon monoxide up to 400°C in the sample has to be correlated with the electrical local fields which are produced by defect charges in the crystal structure and are able to adsorb strongly CO molecules. The increase of the paramagnetic signals with increasing heating rate of the samples under maintained vacuum agrees with the above-mentioned assumption because the readsorption of the carbon is faster now. Furthermore, it can not be excluded that the heating procedure in high vacuum up to 400°C produces coke particles which causes readsorption of CO, leading to an increase in the concentration of paramagnetic centres.

Acknowledgements

This work was supported by the "Consejo de Desarrollo Científico y Humanístico" of the Central University of Venezuela, the CONICIT of Venezuela and by the Volkswagenwerk Foundation in the German Federal Republic. The authors wish to thank Miss Marlene Mendoza for the mass spectra measurements.

References

1. A. Martinez: Ph.D. thesis, Central University of Venezuela, Faculty of Science, to be published.
2. Ch. Poole: Electron Spin Resonance, John Wiley and Sons, 1982, p. 409.
3. J. Babor and J. Ibarz: Química General Moderna, Marín, 1968, p. 666.
4. A. Gutsze and S. Orzeszko: private communication.

Part IV

Structure and Characterization of Surfaces

Surface Extended X-Ray Absorption Fine Structure (SEXAFS) Studies of Various Phases of Chlorine and Caesium Adsorbed on Silver Single Crystals

G.M. Lamble and D.A. King

The Donnan Laboratories, University of Liverpool, Liverpool L69 3BX, UK

A number of submonolayer phases of caesium and chlorine adsorbed on silver single crystal surfaces have been studied using the technique of Surface Extended X-Ray Absorption Fine Structure. A multi-shell curve fitting data analytical procedure was employed.
When chlorine is adsorbed on Ag(111) the nearest neighbour distance is determined to be 2.70 ± 0.01 Å at both 0.4 monolayer and 0.7 monolayer coverages. The insensitivity of bond distance to coverage, for chlorine on silver, was verified by the determination of the chlorine to silver nearest neighbour distance for 0.5 and 0.75 monolayer coverages of chlorine on Ag(110). This distance was determined to be 2.56± 0.04 Å for both phases.
In complete contrast, caesium adsorbed on Ag(111) exhibits a strong dependence of adsorbate-substrate nearest neighbour distance on coverage. The Cs-Ag distance was found to increase from 3.20 to 3.50 Å on increasing the coverage from 0.15 to 0.30 monolayers. The change in distance, with increase in coverage is interpreted as an 'ionic' to 'covalent' transition of the Cs-Ag bond type.
This contradictory behaviour of Cl and Cs adsorbates is discussed in terms of the different electron transfer processes between electronegative and electropositive adsorbates with the substrate.

INTRODUCTION.

The most-established technique currently in use to determine surface structures is Low Energy Electron Diffraction (LEED). However, unlike the analogous bulk technique of X-ray diffraction, kinematic theory is not sufficient to describe the scattering: Thus, analysis of LEED data is notoriously complicated and time-consuming.
In 1976, it was recognised that the technique of Extended X-ray Absorption Fine Structure (EXAFS) could be made surface sensitive (1). It could thus be applied to the problem of determining bond distances and adsorption sites for adsorbates. Although the technique is inherently experimentally more demanding than LEED, and can only be performed with an intense source of radiation (as provided by a synchrotron), surface EXAFS (SEXAFS) is dominated by single scattering. Data analysis is more direct, less tedious and more accurate than the long-range order technique.
Surface sensitivity is achieved by monitoring the adsorbate edge, so that all the information relates to the environment of the surface atoms. Thus the adsorption is measured indirectly from the products of core-hole annihilation, by detecting: 1) the elastic Auger yield, 2) the fluorescence yield, 3) the total secondary electron yield or 4) a partial secondary electron yield. The secondaries consist of mainly inelastic Auger electrons and the cascade resulting from Auger emission. Since the elastic Auger channel remains the same with increasing photon energy, all the channels of the resulting cascade will be the same; the resulting yield modulating with the rate of core-hole creation.

The aim of our surface EXAFS work (2) was to investigate various submonolayer phases of both a strongly electronegative species, chlorine, and a strongly electropositive species, caesium, adsorbed on silver. Silver is a very effective catalyst for the oxidation of ethylene to ehtylene oxide. Adsorbed chlorine is used as a modifier for the catalytic process, increasing the selectivity of the silver catalyst to epoxidation reaction, whilst Cs, along with other alkali metals, is used as a promoter; increasing the activity of the reaction. This indicates that the effect, of both chlorine and caesium, is to change the surface electronic properties and, thus, reactivity. However, due to the difference in nature of the two species, the way in which the surface electronic properties are modified must be by very different mechanisms.

EXPERIMENTAL.

The success of a surface EXAFS experiment is wholly dependent on the performance of the beamline and optics. The very nature of the experiment places heavy demands on the beamline operation and strict constraints on its design. The beamline and monochromator are described in detail elsewhere by MacDowell et al. (3). Only a brief outline is presented here.
The SEXAFS measurements were made on station 3, beamline 6, at the Daresbury Synchrotron Radiation Source (SRS), near Liverpool, England. A schematic of the beamline is shown in figure 1. A double-focussing, gold-coated, toroidal pre-mirror is situated 11 m from the tangent point. This accepts up to 5 mrad (horizontal) by 0.6 mrad (vertical) of synchrotron radiation, at a grazing angle of 0.5^{o}. It acts to demagnify the storage ring source size by 2:1 at the sample. The size of the monochromatic beam, as a result of this demagnification, is 1.0 mm (vertical) by 4 mm (horizontal) FWHM. The horizontal and vertical aperture of the white beam can be adjusted by the sidejaws and vertical apertures situated in front of the prefocussing mirror. A range of carbon, beryllium and aluminium filters, with varying attenuation characteristics, are situated prior to the apertures. The filters act as a low-energy cut-off to eliminate the uv and visible component of the synchrotron radiation, which would otherwise be transmitted by the double-crystal monochromator by normal reflections.
The monochromator is unique insofar as it covers a very large energy range, in order to reach as many absorption edges as possible. The range is from 0.1 to 10 kV. The higher range is covered by using a double-crystal combination; the lower range by a grating/mirror assembly. The double-crystal section consists of three pairs of crystals which include Ge(111), Ge(220) and InSb(111). Since the monochromator is at U.H.V. (10^{-9} mbar), all operations of the optical elements are performed externally by computer-controlled stepper motors via a CAMAC hardware system. A quantitative reference to the incident photon intensity is provided, for the sample under investigation, by a beam intensity monitor. The monitor (1) is situated between the monochromator and the sample chamber. The incident intensity is measured as a function of the drain current leaving a photocathode. The choice of material depends on its suitability to the experiment with regard to the transmission flux and linearity in the energy range of interest. In the experiments presented here, an 0.75 mm Al foil was used.
The I_o monitor is mounted in a manifold section together with a sputter gun (to clean or passify the monitor), a set of calibration foils and filters, a rack of beam-defining apertures and a subsidiary beam monitor ($I_o i$). The beam-defining apertures, prior to the sample reference monitor, can be used to enforce 'pseudo' spatial stability at the sample

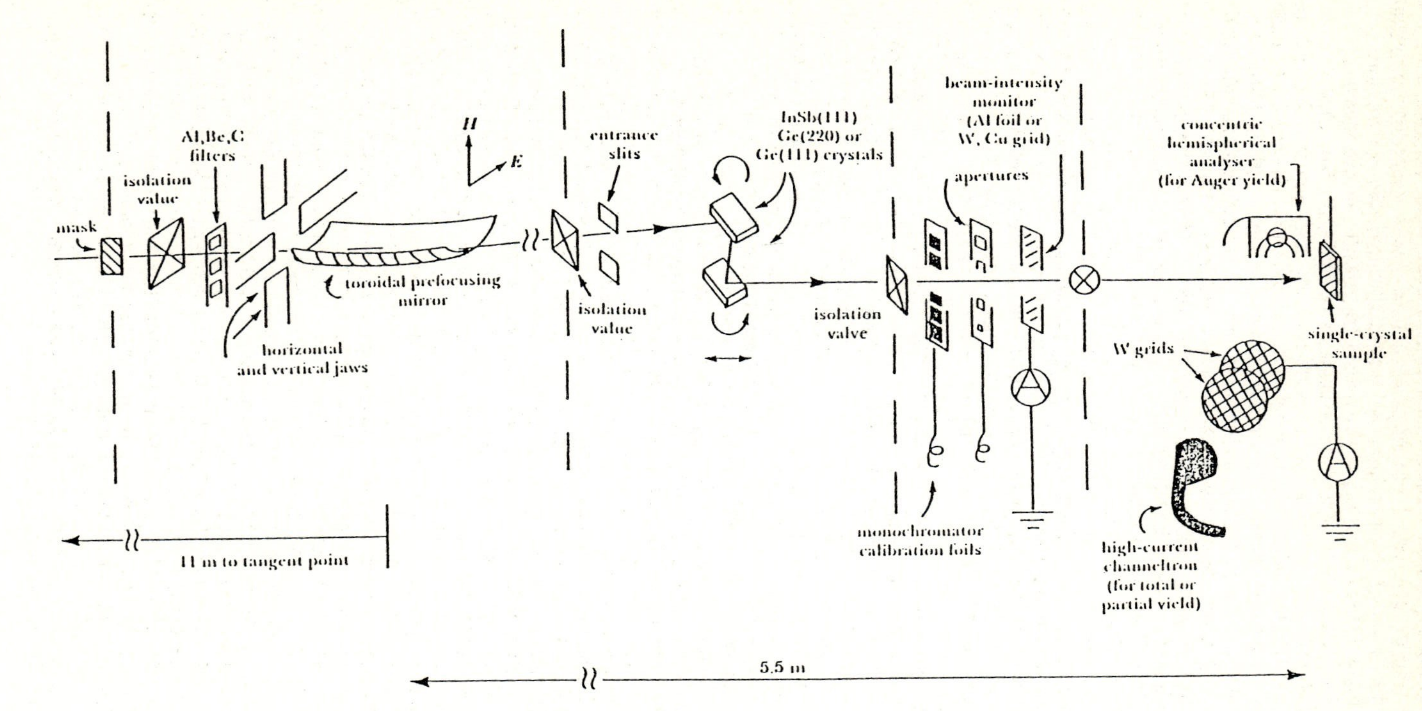

Figure 1 Schematic diagram of the Daresbury Synchrotron Radiation Source beamline for SEXAFS studies, with the light path shown arrowed from left to right

i.e. any vertical deviation causes the beam to move over the aperture which blocks any displaced beam from the sample. This is at the expense of decaying incident intensity but, as far as the sample is concerned, the beam is essentially fixed. Since the sample reference monitor is also beyond the apertures, the reduction of intensity at the sample is correctly normalised.

The SEXAFS experiments were carried out in a conventional stainless steel U.H.V. chamber. It was pumped by a 300 ls^{-1} turbo pump and a titanium sublimation pump. Additionally it had a liquid nitrogen cryopanel, which was used for the experiments requiring vacuum better than 10^{-10} mbar. The best attainable vacuum in the chamber was 5 x 10^{-11} mbar. The gas handling line was pumped by an oil diffusion pump. The chamber has an upper and a lower analysis level. The upper level was used for surface preparation and characterisation. It housed a three-grid retarding field analyser (for LEED and Auger measurements), an argon ion gun (for cleaning the sample), an ionisation gauge and the crystal-dosing sources. All SEXAFS measurements were carried out in the lower level.

The silver single crystals, used for all the surface studies presented, were cut and polished to within 0.5°. The crystals were cleaned by continuous argon ion bombardment (10 μA, 1.5 kV) and anneal cycles, at 800-870 K, until no contaminants were evident in the Auger spectra and sharp (1x1) LEED patterns were observed. The crystals were mounted on a 'cold-probe' manipulator built by Vacuum Science Workshops. Efficient cooling was by virtue of the liquid nitrogen tank which extended into the vacuum chamber. It was in direct thermal contact with two copper feedthroughs from the base of the tank, on which the crystal was mounted via tantalum support wires. Temperatures of 110 K were routinely achieved within 5 minutes. The crystal was heated radiatively by a tungsten filament suspended behind the crystal and supported from another two copper feedthroughs from the base of the liquid nitrogen tank. The temperature measurements were made with a Chromel/Alumel thermocouple. The crystal had no azimuthal rotation mechanism; this was the price paid for efficient cooling and simple mounting. Alteration of the vertical position of the crystal was by means of a pneumatically operated linear drive. Polar rotation of the manipulator probe was via a differentially pumped Wilson seal section.

In choosing the dosing sources, to dose the sample, caution was taken so that contamination of the monochromator and beamline was avoided. For the chlorine experiments, an electrolytic chlorine source consisting of a silver gauze anode and a platinum foil cathode, with an 'electrolyte' composition of 1% $CdCl_2$ and 99% AgCl (by mass) was used. Ionic mobility was encouraged by heating the electrolyte pellet with a tungsten filament to 320-370 K. Using a current of 50 μA through the pellet, a dosing rate of approximately 1 monolayer in 5 minutes was achieved with a source-to-sample separation of approximately 15 cms. Adsorbed chlorine rendered the surface extremely unreactive; the system was stable for several hours at a vacuum of $1x10^{-10}$ mbar, allowing continuous SEXAFS date collection, on the same surface, for several hours. The crystal was dosed with caesium using an SAES getter source. This was mounted on a linear drive. The source was placed at a distance of approximately 5 cm from the sample during dosing. A current of 5.5 A through the source gave a dosing rate of the order of a monolayer in 5 minutes. These dosing conditions produced a homogeneous overlayer as indicated by the Auger Cs/Ag peak height ratio over the surface. For the caesium adsorbate studies, a particularly good vacuum was necessary, on account of the enhanced reactivity of a surface which is induced by alkali adsorption. Thus, a vacuum of $5x10^{-11}$ mbar was maintained throughout these experiments, by filling the liquid nitrogen cryopanel.

In the SEXAFS studies presented here, the measurements were made by collecting the total secondary electron yield as a result of photon absorption. This was achieved by monitoring the drain current leaving a tungsten mesh, placed at a distance of 3 cm from the sample. The mesh was biased by +9V to optimise the yield. The drain current passed through a 'pico-ammeter' to a voltage-to-frequency converter and then to the CAMAC counting system. This sample signal (I_s) was then directly normalised to the incident intensity (I_o) (as measured by the beam monitor). All SEXAFS data were taken at a crystal temperature of 110 K. The essentially uncomplicated manipulator 'foot' on which the crystal was mounted allowed any beam movement off the crystal (during a scan) to be detected by a phosphor screen placed behind the sample.

The bulk data used to obtain correct phase shifts were obtained by mounting a series of model compounds in the same vacuum system, and measuring the bulk EXAFS by the same total yield method. In this case the information received was from the bulk, since the absorber species is, itself, located throughout the bulk and not just residing in the top layer. The series of model compounds included: AgCl, CuCl, CuBr, AgBr, CsCl, CsI, CsBr, NaCl, KCl, RbCl, $CdCl_2$, $CoCl_2$, $MnCl_2$, $LaCl_3$, NiCl and Ag_2S. The samples were doped with 10% (by weight) graphite and pressed into pellets. They were then mounted on a copper block of dimensions 3 cm x 3 cm x 15 cm. The block was attached to the copper feedthroughs from the base of the liquid nitrogen probe of the manipulator. This allowed sufficient thermal conductivity for the model compounds to achieve temperatures of 110 K, as measured by a Chromel/Alumel thermocouple (with the 'hot' junction placed between the block/pellet interface of a pellet mounted at the base of the copper block).

Data Analysis.

The method used to obtain structural information from all the data presented here was by using phase shifts obtained in a semi-empirical way. These were then employed, within the curve fitting method, to deduce structural information. Thus as regards the phase shifts the method was not wholly theory dependent. The phase shifts were initially calculated, for a given standard compound, using the method outlined in the previous section. These were input to the EXAFS analysis program, together with the parameters for the standard compound (known from X-ray diffraction data). The resulting theoretical spectrum was refined against the experimental spectrum, for the particular compound, by variation of only the phase shifts and the E_o. Reasonable modification of the precise values of these parameters were unknown; their variation, within reason, did not affect the phase terms.

Having obtained discrete phase shifts for absorber and backscattering atoms for a number of standard compounds, their accuracy was checked by substituting a particular absorber or backscatterer phase shift into the theoretical EXAFS calculation for a different compound which possessed the same absorber or backscattering atom. The theoretical bond distances of the compound were then allowed to vary and the amount by which the distance'moved', from its true position, gave a measure of the accuracy of the particular phase shift in real space terms.

A computer program called EXCURVE (4) was used to analyse the SEXAFS and EXAFS data presented. This is based on the original theory of Lee and Pendry (5). It uses a curved wave description of the photoelectron wave. Use of the more accurate 'curved wave' description means that a longer range of the spectrum can be analysed, by extending the calculation to a lower energy. This is particularly useful for weak scatterers which only contribute to the low energy part of the spectrum. The program, EXCURVE, is a rapid version of the curved wave calculation in the Daresbury EXAFS program. The latter is, in turn, a modification of the original version written by Lee and Pendry at Bell Laboratories. The price paid for its rapid execution was to exclude polarisation considerations. The simplification of the calculation, by an angle averaging process, is of no consequence for polycrystalline or amorphous samples. However, the inclusion of polarisation effects is particularly useful for single crystal studies. So, for the surface EXAFS data, analysed by this program, effective coordination numbers at a given angle of incidence, were regarded as being somewhat qualitative. In particular, for the caesium adsorption, where an L_3 edge was monitored, the anisotropy in the latter situation was less marked but more complex.

RESULTS

CHLORINE ADSORBED ON Ag(111)

SEXAFS data was taken from Cl adsorbed on Ag(111) at coverages of 0.4 and 0.7 monolayers. A monolayer is defined as the same surface concentration as a close-packed Ag(111) plane. A complete structural determination for the higher coverage phase is published elsewhere (6).
Serious anharmonic effects occur at room temperature for the Cl/Ag system (2, 7); thus, all SEXAFS and EXAFS measurements were made at both room temperature and at 100 K, where the deviation from the simple harmonic approximation was negligible.
Raw SEXAFS data taken from the high coverage phase are shown in fig 2. The k^3-weighted back-transformed experimental frequency spectrum (with all real space components above 5.0 Å filtered out) is shown in figure 3a), together with the theoretical fit. The corresponding Fourier transforms are shown in figure 3b). The first nearest neighbour distance is determined to be 2.70 ± 0.01 Å.

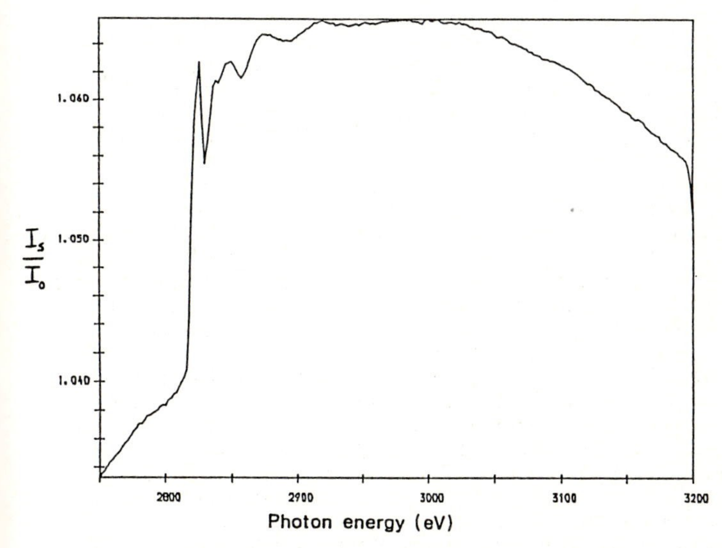

Figure 2 A normalised raw surface EXAFS spectrum, of the chlorine K edge, from the 0.7 monolayer phase on Ag(111).

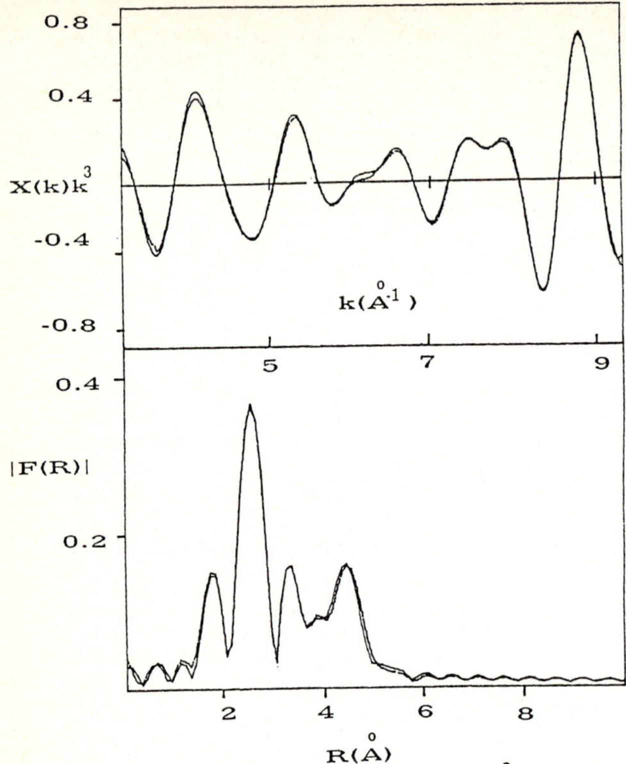

Figure 3 (a) The full line shows the k^3-weighted, back-transformed experimental frequency spectrum (to include real space components up to 5.0 Å) together with the theoretical multishell fit (broken line), of the phase observed at a coverage of 0.7 monolayers of Cl/Ag(111).

(b) Experimental and theoretical Fourier transforms of a).

The raw data from the low coverage phase is shown in figure 4. The back-transformed and k^3-weighted surface EXAFS data are shown in figure 5a). Again, all real space components above 5.0 Å are filtered from the raw data. The corresponding Fourier transform is shown in figure 5b). The nearest neighbour distance was determined to be 2.70 ± 0.01 Å, i.e. identical to that determined for the higher coverage phase.
A LEED I (V) analysis is not a viable structure technique for either of these surface phases, since no long range order exists in the low coverage phase whilst there is only weak order (8) (9) for the higher coverage phase. Thus SEXAFS is the only applicable method for the determination of bond distances in these systems.

CHLORINE ADSORBED ON Ag(110)

The measurements were made at two different chlorine coverages: 0.5 monolayers and 0.75 monlayers. Long range order is observed for both of these phases (8) (10), using the technique of Low Energy Electron

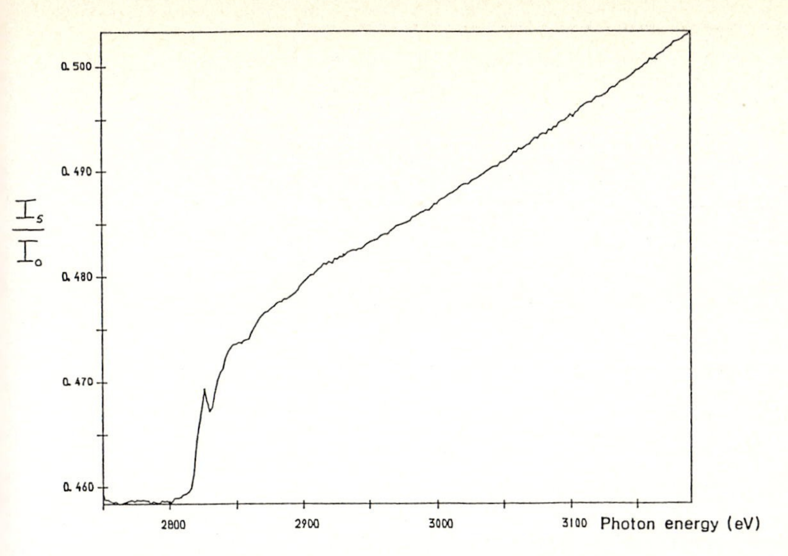

Figure 4 The raw SEXAFS spectrum for the 0.4 monolayer phase of Cl on Ag(111)

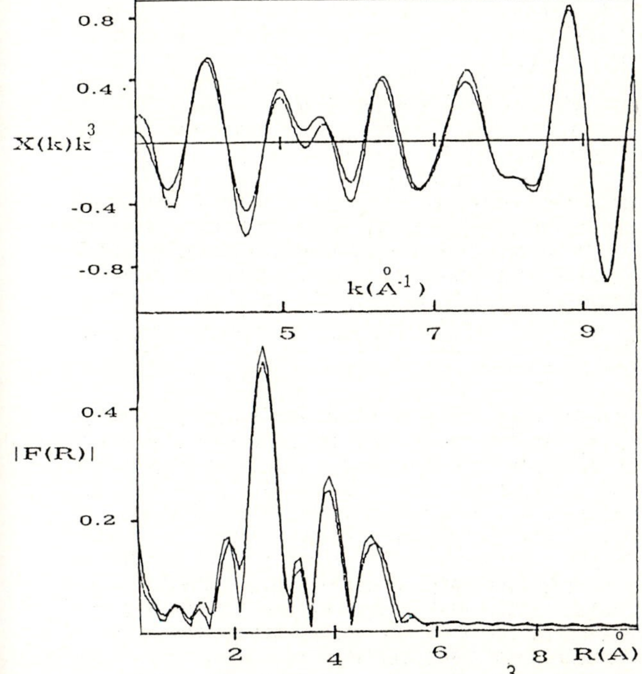

Figure 5 (a) The full line shows the k^3-weighted, back-transformed experimental frequency spectrum (to include real space components up to 5.0 Å) together with the theoretical multishell fit (broken line), of the phase observed at a coverage of 0.4 monolayers of Cl/Ag(111).

(b) Experimental and theoretical Fourier transforms of a).

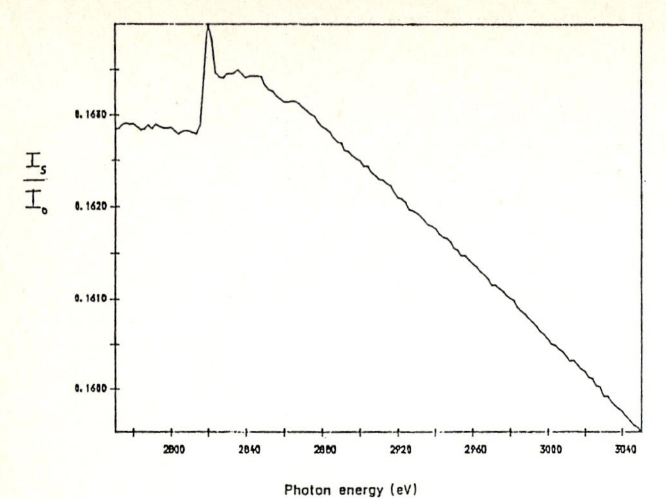

Figure 6 The raw SEXAFS data obtained from the 0.5 monolayer phase of chlorine adsorbed on Ag(110).

Diffraction (LEED). At a coverage of 0.5 monolayers a p (2x2) LEED pattern is observed, in which chlorine exists as a simple chemisorbed overlayer. At the higher coverage a different 'incommensurate' LEED pattern exists. A structural model for this has been determined from a combination of the SEXAFS results obtained and a laser simulation study. This is reported elsewhere (11). Here, only the first nearest neighbour distance determinations are reported for both coverages. The aim of this work was to investigate whether there was a similar insensitivity of the first nearest neighbour bond distance to coverage as with Cl/Ag(111). The raw data from the 0.5 monolayer phase are shown in figure 6. Figure 7a) shows the Fourier filtered and back-transformed data together with the theoretical fit. The corresponding experimental and theoretical Fourier transforms are shown in figure 7b). A first nearest neighbour distance is determined to be 2.56 ± 0.04 Å for this p (2x2) structure. The raw spectrum from the 0.75 monolayer phase is shown in figure 8. The Fourier filtered and back-transformed frequency spectrum is shown with the calculated spectrum in figure 9a). The theoretical and experimental Fourier transforms of these are featured below in figure 9b). The Cl-Ag nearest neighbour distance for this higher coverage phase was found to be 2.56 ± 0.04 Å ; i.e. the Cl-Ag distance is invariant with coverage.

CAESIUM ADSORBED ON Ag(111)

A great deal of effort has been invested in alkali metal adsorption systems since the first observations of drastic changes in the electronic properties of the surface resulting from the uptake of these species (12). These original studies showed that the adsorption of caesium on a tungsten surface greatly enhanced its thermionic electron emission. This phenomenon has obvious technical applications but is still, as yet, not fully understood at the fundamental level. The numerous experimental studies during the 50 years subsequent to this discovery reach the general conclusion that, as the alkali metal coverage increases, the work function of the substate initially exhibits a rapid non-linear decrease. This

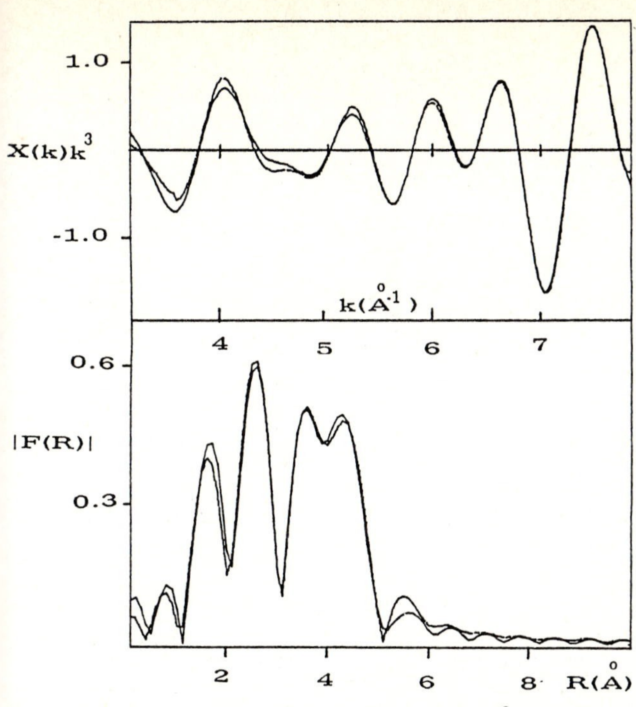

Figure 7 (a) The full line shows the k^3-weighted, back-transformed experimental frequency spectrum (to include real space components up to 6.0 Å) together with the theoretical multishell fit (broken line), of the phase observed at a coverage of 0.5 monolayers of Cl/Ag(110).

(b) Experimental and theoretical Fourier transforms of a).

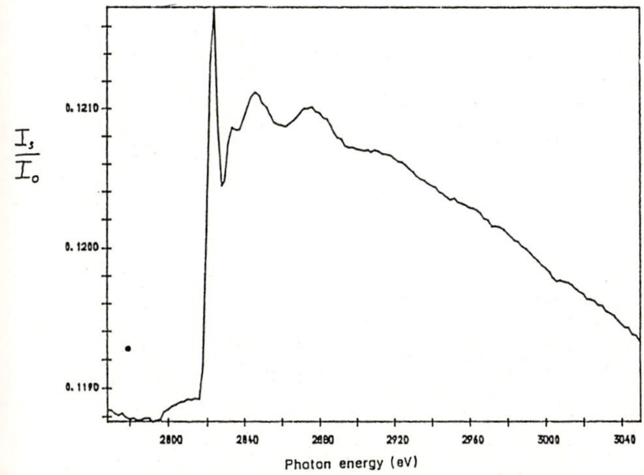

Figure 8 The raw SEXAFS data obtained from the 0.75 monolayer phase of chlorine adsorbed on Ag(110).

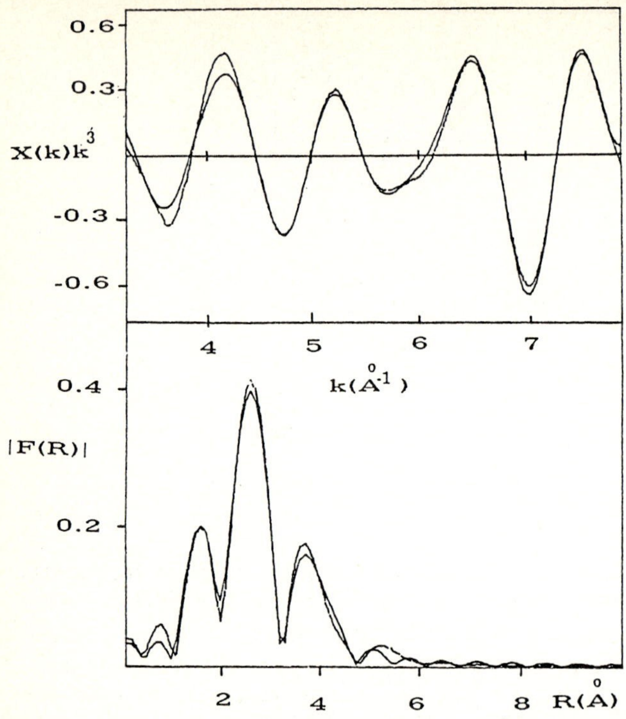

Figure 9 (a) The full line shows the k^3-weighted back-transformed experimental frequency spectrum (to include real space components up to 6.0 Å) together with the theoretical multishell fit (broken line), of the phase observed at a coverage of 0.75 monolayers of Cl/Ag(110).

(b) Experimental and theoretical Fourier transforms of a).

reaches a minimum at less than a monolayer coverage and then increases slightly, to a plateau, with further increase in coverage. This plateau is approximately equal to the work function of the bulk alkali metal. It is also generally found that the binding energy of the adsorbate decreases with increase in coverage, as first illustrated by Taylor and Langmuir (13). Numerous theoretical interpretations, using different approaches, have been proposed to account for the changes in electronic structure occurring during the adsorption process and a lot of these have proved to be successful, regardless of their varying complexity. It is generally accepted that electron donation occurs from the adsorbate to the substrate, the extent of which is greatest at low coverage and decreases as the coverage increases. In this work we present the first direct observation of a substrate-adsorbate bond distance increase of 0.3 Å ± 0.06 Å, which accompanies the coverage change from 0.15 to 0.3 monolayers. We believe this structural change is a result of the transition from a mainly ionic to a mainly covalent state of the adsorbate-substrate bond. The determination illustrates the great advantage of the technique in its ability to monitor subtle bond length changes. The SEXAFS was monitored above the Cs L_3 edge. All data were taken with the crystal at 110 K.

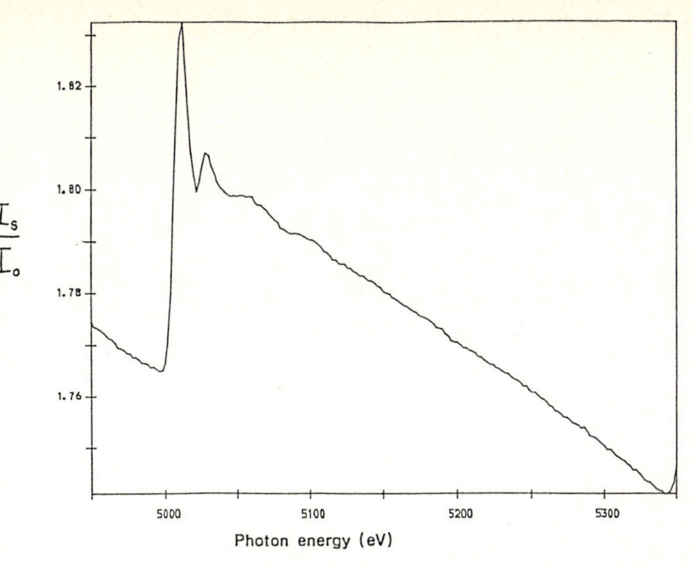

Figure 10 A normalised raw surface EXAFS spectrum, of the caesium L_3 edge, from the 0.3 monolayer phase on Ag(111).

Using experimental data from: Ag metal, AgBr, CsBr, CsI and AgCl, a total error bar was obtained, to account for the combined Cs and Ag phase shifts of $\pm$ 0.03 Å (phase shift transferability being assumed).
The raw data obtained for the highest coverage phase studied are shown in figure 10. The surface concentration here is 0.3 monolayers, where one monolayer is equal to the same atom concentration as that of the (111) plane in silver. These data are considered to be of particularly high quality, considering the dilution of the adsorbate layer. The Fourier filtered and k^3-weighted background subtracted spectrum is shown with the theoretical fit in figure 11a). This includes all real space components out to 8.5 Å; the corresponding Fourier transforms are shown in figure 11b).
The raw data obtained for the lowest coverage phase studied are shown in figure 12. The surface coverage here is 0.15 monolayers. The Fourier filtered and k^3-weighted frequency spectrum, with all components out to 8.2 Å included, is depicted in figure 13a) together with the calculated spectrum. The corresponding experimental and theoretical Fourier transforms are shown in figure 13b).
The first nearest neighbour distance for the higher coverage, more 'covalent' phase is calculated to be 3.50 $\pm$ 0.03 Å; whilst that for the lower coverage, more 'ionic' phase is calculated to be 3.20 $\pm$ 0.03 Å. A full multishell structural analysis for this phase is reported elsewhere (14). Figure 14 provides a visual appreciation in frequency terms of the observed bond distance increase. It shows the superimposition of the first shell contributions from both the high and low coverage phases (i.e. the first nearest neighbour contributions in each case). The overlayed frequency curves are those extracted from the theoretical spectra. These contributions are separated from the sinusoidal sum of the total calculated spectrum, since a back-transform of the first R-space component in the experimental spectrum would include the interfering Ramsauer-Townsend resonance (15) of the second

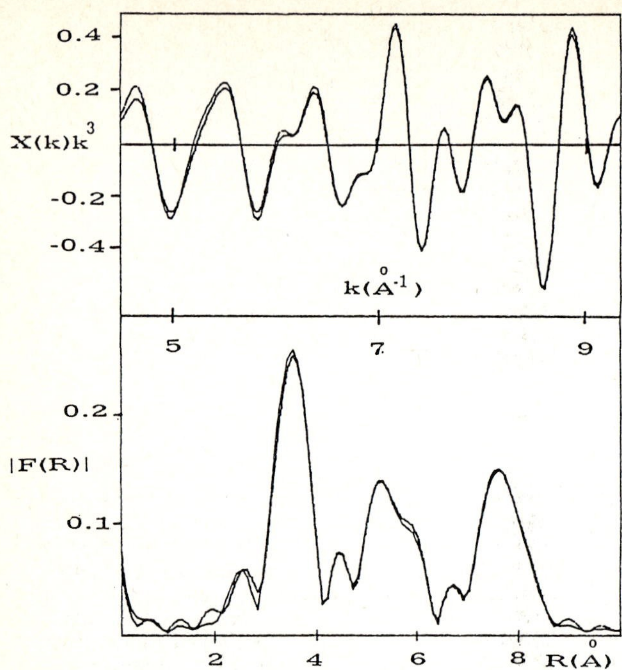

Figure 11 (a) The full line shows the k^3-weighted, back-transformed experimental frequency spectrum (to include real space components up to 8.5 Å) together with the theoretical multishell fit (broken line), of the phase observed at a coverage of 0.3 monolayers of Cs/Ag(111).

(b) Experimental and theoretical Fourier transforms of a).

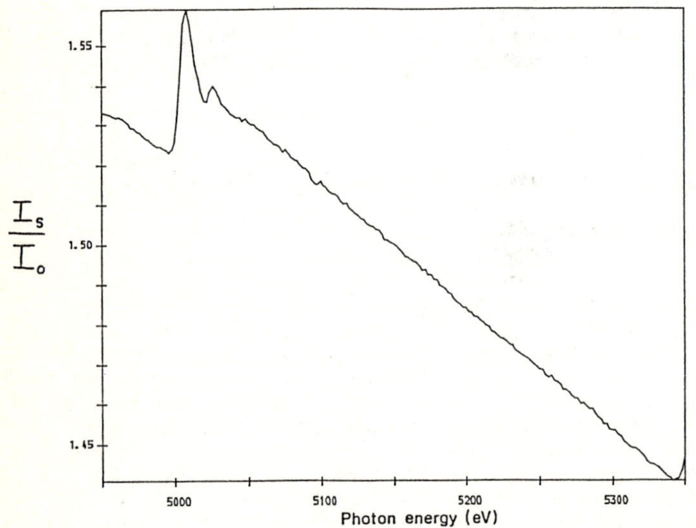

Figure 12 A normalised raw surface EXAFS spectrum, of the caesium L_3 edge, from the 0.15 monolayer phase on Ag(111).

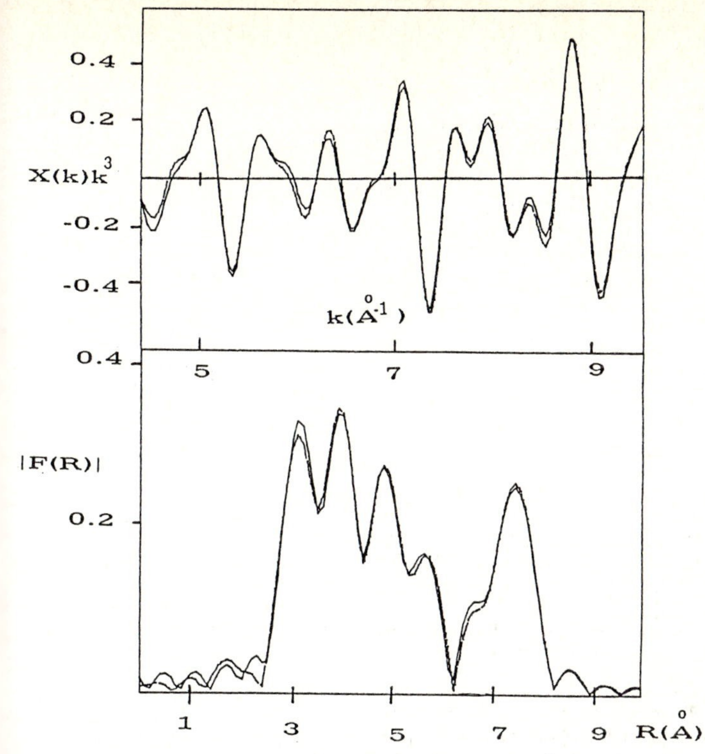

Figure 13 (a) The full line shows the k^3-weighted, back-transformed experimental frequency spectrum (to include real space components up to 8.2 Å) together with the theoretical multishell fit (broken line), of the phase observed at a coverage of 0.15 monolayers of Cs/Ag(111).

(b) Experimental and theoretical Fourier transforms of a).

nearest neighbour shell. This resonance arises from the periodicity of the energy dependent backscattering factor which occurs generally for the heavier atoms. The existence of this extra frequency term is one reason why the Fourier transform (FT) method of data analysis is less suitable than theoretical curve fitting (CF) methods for cases involving high atomic weight backscattering atoms; the back-transform of a real space component, as employed by the FT method, is likely to include the interfering resonance of a higher shell component.

No account, in the EXAFS calculation, is taken of the polarisation effects arising from the L_3 edge absorption, although it has been suggested that errors associated with this are small (16). A conservative estimate of the error in the absolute distance determinations is 0.03 Å. The observed increase in the adsorbate-substrate interatomic spacing as the coverage is increased from 0.15 to 0.30 monolayers is 0.30 ± 0.06 Å.

At this point it may be illuminating to consider the sum of the metallic silver radius with the ionic, covalent and metallic radii of caesium: these are 3.09, 3.69, and 4.07 Å respectively (17). This serves only to

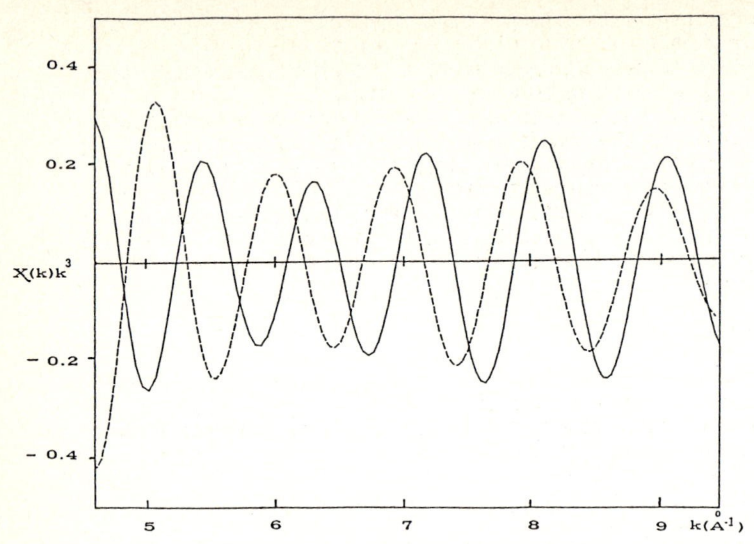

Figure 14 Superimposition of the first shell contributions from both the high coverage (covalent) and low coverage (ionic) phase of Cs/Ag(111). The obvious difference in frequency provides a visual demonstration of the change in bond distance with increasing coverage.

illustrate the "ball-park" limits and to indicate the order of magnitude of bond distance variation which we may expect to observe; our knowledge of the precise description of the adsorbate-substrate bonding is, as yet, limited at any stage of the adsorption process. At a qualitative level, this picture adequately suffices. Indeed, simple theoretical descriptions based on a classical interpretation, with emphasis on the local aspects of the bonding, have been shown to be qualitatively successful (within given coverage limits (18) (19) (20)). Jellium type calculations have also achieved similar success (21) (22).

Quantum mechanical models have been developed. Using an LCAO method, Wojciechowski (23) found that the transition from an ionic state to a metallic state, with increase in coverage, gave a valid description.

The aforementioned quantum mechanical treatment, along with those applied subsequently, only considered the sp electrons of the substrate. This seemed reasonable for simple and noble metals, where the d-band lies well below the Fermi-level. A total electron calculation of the electronic structure of a caesium overlayer on W(100) (24) (25) (using the FLAPW method) showed that, for this case, the valence charge density in the surface region is strongly dominated by the d-like surface states. Interaction between the Cs 6s electrons and these states resulted in the polarisation of the valence electrons. Additionally, it was found that the Cs 5p also participated in the bonding, becoming 'counter polarised' as a result of the 6s polarisation. Thus here, a different picture was presented which interpreted the Cs/W(100) overlayer as a 'polarised' metallic rather than ionic system. The authors posed the question as to whether this is the electronic description, for alkali metal adsorption or whether tungsten was a special case, on account of its dominating d-like surface states. Certainly for Cs on Ag this would

seem to be an inappropriate interpretation, as the d-band is of little importance with regard to the bonding mechanism of this metal.
The results presented here strongly support the electronic description of a transition from an ionic to a covalent state.
At the time of completion of this work, the examination of a variation in the substrate-adsorbate bond distance as a function of coverage, had never been treated with either a classical or quantum mechanical approach. However, independant theoretical treatment (29) has indeed predicted that increase in alkali metal coverage would lead to such variations. It is thus suggested that much rigorous effort in this field is now required for a more precise understanding of these electronic changes.

SUMMARY AND CONCLUSIONS

The unique capacity of SEXAFS to elucidate surface structures with only short range order was demonstrated for Cl/Ag(111), by the study of: 1) a partially ordered phase observed at a 0.7 monolayer coverage and 2) a disordered phase observed at a 0.4 monolayer coverage. For these two cases, LEED is unsuitable and not applicable, respectively. In this work, highly accurate bond distance determinations were made of 2.70 ± 0.01 Å for both phases.
A particularly significant observation from this study was that the nearest neighbour distance did not vary with increase in chlorine coverage. The latter was also found to be true in the SEXAFS study of the two ordered phases of Cl/Ag(110). These results were supportive of a general conclusion about the adsorption of electronegative species.
In complete contrast to the adsorption of chlorine it was found that the adsorption of the electropositive Cs species on Ag(111) showed a marked coverage dependence of adsorbate-substrate distance with coverage. The bond distance was found to increase from 3.20 Å to 3.50 Å as the coverage was increased from 0.15 to 0.30 monolayers. This work demonstrated the sensitivity of SEXAFS to a subtle bond distance variation.
The question arises as to how the electronegative and electropositive species are interacting with the substrate and how the surface properties are affected in each case: this being of fundamental importance with regard to the role of these species in catalysis. A particularly relevant industrial case, to which all the surface studies presented in this work are pertinent, is that of the silver catalysed oxidation of ethylene to ethylene oxide. Selectivity in the required ethylene oxidation is reduced mainly by the over-oxidation of the ethylene oxide product. This is improved by the addition of gas phase organo-chloro compounds. The reason is believed to be that sites, on the supported silver catalyst, are blocked by adsorbed Cl atoms; this inhibits the adsorption of unfavourable states of oxygen. However, the adsorption of the chlorine atoms also decreases the total activity of the catalyst.
The activity of the silver catalyst is improved thus by the addition of small amounts of alkali metal 'promoters'. This also further increases the selectivity of the catalyst.
With regard to the different roles of an electropositive and an electronegative species in the catalysis of a reaction, Feibelmann and Hamman (26) have performed theoretical calculations for the adsorbate species; Cl, S, P and Li on Rh(100). The study shows that the perturbations to the valence charge by the electronegative species (Cl, S, and P) are large but local; they do not extend much beyond the first nearest neighbour Rh distance. This is attributed to the efficient screening of the charge density by metal. Thus, the rest of the substrate is macroscopically unchanged by the adsorption of the

electronegative species. However, the local density of states at the Fermi level (the magnitude of which is a measure of the ability of the surface to respond to reactants) is depleted, and this effect extends beyond the first nearest neighbour. This would explain why the activity of silver in the ethylene oxide reaction is reduced by the adsorption of chlorine. In complete contrast, by the adsorption of small amounts of Li, the electronic charge density increases over most of the surface, whilst the local density of states at the Fermi level is increased. Thus small amounts of adsorbed alkali atoms will enhance the reactivity of the whole substrate. This explains the 'promoting' effect (i.e. activity enhancement) by adsorbed alkali atoms.

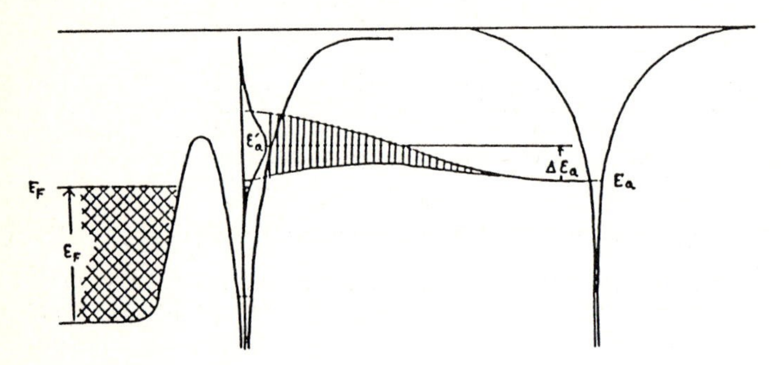

Figure 15 Change in adsorbate energy level on approaching a metal surface.

It may be instructive to consider how the electronic charge transfer, to or from the substrate, is different for electropositive and electronegative species. The case of an alkali atom being brought up to the surface is illustrated in figure 15. The valence energy level of the isolated adsorbate atom, E_A, moves up in energy as the atom moves closer to the surface. When the wave function of the adsorbate atom is close enough to the metal surface for overlap to occur, mixing of the wave functions takes place which broadens the previously sharp atomic level into a resonance. If the resonance lies above the Fermi level then an electron will be transferred from the adsorbate to the substrate. This is the situation for a single adsorbed alkali metal atom. Thus, a large dipole moment exists on the surface, and correspondingly, the work function initially decreases. With further increase in coverage (beyond approximately half of saturation coverage) the dipoles deactivate each other by depolarisation. Thus, the energy level E_A moves downwards across the Fermi level and the substrate-adsorbate bond distance increases as electrons flow back from the substrate to the adsorbate. This reneutralises the alkali species and the work function passes through a minimum. The occupation of the resonance with increase in coverage has, thus, increased and the originally high ionic character of the adsorbate-substrate bond becomes more covalent.
One might suppose that the behaviour of the chlorine adsorbate would be opposite, but analogous, in terms of the charge transfer process. However, the behaviour is quite different. In particular, the work function is seen to slightly increase with increase in chlorine coverage in a simple linear manner.

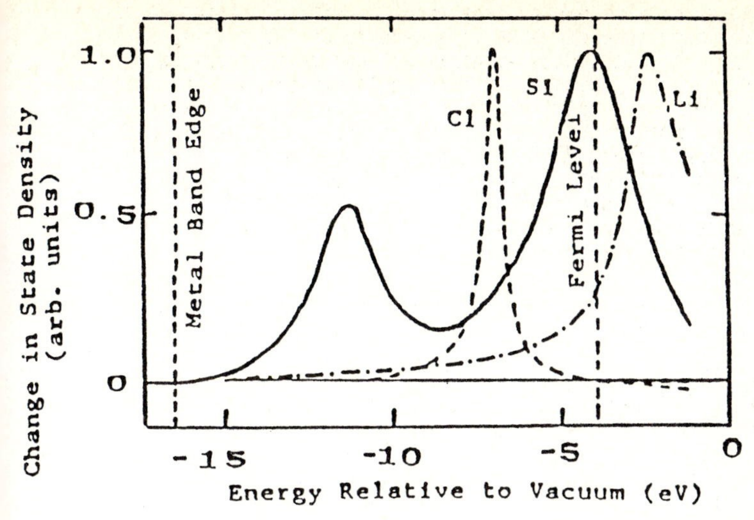

Figure 16 The local density of states at the adsorption site for an electronegative, electropositive and covalent species (from reference 12).

Lang and Williams (27) (28) have studied the adsorption of a strongly electropositive, a strongly electronegative and a covalent species (Li, Cl and Si respectively), adsorbed on a jellium substrate. The calculated local density of states for the three systems are shown in figure 16. For lithium, the adsorbate resonance is initially mostly unoccupied, since it lies mainly above the Fermi level; whereas the chlorine resonance is totally filled as it lies completely below the Fermi level. The silicon resonance lies almost exactly halfway across the Fermi level, exhibiting covalent character, with equal amounts of bonding and anti-bonding character.
Considering the alkali metal case, as the coverage increases the resonance moves down over the Fermi level and the occupation of its resonance increases i.e. the electron transfer is back to the adsorbate atom. The alkali-substrate bond then becomes more covalent, resembling that illustrated for Si in figure 16. However, the occupation of the chlorine resonance cannot vary with the increased coverage as it is fully occupied from the initial stages of adsorption. This is illustrated by its work function behaviour; the dipole moment for each adsorbate-substrate bond is the same, thus, there is merely a linear increase in the total dipole moment of the surface with increase in number of atoms contributing to this total dipole moment.
This fundamental difference in electronic behaviour may help to explain why there is a variation in bond length with coverage for Cs/Ag(111); whilst for Cl/Ag(111), and Ag(110), the bond distance remains the same regardless of the coverage. Since the occupation of the Cs resonance is varying with coverage, the equilibrium distance (for minimum energy of the adsorbate) will also vary. However, since the occupation of the Cl resonance is invariant with coverage, nothing is to be gained (in terms of energy) by altering its distance from the surface.

ACKNOWLEDGEMENTS

We acknowledge the SERC (UK) for grants that enabled the development of the SEXAFS station at Daresbury, and for a studentship to G.L. We are particularly grateful to all our colleagues who have participated in this work, including those based at Liverpool (R. Brooks, J.C. Campuzano, S.Ferrer, D. Holmes), and at Daresbury (D. Norman, A. MacDonald), with special thanks to N. Binsted for assistance with computation.

REFERENCES

1. P. A. Lee, Phys. Rev. B13 5261 (1978)
2. G. M. Lamble and D. A. King, Phil. Trans. Roy.Soc. Lond.A318 (1986) 203.
3. A. A. MacDowell, D. Norman, J.B. West, J.C. Campuzano and R.G. Jones, Nucl. Instrum.Meth. (in press): A.A. MacDowell, D.Norman and J..B. West, (to be published). Rev.Mod. Phys.
4. S. J. Gurman, N. Binstead, and I. Ross, J. Phys. C17 143 (1984)
5. P. A. Lee and J. B. Pendry, Phys. Rev. B 11, 2795 (1975)
6. G. M. Lamble, R.S. Brooks, D. A. King and D. Norman, Phys.Rev. B, 1986.
7. G.M. Lamble, R.S. Brooks and D.A. King, (in progress).
8. G. Rovida and F. Pratesi,Surf. Sci. 51, 270 (1975).
9. M. Bowker and K.C. Waugh, Surf.Sci. 134, 639 (1983).
10. M. Bowker and K.C. Waugh, Surf.Scu. 155 1 (1985).
11. D.J. Holmes, F. Della Valle, C.J. Barnes, N. Panagiotides, G.M. Lamble and D.A. King, Journal of Vac.Sci. Technol., in press (1986 Baltimore International Congress Proceedings).
12. K. H. Kingdon and I. Langmuir, Phys.Rev. 21, 280 (1923).
13. J.B. Taylor and I. Langmuir, Phys. Rev. 44, 423 (1933).
14. G.M. Lamble, R.S. Brooks, D. Norman and D.A. King (in progress).
15. N.F. Mott and H.S.W. Massey, 'The Theory of Atomic Collisions', 3rd edn. (Clarendon Press, Oxford, 1965), p562.
16. P.H. Citrin, Phys. Rev. B31, 700 (1985).
17. L. Pauling, 'The Nature of the Chemical Bond', (Cornell Univ. Press, 1939).
18. N.S. Rasor and C. Warner, J. Appl. Phys. 35, 2589 (1964).
19. J.W. Gadzuk and E.W. Carabeates, J. Appl. Phys. 35, 357 (1965).
20. J.R. MacDonell and C.D. Barlow, J. Chem. Phys. 44, 202 (1966).
21. N.D. Lang, Phys. Rev. B4, 4234 (1971).
22. H. Yamauchi and V. Kawabe, Phys. Rev. B14, 2687 (1976).
23. K.F. Wojciechowski, Surf.Science. 55, 246 (1976).
24. E. Wimmer, A.J. Freeman, J.R. Hiskies and A.M. Karo, Phys. Rev. B28, 2074 (1983).
25. P. Soukiassian, R.Riwan, J. Lecante, E. Wimmer, S.R. Chubb and A.J. Freeman, Phys. Rev. B31, 4911 (1985).
26. P.J. Feibelman and D.R. Hamman, Surf. Sci. 149, 48 (1985).
27. N.D. Lang and A.R. Williams, Phys. Rev. Lett. 34, 531 (1975).
28. N.D. Lang and A.R. Williams, Phys. Rev. Lett 37, 212 (1976).
29. J.P. Muscat and I.P. Batra, Phys. Rev. B (1986).

Improved Spectrometer for Auger Appearance Potential Fine Structure Studies

J.L. del Barco and R.H. Buitrago

Instituto de Desarrollo Tecnológico para la Industria Química, Universidad Nacional del Litoral, Consejo Nacional de Investigaciones Científicas y Técnicas, Güemes 3450, 3000 Santa Fe, Argentina

1. Introduction

The primary goal of the extended fine structure spectroscopies is to provide information about the spacing of nearest neighbors in ordered and disordered systems. They have become important tools for the analysis of local structure and other bulk and surface parameters [1]. In particular, Extended X-ray Absorption Fine Structure Spectroscopy (EXAFS) [2] has emerged as a powerful technique since intense synchrotron x-ray sources were available, allowing for both high signal-to-noise ratios and rapid data collection.

A fine structure that extends above the appearance potential threshold has been observed when a material is bombarded with electrons. Recently, it has been demonstrated that it might be possible to use this extended fine structure in a similar manner to that of EXAFS [3,4], to obtain nearest neighbor distances for a local arrangement of atoms in solid surfaces. The fine structure originates from the interference of an outgoing wave of a scattered electron with backscattered components from neighboring atoms, and has been named Extended Appearance Potential Fine Structure (EAPFS) [6]. The potential capability of EAPFS is well known [4] and provides motivation for dealing with its complications, namely those arising from physical processes involved in Appearance Potential Spectroscopy (APS).

2. Experimental Arrangement

The appearance potential fine structure spectroscopy which measures the total current of the secondary electrons is called Auger Electron Appearance Potential Fine Structure Spectroscopy (AEAPFS). In order to do AEAPFS we followed the method of Cohen et al. [3], which provide the best signal-to-noise ratio. Electrons from an electron gun strike the target, with an energy which has been modulated by adding a sinusoidal voltage to the accelerating potential. The target current, which is the difference between primary and the secondary current leaving the probe, is fed into a lock-in amplifier via a tuned tank circuit, which works as a tuned current to voltage converter. In order to obtain the second derivative of the target, current, the modulation frequency is a half of that required for the resonance of the tuned circuit.

In AEAPFS technique it is a matter of utmost importance to avoid irradiation damage during the measurement. To obtain simultaneously a high signal-to-noise ratio, for rapid data collection, and a relatively good spatial resolution, there is no alternative but to make the measurements with low current intensity and lock-in time constant. Consequently, under these imposed conditions an improvement in the signal-to-noise ratio of the circuit that feeds the lock-in amplifier input is required.

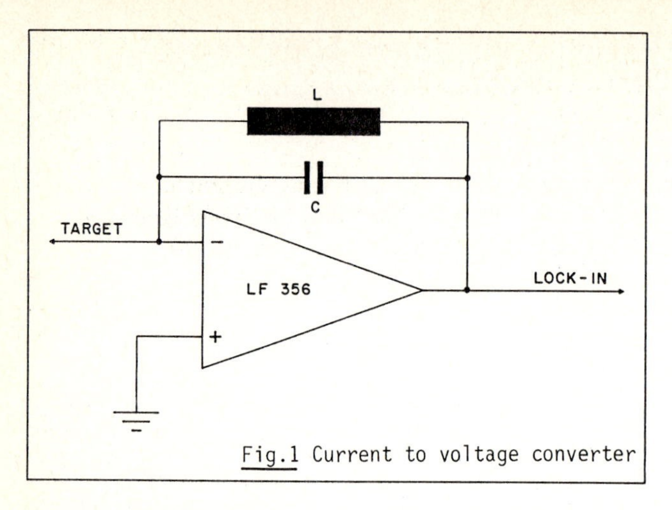

Fig.1 Current to voltage converter

3. Experimental Results

As a performance test, the oscillation above the threshold in polycrystalline clean and oxidized cobalt, iron and titanium have been measured. The current of the primary bombarding, the recording speed and the time constant used was optimized for each sample. Figure 2 shows a clean polycrystalline iron AEAPFS spectra obtained using a primary current of approximately 80 μA, a lock-in time constant of 1 sec. and a recording time of the extended fine structure spectrum of 10 min. The recording speeds are considerably increased compared to the those reported previously [10].

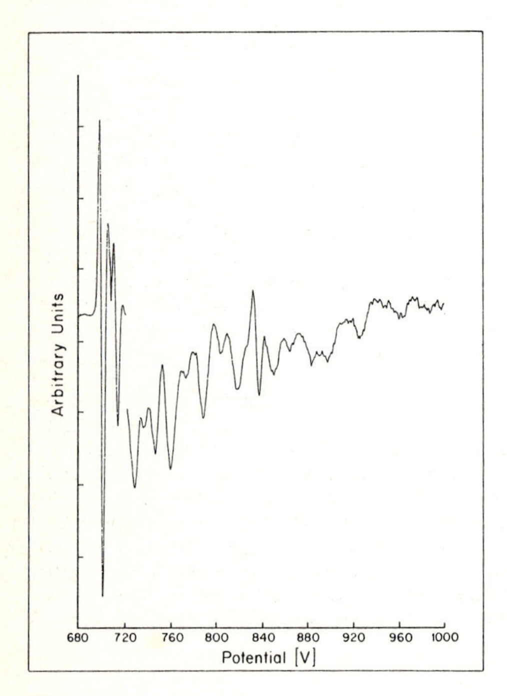

Fig.2 Polycrystalline iron AEAPFS spectrum showing the 2p3/2, the 2p1/2, the 2s core edges and the extended fine structure

The output signal of the lock-in amplifier was digitalized using a 12 bit analog-to-digital converter and transferred to a computer. The data was analyzed following a formalism and computational technique similar to that usually employed in EXAFS [2,10]. The Fourier transform extracted from the iron AEAPFS spectrum is shown in Fig. 3. The first peak position is centered at 2.17 Å and since the phase shift correction is about 0.22 Å [10] the measured nearest neighbor spacing become 2.39 Å. Therefore, the comparison with reported measurements [10-12] confirms the good performance of the constructed spectrometer.

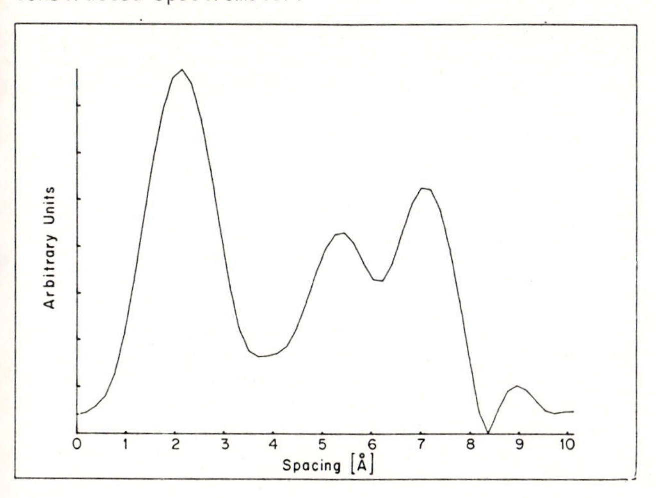

Fig.3 Magnitude of Fourier transform of the AEAPFS polycrystalline iron fine structure starting above 2s edge and extending to about 1100 eV.

We have constructed a simple spectrometer including two major improvements which allow for a high quality AEAPFS spectra. Firstly, a built in-house electron gun was designed to provide electron beam energies up to 3 keV, and a current up to 100 μA, with a spot size lower than 2 mm according to the operating conditions. The arrangement consists of an electron gun with tetrode electrostatic optics as in a typical cathode ray tube gun [7], with a hairpin tungsten filament. The control electrode is a cylinder slightly larger than the filament, with a small aperture at the cap end. Opposite the aperture is the accelerating electrode which is interposed between the control electrode and the focusing electrode. The accelerating electrode is connected to the last electrode to form the focusing lens.

All metal parts and lenses were made out of a non-magnetic stainless steel and all insulators in the assembly were made of a low porosity ceramic. The gun was constructed on a standard 2.75" flange with four electrical feedthroughs and shielded from external electrostatic influences by a non-magnetic stainless steel tube. A programable high voltage low ripple power supply was used to provide the beam voltage. The electrode control and the focus electrode voltage was obtained with a divider network . The filament current was supplied by a floating current regulated power supply. The gun with an aligner bellows systems was mounted in the vacuum chamber of an Auger Spectrometer. In this way one profits from a combination of the two techniques, the ion gun and the specimen stage facilities.

The second improvement has its value in the fact that we have obtained ar improved simplified tuned current-to-voltage converter circuit, with a high quality factor. The circuit of the converter proposed is shown in Fig. 1. This circuit uses a similar LC arrangement of the converter used in reference [2] but connected as a feedback network of an operational amplifier. This new design requires a few additional electronic components of low cost; an operational amplifier and a low noise power supply. We selected an LF356 operational amplifier [8], a battery pack as power supply, an N28 ferrite 47 x 28 pot core [9], and a polyestyrene capacitor. The use of the operational amplifier eliminates the problems associated with the dependence of the converter electrical parameters on the electron gun-target system. Therefore, the circuit makes the best use of the inductor and capacitor characteristics. With this arrangement the portion of the target current signal corresponding to the second harmonic of the modulation frequency is more selectively transmitted, while the noise that could overdrive the lock-in amplifier is reduced. The target current signal pass through the resonant band of the converter, the bandwidth being determined by the time constant of the lock-in amplifier, and is demodulated as an AEAPFS spectrum. When the recording spee of the signal is increased, the time constant of the lock-in amplifier must be reduced, that is, the corresponding bandwidth becomes wider. Therefore, t signal-to-noise ratio of the AEAPFS spectrum is determined by the narrower bandwidth of the two: the converter and the lock-in amplifier, being the bandwidth of the converter more important in such case.

The operation frequency is conditioned by the circuit stability. By a suitable election of the operational amplifier, inductor and capacitor value it is possible to obtain an adequate frequency of operation. Thus the noise problems associated with the electrical interference of the line frequency are avoided. The above-described converter was operated satisfactorily at 10.631 Hz.

4. References

1. G.S. Knapp, F.V. Fradin: In Electron and Positron Spectroscopies in Materials Science and Engineering, ed. by O. Buck, J.K. Tien and H.L Marcus, (Academic, London 1979), pp. 243
2. P.A. Lee, P.H. Citrim, P. Eisenberg and B.M. Kincaid, Re. Mod. Phys. 53, 769 (1981)
3. P.I. Cohen, T.L. Einstein, W.T. Elam, Y. Fukuda and R.L. Park, Appl. Sur Sci. 1, 538 (1978)
4. G.E. Laramore, Phys. Rev. B18, 5254 (1978)
5. T.L. Einstein, Appl. Surf. Sci., 11/12, 42 (1983)
6. T.L. Einstein, M.J. Mehl, J.F. Morar, R.L. Park and G.E. Laramore: In EXAFS and Near Edge Structure, ed. by A. Bianconi, L. Incoccia and S. Stipcich, (Springer-Verlag, Berlin, 1983), pp. 391-393
7. J. Arol Simpson: In Methods of Experimental Physics, ed. by L. Marton an C. Marton (Academic, New York, 1967) vol. 4A, pp. 93
8. National Semiconductor Corp., 2900 Semiconductor Dr., Santa Clara, CA 95051, U.S.A.
9. Siemens AG, Balanstrasse 73, D-8000 Müchen 80, Federal Republic of Germa
10. W.T. Elam, Extended Appearance Potential Fine Structure Analysis of Titanium, Vanadium and Iron Surfaces, Ph.D. thesis, University of Maryla (U.S.A.), 1979
11. T. Ishida, H. Maeda, N. Kamijo, K. Umeda, J. Appl. Phys. Japan, 52, 9, 29 (1983)
12. S.W. Schulz, H.U. Chun, Solid State Commun., 49, 4, 399 (1984)

Characterization of Heterogeneous Surfaces Using Photoemission Techniques

K. Wandelt

Fritz-Haber-Institut der Max-Planck-Gesellschaft, Faradayweg 4–6, D-1000 Berlin 33

1. INTRODUCTION

The physics and chemistry of solid surfaces are strongly effected - if not dominated - by structural and chemical heterogeneities. Defects and dopant atoms create impurity states in the band gap of semiconductors, which determine the electronic properties of semiconductor devices. "Extrinsic" surface sites influence the growth mode and final structure of epitaxial films by heterogeneous nucleation. Composition and topography are also decisive parameters in the reactivity and catalytic activity of solid surfaces. After two decades of detailed model studies with well prepared single crystal samples, surface heterogeneities begin to dictate a new trend in surface science. Several new analytical techniques are at different stages of development, which can provide information about structural and chemical surface disorder on different scales of resolution, such as Low Energy Electron Reflection Microscopy (LEERM) /1/, Ultra-electronmicroscopy /2/, Photoelectronmicroscopy /3/, Scanning Tunneling Microscopy (STM) /4-6/ and Photoemission of Adsorbed Xenon (PAX) /7-9/. All these techniques can be applied to extended heterogeneous surfaces and do not require fine tips as the Field Ion Microscopy does /10/.

The present paper explores the capability of photoelectron emission techniques to provide information about the structural and chemical properties of heterogeneous surfaces as well as the related electronic structure. These properties will here only be discussed for the bare substrate surfaces; the breadth of information obtained from photoemission of chemisorbed gases has been reviewed at length in Ref. 6. After a brief recollection of the characteristic features of a photoemission experiment in part 2, the selected examples are divided into two classes. In part 3 we discuss data based on the photoemission of the surface constituent atoms themselves, while part 4 presents photoemission results from adsorbed Xenon probe atoms which give information about the nature of the respective adsorption site even on an atomic scale.

2. PHOTOEMISSION EXPERIMENT

Typical set-ups for a photoemission experiment are illustrated in Fig. 1. The sample, mounted in ultra-high vacuum, is irradiated by unpolarized or polarized light in the X-ray or UV-regime (XPS, UPS) from a given direction of incidence using conventional light sources or synchrotron radiation. Since the differential cross-section σ for photoionization is given by

$$\frac{d\sigma}{d\Omega dE}\ (\varepsilon_f,\vec{k}_f,h\omega,\vec{A}) \sim \sum_i |\langle f|\vec{A}\cdot\vec{p}|i\rangle|^2 \delta(\varepsilon_f - \varepsilon_i - h\omega), \quad (1)$$

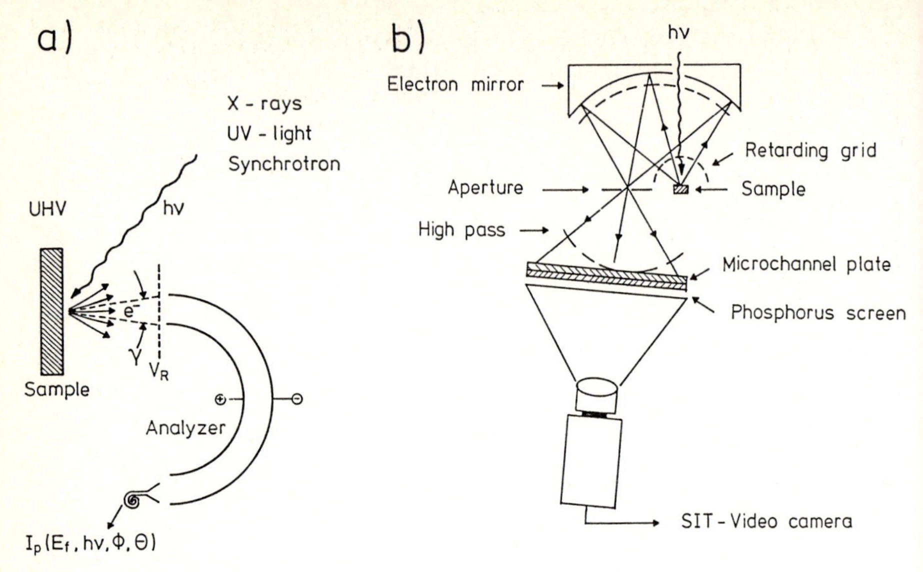

Fig. 1: Schematics of two types of photoelectron spectrometers, a) hemispherical analyzer and b) two-dimensional electron-mirror-display-analyzer (EMDA) /11/

analysis of the detected photoemission current I_p with respect to its intensity, kinetic energy (ε_f), emission direction $\vec{k}_f$ also as a function of photon energy $h\omega$ and polarization $\vec{A}$ yields information about surface concentrations, characteristic electron binding energies ε_i and, in particular, their "environmental" shifts, orbital symmetries, structure related energy and intensity losses etc. Figure 1 displays two different types of electron analyzers, a) a conventional hemispherical analyzer (HA) and b) a two-dimensional electron-mirror-display-analyzer (EMDA) /11/. While the first type collects the photocurrent I_p into a small acceptance cone of solid angle γ and is movable in space ($\emptyset$ = azimuth angle, θ = polar angle) the EMDA typ collects all photoelectrons within a solid angle of 88° and projects them angular-resolved onto a multi-channel plate. Angular anisotropies in the photoemission intensity become immediately visible on a phosphorescent screen by use of a SIT-video camera /11/. In the current-integrating mode the EMDA yields large intensitites at high energy resolution due to the large collection angle. The examples discussed in this work will take advantage of both detection geometries.

Another important fact is the pronounced energy dependence of the inelastic mean free path (impf) length of electrons in solids /12/, providing indepth resolution of the distribution of a special kind of atoms normal to the surface.

3. PHOTOEMISSION FROM SURFACE CONSTITUENTS

3.1 Surface Composition

The determination of surface composition is one of the primary goals of surface investigations. By now it is a well established matter of fact that the composition even of clean and equilibrated surfaces of binary

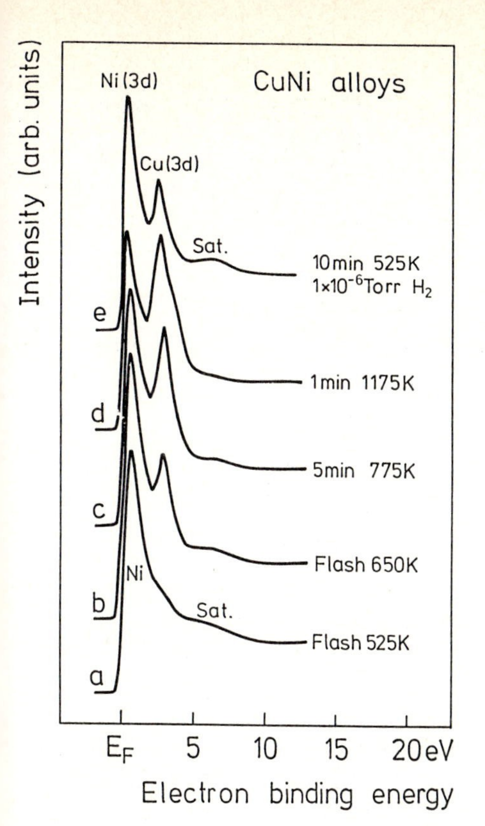

Fig. 2: UV(HeII) photoemission valence band spectra from a 50%Cu 50%Ni alloy as a function of surface treatment

alloys may not be the same as in the bulk of the same sample. This is demonstrated in Fig. 2 with the "classical" system of a 50% Cu 50% Ni bulk alloy. Spectrum a) shows the HeII excited UPS valence band spectrum of the (110) surface of this alloy after sputter-cleaning and a brief flash to 525 K. Besides the d-band emission close to E_F and a satellite peak around 6 eV typical for Ni, the spectrum shows hardly any emission from the Cu(3d) band at ∿ 2 eV below E_F. The latter one appears only after extended annealing at higher temperatures as made clear by spectra b)-d). Analysis of the final intensities (areas) under both d-band peaks (after subtraction of a secondary electron background) yields a final surface composition of approximately 80% Cu 20% Ni.

Due to the very short escape depth of the involved electrons of ∿ 35 eV kinetic energy this information pertains to the first two atomic layers only, and indicates substantial surface enrichment with Cu over the bulk composition. This finding is in agreement with current theories of surface segregation /13/ which are based on bond energy (ΔH) (surface tension) and lattice mismatch (elastic strain) considerations. For the same reasons the amount of segregation is expected to depend on the crystallographic orientation, namely on the atomic density, of the surface, which could also be verified with CuNi (fcc) alloys by means of photoemission. Cu segregation is more pronounced on the (111) face than on the (100) face /14/. The role of lattice vibrational entropy in the context of surface segregation has been studied recently with Pt-Rh alloys /15/.

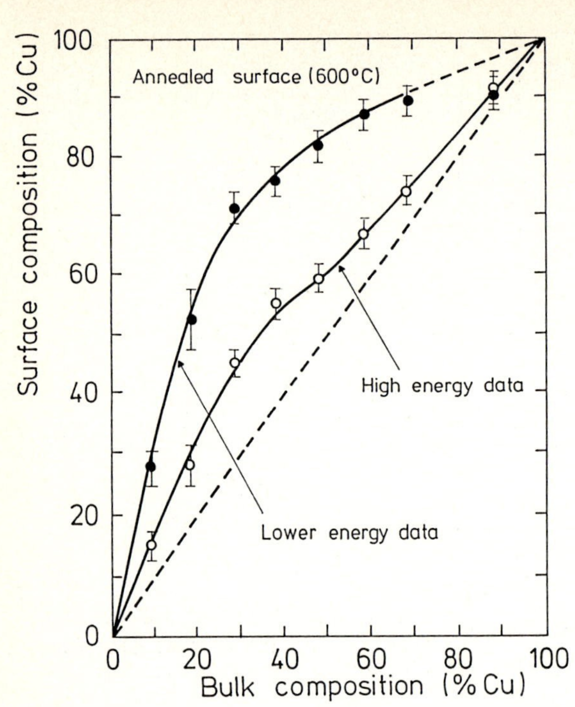

Fig. 3: Cu surface concentration as a function of Cu Ni bulk concentration as deduced from low and high energy electrons /38/

Deviations between surface and bulk concentration due to segregation lead to the formation of concentration gradients normal to the surface. The indepth profile of a species can be determined taking advantage of the energy dependence of the inelastic mean free path lengths and, hence, of the probing depths of electrons of different kinetic energy, by monitoring electrons of different binding energy ε_i excited with monochromatic light or by detecting electrons of constant binding energy ε_i excited with photons of variable energy (synchrotron radiation). As a result of this kind of studies Fig. 3 shows the surface concentration of Cu as a function of the Cu bulk concentration in CuNi alloys as determined from low (see Fig. 1) and high kinetic energy electrons. Other techniques for detecting concentration gradients are the polar-angle variation method (see following paragraph) or sputter-profiling in conjunction with photoemission.

The thus determined surface concentrations are still integral values over the relatively large surface area illuminated by the photon beam. A scanning technique would be reguired to obtain also information about the lateral distribution (possibly due to lateral segregation) of surface species. Since available scanning modes of XPS and Auger electron spectroscopy do not provide lateral resolution better than a few hundred Ångstroms we refer here to paragraph 4.3 which describes a completely different approach to study the lateral distribution of surface species.

3.2 Growth Mode of Metal Deposits

Thin evaporated metal films play an important role in microelectronic devices; ultra-thin metal deposits are often used as model systems for supported catalysts. Their initial growth structure and distribution can also be deduced from photoemission experiments.

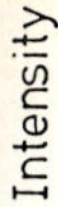

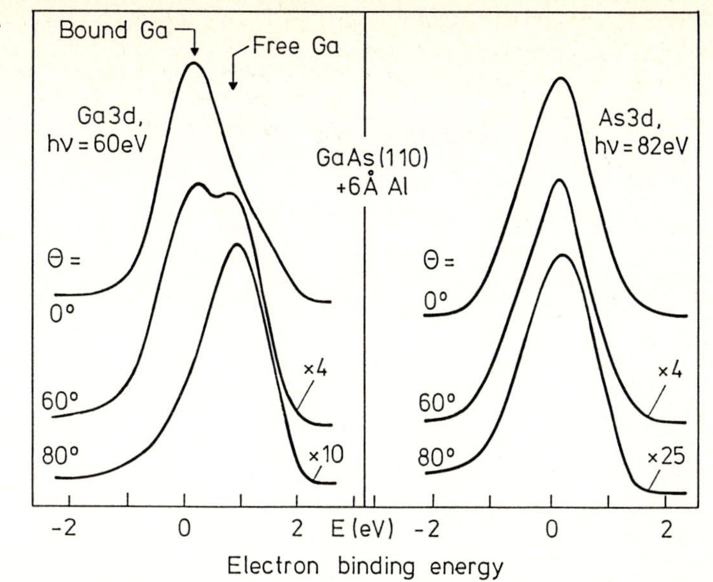

Fig. 4: Ga(3d) and As(3d) core level spectra from a 6 Å Al covered GaAs(110) surface with variable photon energy so that the kinetic electron energy is the same in both cases (36 eV) /16,17/. θ = polar angle

STOFFEL et al. /16,17/ have studied the morphology of thin 6-Å Al overlayers on GaAs(110) by means of polar-angle-resolved photoemission (PARP) from the As(3d), Ga(3d) and Al(2p) core levels, respectively. The photon energy of the synchrotron radiation was selected in each case such as to result in the same kinetic energy of the photoelectrons, namely 36 eV, and, hence, in the same escape depth for all three core levels. The Ga(3d) and As(3d) core level spectra shown in Fig. 4 for different polar angles θ can be interpreted very straightforwardly in terms of geometrical shadowing effects illustrated in Fig. 5.

In normal emission (θ = 0°) both core level peaks (Fig. 4) are attenuated due to the 6-Å Al overlayer only by ∿ 50 %. At a collection angle of θ = 80° the Ga(3d) intensity from the Al covered surface is attenuated by roughly the same factor of 2 as compared to the freshly cleaved GaAs surface, while that of the As(3d) core level is attenuated by almost a factor of 5. This immediately indicates a strong change in the spatial distribution of Ga and As atoms before and after Al deposition. Furthermore, the angular dependence of the Ga(3d) line shape suggests Ga atoms in two different chemical environments (Fig. 4). The authors' conclusions from these data are summarized in Fig. 5. The Al is suggested to form islands on the surface, because based on the imfp of 36 eV electrons in Al, a uniform Al film should suppress the substrate emission lines much more than just by a factor of 2 as experimentally observed. The strong angular variation of the attenuation of As vs. Ga intensity from the Al covered surface is interpreted as As staying at the GaAs/Al interface, while Ga diffuses through the Al to the surface. This is supported by the presence of two chemically different Ga species (Fig. 4), namely Ga bound to As and free Ga at the Al surface, in contrast to the uniform As(3d) signal, as well as

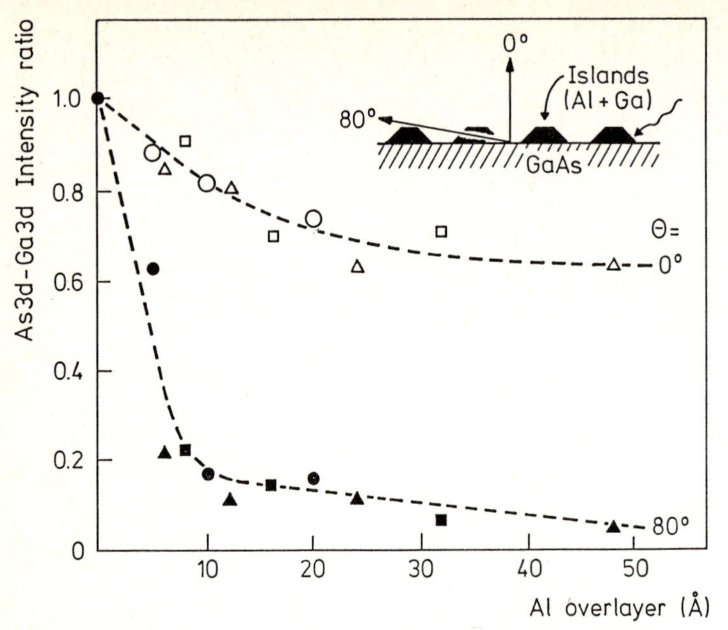

Fig. 5: Al coverage dependence of the As(3d)/Ga(3d) core level intensity ratio as a function of Al coverage for two polar angles /16,17/

by the Al coverage dependence of the As(3d)/Ga(3d) intensity ratio plotted in Fig. 5. For $\theta = 0°$, the photoelectron spectra probe mainly the uncovered stoichiometric GaAs substrate surface region; the ratio between the partial intensity of the As-bonded Ga(3d) emission and the As(3d) intensity, in fact, stays close to unity for all Al coverages shown. For $\theta = 80°$, the Ga, which preferentially diffuses through the Al islands, dominates. Of course, a distribution in island width, height and separation explains why the complete blocking of the substrate emission does not occur abruptly with coverage at any given collection angle.

The same approach, in fact, has been taken quite early /18/ using XPS, in order to obtain the size and distribution of small transition metal clusters on oxidic substrates from intensity shadowing effects. These systems are sometimes regarded as models of supported catalysts.

Much more information about these kinds of metal deposits, however, can be obtained by taking advantage of synchrotron radiation and measuring surface core level shifts (SCS). Surface core level shifts have been detected for a great number of bulk metals and provide a powerful tool for surface characterization /19/. Fig. 6 shows the first evidence for SCS in small deposited metal clusters, namely $4f_{7/2}$ core level spectra of Ir evaporated onto SiO_2 /20/, monitored using synchrotron radiation of photon energy $h\nu = 130$ eV (BESSY) and the EMDA analyzer displayed in Fig. 1b. For the purpose of reference Fig. 6a shows the $4f_{7/2}$ spectrum (data points) of a thick overlayer (the SiO_2 substrate was no longer detectable) together with the best fit (full line) with two equal Doniach-Sunjic lines with an asymmetry factor of $\alpha = 0.12$ (linear background substracted is also shown). The two lines are separated by 0.4 eV and correspond to $4f_{7/2}$ core level emission from the Ir bulk (at 60.7 eV) and the topmost surface (at

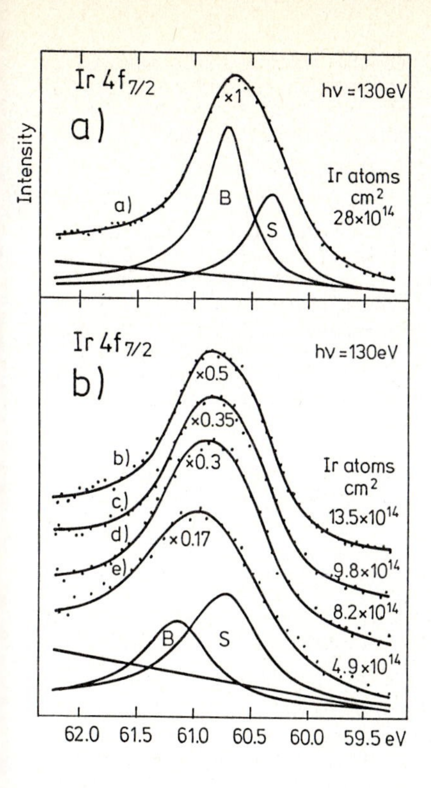

Fig. 6: Ir $4f_{7/2}$ core level spectra of a) a thick Ir film and b) small Ir clusters on SiO_2. The spectra are decomposed into surface (S) and bulk (B) emission.

60.3 eV), respectively. Their intensity ratio is I_S/I_B = 0.58. These results are very similar to those found for open Ir single crystal surfaces /21/. Fig. 6b then shows experimental spectra (data points) for decreasing iridium coverages (normalized to the same height as spectrum 6a). The coverages given in Ir atoms/cm^2 are estimated assuming that the intensity of spectrum 6a corresponds to 1.6 complete Ir(100) layers as estimated on the basis of imfp considerations. In each case the full line through the data points again corresponds to the best fit with a surface and bulk peak (α = 0.12) separated by 0.4 eV but variable width and relative intensity. For the sake of clarity only the two component peaks of spectrum e) are shown. The shift of the $4f_{7/2}$ emission as a whole to higher binding energy as the Ir coverage decreases is in full accord with all other metal systems studied to date /22,23/. More important here is the fact, that the intensity of the surface emission (lower binding energy component peak) provides a measure of the true, free metal surface, which is the relevant quantity when evaluating the catalytic activity of such a supported catalyst. As expected the I_S/I_B intensity ratio increases as the total Ir load decreases. Furthermore, assuming hemispherically shaped Ir-particles on the SiO_2 surface the average size of the particles could be estimated from the I_S/I_B ratio and the imfp of ~ 70 eV electrons in Ir.

4. PHOTOEMISSION OF ADSORBED XENON (PAX)

In the previous section important conclusions about the composition and structure at surfaces could be drawn from the photoemission spectra of the surface constituents themselves. In this section a novel technique will be described, which is based on the fact, that the photoemission spectra of adsorbed Xe probe atoms reflect variations of the local surface potential and, hence, the local surface charge density /7/. The latter one is directly determined by the local surface geometry and composition.

4.1 The PAX Method

Below $\sim$ 80 K the rare gas xenon can be "physisorbed" on any solid surface. The UV (HeI) photoemission spectrum of Xe adsorbed on a Ru(001) surface is displayed in Fig. 7. The two sharp extra peaks (above the Ru background, dashed line) between 5 and 7 eV below E_F arise from the $5p_{3/2}$ and $5p_{1/2}$ photoemission final states of adsorbed Xe atoms. In the following we will concentrate on the $5p_{1/2}$ signal only, because its physical structure is somewhat simpler than that of the $5p_{3/2}$ signal /7-9/. It is the energy position, intensity and shape of this signal, that will furnish us important surface structural information.

In a number of publications /7-9,24/ it has been shown that the ionization potential of adsorbed Xe atoms (with respect to the vacuum level V) is rather independent of the substrate:

$$E_B^V(5p_{1/2}) = E_B^F(5p_{1/2}) + \varphi = \text{const.} \qquad (2)$$

This correlation has, in fact, been verified by now for more than 25 single crystal substrates of quite different nature, including transition metals (Pd, Pt, Ir, Ni etc.), noble metals (Cu, Ag, Au), alkali metals (K, Cs), semiconductor surfaces (Si, ZnO) as well as oxides (TiO_2, ZnO). There are two physical reasons for this surprising invariance of $E_B^V(5p_{1/2})$. Firstly, the Xe-substrate interaction is very weak in all cases; typical Xe physisorption energies are smaller than $\sim$ 8 kcal/mole. As a consequence

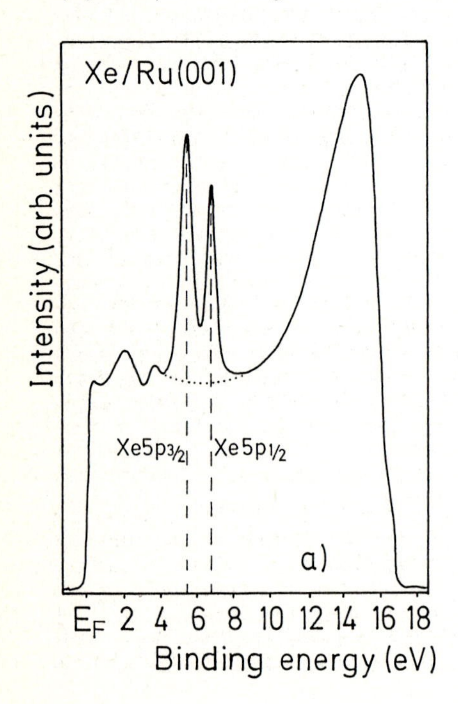

Fig. 7: He I excited Xe $5p_{3/2,1/2}$ photoemission of adsorbed xenon (PAX) spectra

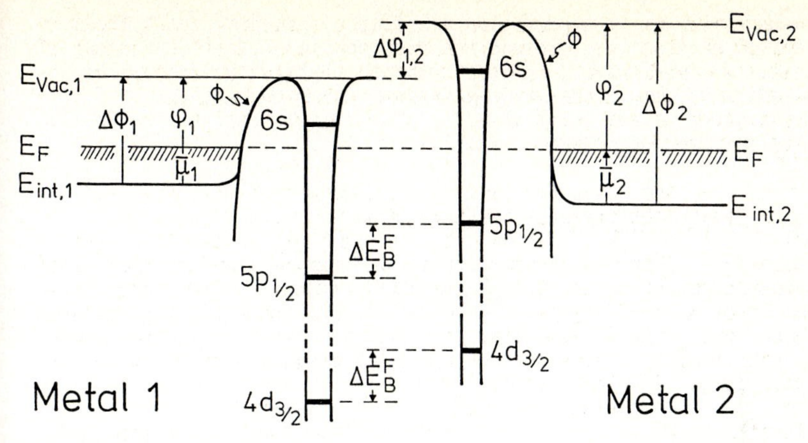

Fig. 8: Energy level diagram for xenon adsorbed on two metals of different work function φ, illustrating the PAX mechanism based on equ. (3)

initial state bonding shifts, and in particular their changes between different substrates, are negligibly small (< 0.2 eV). Secondly, a Xe atom is rather big (diameter 4,5 Å). Therefore the center of an adsorbed Xe atom is located outside of the steeply varying electrostatic surface potential ϕ (Fig. 8), which arises from the surface dipole layer. As a consequence the potential well of an adsorbed Xe atom is "pinned" to the vacuum potential outside the surface dipole barrier and "floats" up and down when the work function φ of the substrate is changed, for whatever reason, e.g. when changing the crystallographic orientation of the substrate surface, when changing the substrate material or when preadsorbing some other adsorbate. This "floating" model is illustrated in Fig. 8 and has received strong theoretical /25/ as well as experimental support. For instance, the same invariance as expressed in equ. (2) could be verified for the Xe(4d) and Xe(3d) core levels /9/ as well as for the unoccupied Xe(6s, 6p, 5d)-states /26/.

An immediate consequence of equ. (2) is:

$$\Delta E_B^F(5p_{1/2}) = -\Delta\varphi \ , \tag{3}$$

as becomes clear from Fig. 8. The crucial point now is, that this equation holds also for heterogeneous surfaces: Xe atoms being adsorbed on two "patches" of the same surface which have different local work functions, differ in their $E_B^F(5p_{1/2})$ binding energy values (with respect to the Fermi level E_F) accordingly, and two Xe (5p) spectra appear simultaneously, shifted by

$$\Delta E_B^F(5p_{1/2}) = -\Delta\varphi_{local} \tag{4}$$

with respect to each other. Equation (4) is the basis of PAX. Considering further, that the adsorption energy E_{ad} of Xe also depends on the chemical nature and coordination of the specific adsorption site, different surface sites can be populated successsively in the sequence of increasing E_{ad} and can thus be characterized by PAX separately on an atomic scale /7,24/. The role of the adsorbed Xe atoms is nothing more than to deposit a "test electron" (bound in the 5p level) at a particular surface site (controlled by E_{ad}) which then via photoemission provides information about the surface potential at this site. Beyond the distinction of different kinds of surface sites by their $5p_{1/2}$ electron binding energy (qualitative analysis) evaluation of the corresponding partial intensities yields the surface concentration of each kind of site (quantitative analysis). The following sections present selected examples.

4.2 Characterization of Surface Structure Defects

The sensitivity of PAX towards structural surface defects is illustrated best by Xe adsorption on a vicinal Ru(001) surface. Vicinal metal surfaces, cut under a small angle (< 10°) off a low-index plane, acquire a more or less regularly stepped structure after cleaning and annealing in UHV. The step height, step direction and step density can be determined very accurately by LEED /27/, so that the relative abundance of (high-coordination) step sites versus flat terrace sites is known.

Figure 9 shows on an expanded energy scale a series of Xe($5p_{3/2,1/2}$) spectra from a regularly stepped Ru(001) surface /28/. The successive population of two Xe states is clearly seen as the total Xe coverage increases. The Xe(S) state saturates after 1L Xe exposure (1 L = 10^{-6} Torr sec); its $5p_{1/2}$ electron binding energy is $E^F_{B,S}$ = 7.60 eV. The Xe(T) state emerges at higher coverages at $E^F_{B,T}$ = 6.90 eV. After completion of the first Xe monolayer on this stepped Ru(001) surface the Xe(T)/Xe(S) intensity ratio is 10 : 1. This intensity ratio, which agrees quantitatively

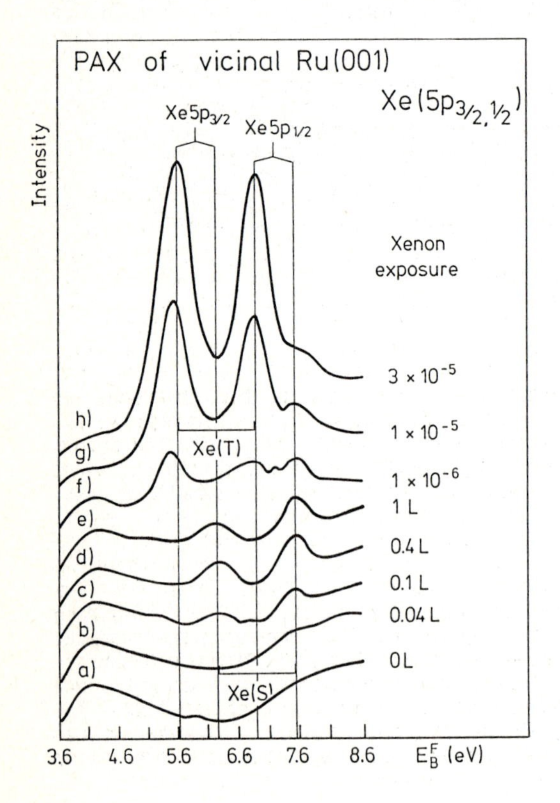

Fig. 9: Xe($5p_{3/2,1/2}$) photoemission spectra of Xe adsorbed on a vicinal (stepped) Ru(001) surface. Xe(S) = Xe atoms at step sites: Xe(T) = Xe atoms at terrace sites

with that of the available terrace and step sites (as deduced from LEED) as well as the sequence of site population (higher coordinated step sites first) lead to the assignment of Xe(S) arising from Xe atoms at step sites and Xe(T) originating from Xe atoms at terrace sites /28/. Indeed, the Xe(S) state is absent on a perfectly flat Ru(001) surface. According to equ. (4) the higher $E_{B,S}^F$ value suggests the "local work function" at the step sites to be lower by ~ 700 meV than on the terraces. This finding is in perfect agreement with predictions based on the so-called Smoluchowski-electron-smoothing effect /29/. On Pt(111) the local work function at step sites is lower by even 1 eV compared to the (111) terraces. It is this huge change in local surface potential, and, hence, in local charge distribution, which may be the cause of the strongly enhanced catalytic activity of step (defect) sites, and not the one which is deduced from averaging work function techniques.

This example of Xe photoemission from a stepped Ru(001) surface demonstrates the capability of PAX to provide the local surface potential at step (defect) sites as well as the number of surface defects. The detection limit is better than 1% step defects per cm^2. A summary of all kinds of surface roughness effects that have been studied with PAX is contained in Ref. 30.

4.3 Bimetallic Surfaces

PAX can also provide quantitative information about the concentration and lateral distribution of heteroatoms on a metal surface, as well as the electrostatic surface potential in their immediate vicinity.

Figure 10 shows a series of Xe($5p_{3/2,1/2}$) spectra from a perfect Ru(001) surface, which prior to Xe adsorption, was covered with ~0.4 monolayers of copper by vapor deposition and well annealed at 520 K. Up to an exposure of 3 L Xe the spectra exhibit two $5p_{1/2}$ states at 6.75 eV and ~7.1 eV, respectively. The dominant peak A at 6,75 eV is close to the position as on clean Ru, and is therefore assigned to Xe atoms on bare Ru patches; Xe(Ru). Above 3 L Xe exposure a new $5p_{1/2}$ peak B emerges at ~ 7.4 eV; its $5p_{3/2}$ counterpart grows between the $5p_{3/2}$ and $5p_{1/2}$ signals of the Xe(Ru) state. Signal B is very close to the position as on pure Cu(111) and is therefore assigned to Xe atoms on the deposited Cu; Xe(Cu). The fact, that these Cu sites are populated with Xe after the Ru sites is in agreement with the lower Xe adsorption energy on Cu compared to Ru /31/. The Xe(Cu) $5p_{3/3}$ signal (between the Xe(Ru) peaks) is clearly split into two peaks. It is generally accepted, that the origin of this splitting is the formation of a two-dimensional (2D) electronic band structure within the adsorbed Xe /31,32/, which in the present case is only conceivable, if the Cu deposit forms flat islands so that the Xe atoms on top can also form a densely packed overlayer. Hence, the structure of the Xe(Cu) spectrum itself, namely the $5p_{3/2}$ splitting, conveys the unambiguous information that submonolayers of Cu on Ru(001) form islands, which should be monoatomically thick, because the Cu-Ru interaction is stronger than the Cu-Cu interaction /33/. This together with the fact, that Cu and Ru are not miscible, and that therefore Cu forms 2D islands on the Ru surface leads to the assignment, that the peak C in Fig. 10 corresponds to Xe atoms at the (mixed) Cu/Ru step sites along the Cu island boundaries; Xe(Cu/Ru) (see insert of Fig. 10). As has been demonstrated earlier with the similar Ag/Ru system /30/ the partial intensities of all three Xe states, Xe(Ru), Xe(Cu) and Xe(Cu/Ru), are a quantitative measure of the relative abundance of these three kinds of surface sites, and the local surface potential at the mixed Cu/Ru boundary sites is intermediate be-

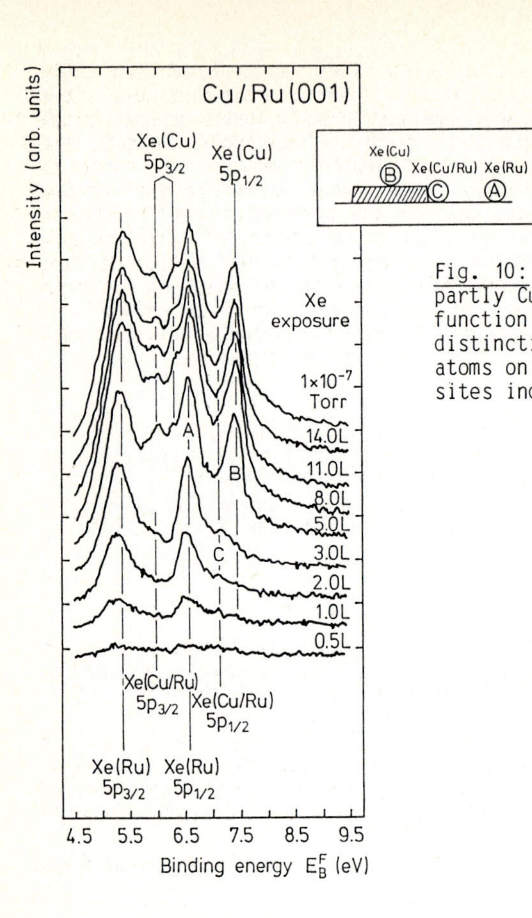

Fig. 10: Xe($5p_{3/2,1/2}$) spectra from a partly Cu-covered Ru(001) surface as a function of Xe coverage. Note the clear distinction of the emission peaks from Xe atoms on the three possible adsorption sites indicated in the insert

tween those on Ru and Cu patches, respectively. The assignment of peak C being due to Xe atoms at mixed Cu/Ru sites, by the way, is strongly supported by PAX spectra from a Cu covered Ru(001) surface, which prior to Cu deposition was slightly sputtered. In this case, formation of 2D Cu islands is prevented due to the high concentration of sputter-induced defect sites which act as heterogeneous nucleation centers /34/.

This Cu/Ru example shows that PAX cannot only distinguish qualitatively between the three possible surface sites for Xe adsorption, namely Ru, Cu and Cu/Ru boundary sites, but also provides a quantitative measure of their relative surface concentration, their local surface potential by virtue of their $5p_{1/2}$ electron binding energies using equ. (4) as well as the distribution of the Cu deposit, namely 2D island formation on the flat Ru(001) substrate versus nearly atomic dispersion on a sputtered Ru substrate /34/. Even more spectacular PAX results could be obtained from a Ru(001) surface covered with 0.05 monolayers of potassium. In this case the electrostatic potential around individual K^+-ions on the surface could be determined /35/ in remarkable agreement with theoretical predictions /36,37/.

5. SUMMARY

Both initial state and final state (loss) effects in the photoelectron spectroscopy from atoms in solid surfaces provide information not only about their chemical nature but also about their geometrical environment. This is particularly facilitated using synchrotron radiation in conjunction with angular resolved detectors and surface core level shift measurements. Quite another possibility is to probe variations of the electrostatic surface potential on an atomic scale by means of a "test-electron" which is brought very close to the surface. This can be achieved by Photoemission of Adsorbed Xenon (PAX). Combinations of these techniques are increasingly applied to heterogeneous surfaces as demonstrated in the present article.

6. LITERATURE

1 W. Telieps, E. Bauer: Surf. Sci. 162, 163 (1985); see also: Ultramicroscopy 17, 51+57 (1985)
2 Consult the journal Ultramicroscopy for ultra-high resolution electron microscopy work
3 H. Bethge, Th. Krajewski, O. Lichtenberg: Ultramicroscopy 17, 21 (1985)
4 G. Binnig, H. Rohrer: Surf. Sci. 126, 236 (1983), and G. Binnig, H. Rohrer: Surf. Sci. 152/153, 17 (1985)
5 R. J. Behm, W. Hösler: In Physics and Chemistry of Solid Surfaces VI, ed. by R. Vanselow and R. Howe (Springer, Berlin, Heidelberg, 1986)
6 K. Wandelt: Surf. Sci. Rep. 2, 1 (1984)
7 K. Wandelt: J. Vac. Sci. Technol. A2, 802 (1984)
8 K. Wandelt, J. Hulse: J. Chem. Phys. 80, 1340 (1984)
9 R. J. Behm, C. R. Brundle, K. Wandelt: J. Chem. Phys. (1986) in press
10 E. W. Müller, T. T. Tsong: Field Ion Microscopy (Elsevier, New York, 1969)
11 D. Rieger, R. D. Schnell, W. Steinmann, V. Saile: Nucl. Instr. Meth. 208, 777 (1983)
12 G. Ertl, J. Küppers: Low Energy Electrons and Surface Chemistry (Verlag Chemie, Weinheim, 1985)
13 F. F. Abraham, C. R. Brundle: J. Vac. Sci. Technol. 18, 506 (1981)
14 K. Wandelt, C. R. Brundle: Phys. Rev. Lett. 46, 1529 (1981)
15 A. D. van Langeveld, J. W. Niemantsverdriet: Surf. Sci. (Proc. ECOSS 8, 1986) in press
16 N. G. Stoffel, M. K. Kelly, G. Margaritondo: Phys. Rev. B27, 6561 (1983)
17 N. G. Stoffel, M. Turowski, G. Margaritondo: Phys. Rev. B30, 3294 (1984)
18 D. J. Hnatowich, J. Hudis, M. L. Perlmann, R. C. Ragaini: J. Appl. Phys. 42, 4883 (1971)
19 D. Spanjaard, C. Guillot, M.-C. Desjonquères, G. Tréglia, J. Lecante: Surf. Sci. Rep. 5, 1 81985)
20 K. Wandelt, R. Miranda, D. Rieger, R. D. Schnell: BESSY-Jahrbuch 1984
21 J. F. van der Veen, F. J. Himpsel, D. E. Eastman: Phys. Rev. Lett. 44, 189 (1980)
22 M. G. Mason: Phys. Rev. B27, 748 (1983) and references therein
23 P. H. Citrin, G. K. Wertheim: Phys. Rev. B27, 3176 (1983)
24 J. Hulse, J. Küppers, K. Wandelt, G. Ertl: Appl. Surf. Sci. 6, 453 (1980)
25 N. D. Lang, A. R. Williams: Phys. Rev. B25, 2940 (1982)
26 K. Wandelt, W. Jacob, N. Memmel, V. Dose: Phys. Rev. Lett. (1986) submitted

27 H. Wagner: Springer Tracts in Modern Physics, Vol. 85 (Springer, Berlin, Heidelberg, 1979)
28 K. Wandelt, J. Hulse, J. Küppers: Surf. Sci. 104, 212 (1981)
29 R. Smoluchowski: Phys. Rev. 60, 811 (1941)
30 A. Jablonski, S. Eder, K. Wandelt: Appl. Surf. Sci. 22/23, 309 (1985)
31 S. Eder, K. Markert, A. Jablonski, K. Wandelt: Ber. Bunsenges. Phys. Chem. 90, 225 (1986)
32 M. Scheffler, K. Horn, A. M. Bradshaw, K. Kambe: Surf. Sci. 80, 69 (1979)
33 K. Christmann, G. Ertl, H. Shimizu: Thin Solid Films 57, 247 (1979) and J. Catalysis 61, 397 (1980)
34 K. Markert, S. Eder, K. S. Kim, K. Wandelt: Surf. Sci. to be published
35 K. Markert, K. Wandelt: Surf. Sci. 159, 24 (1985)
36 J. K. Nørskov, S. Holloway, N. D. Lang: Surf. Sci. 137, 65 (1984)
37 N. D. Lang, S. Holloway, J. K. Nørskov: Surf. Sci. 150, 24 (1985)
38 K. Watanabe, M. Hashiba, T. Yamashina: Surf. Sci. 61, 483 (1976)

XPS and AES Characterization of Cu-Cr Oxide Catalysts

G.R. Castro[1], *P. Tacconi*[1], *S. Yunes*[2], *F. Severino*[3], *and J. Laine*[3]

[1]Escuela de Física, U.C.V. Apdo. 21201, Caracas 1020-A, Venezuela
[2]Catálisis Aplicada, INTEVEP S.A. Apdo. 76343, Caracas 1070-A, Venezuela
[3]Centro de Química, I.V.I.C. Apdo. 21827, Caracas 1020-A, Venezuela

1. INTRODUCTION

Copper-chromium mixed oxides are among the best noble metal substitutes for catalyzing carbon monoxide oxidation, with possible application for automobile emission control. Previous works on oxidation of CO with various base metal oxides /1-3/ have reported that copper chromite, $CuCr_2O_4$, is the most active catalyst, as compared with the single oxides of the first-transition metal series.

Based on catalytic activity and reduction measurements /4/, copper in copper chromite is held to be the main active species, and chromium a promoter that inhibits excessive catalyst reduction. However, prereducing copper chromite was reported to enhance catalytic activity /4/, but there was not a clear interpretation of this effect. A more comprehensive understanding of these results should be obtained by means of surface analysis techniques, such as AES and XPS. These have been employed in the present work to examine CuO and $CuCr_2O_4$ oxides in order to complement previous activity and reduction studies.

2. EXPERIMENTAL

CuO and $CuCr_2O_4$ samples were the same employed in the previous report /4/. Spectra were taken on fresh and CO prereduced samples. Prereduction was carried out employing a flow of CO, at atmospheric pressure and 300°C for 3 hours.

XPS spectra were measured in a Leybold-Heraeus apparatus, equipped with a pretreatment chamber, using X-ray excitation from Mg Kα (1253.6 eV) and the carbon 1s signal as reference. Samples were placed under Ultra High Vacuum (10^{-6} torr) at 120°C overnight before analyses. These were carried out at 10^{-8} torr.

Auger spectra were obtained in an all-metal system (Varian) equipped with a mass spectrometer. A study in the glancing gun mode was carried out employing a primary current of 5 μA and a beam energy of 3 keV. The base pressure was 10^{-10} torr. The detector was a Cylindrical Mirror Analyzer. In order to get quantitative results, experimental sensitivity factors for Cu and Cr were previously obtained from CuO and Cr_2O_3 samples. Reproducibility was checked by repeating spectra from different samples.

Catalyst activity for CO oxidation was measured as described previously /4/. More experimental details are given below.

3. RESULTS AND DISCUSSION

3.1 CuO

XPS spectra in the Cu 2p region of CuO samples are shown in Fig. 1. The fresh sample shows typical Cu^{2+} peaks /5/, but the CO pretreated samples has the Cu 2p $_{3/2}$ signal displaced from 935.7 to 932.0 eV, as well as missing satellite peaks. This indicates that CO pretreatment has extensively reduced the surfac although it is not possible to differentiate between Cu^0 and Cu^+ by XPS /5/. Prereducing CuO under those conditions has also been found to lead to catalys deactivation /4/. Therefore, oxidized copper rather than reduced copper shoul be ascribed as the active species on CuO surface.

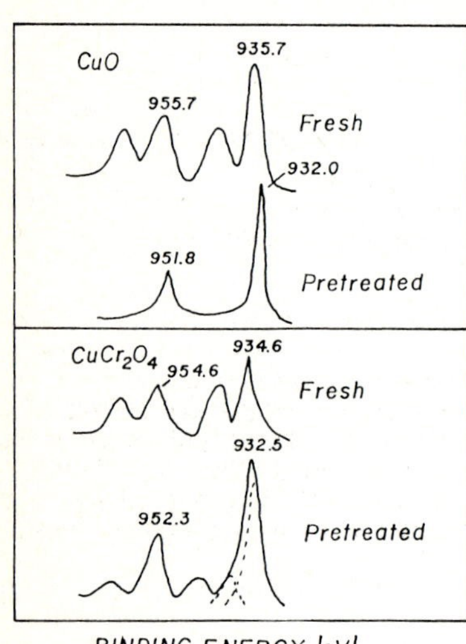

Figure 1. XPS spectra of CuO and $CuCr_2O_4$ in the Cu 2p region

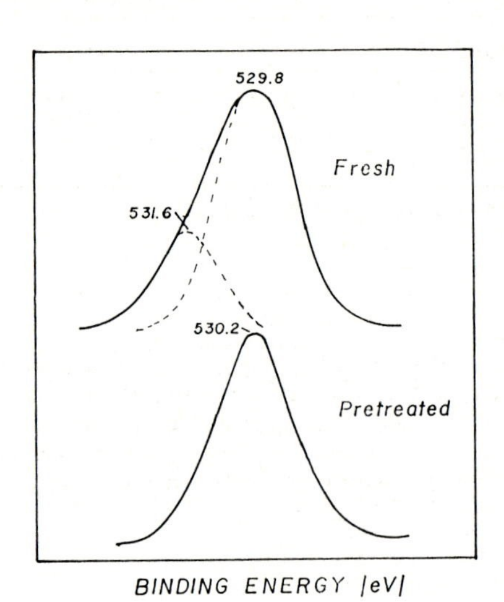

Figure 2. XPS spectra of $CuCr_2O_4$ in the O 1s region

AES of CuO (not shown) showed a surface contamination with carbon, which w eliminated by heating in situ. During thermal treatment the analysis chamber atmosphere was analyzed by mass spectrometry. Heating was interrupted at chos times and Auger spectra were recorded. Remarkably, CO and CO_2 evolution could be observed as long as surface carbon could be detected, but after the carbon AES signal disappeared there was increasing gaseous oxygen and no more carbon oxides evolution. O_2 evolution persisted for a long time, without a noticeabl change of the Cu and O AES peaks. These results suggest that oxygen may be diffusing out from the bulk, and that it may be reacting with the carbon deposits. This diffusion phenomenon is probably in connection with the loss of electrons in the sample due to the Auger effect, thus making the solid behave as a p-type semiconductor.

3.2 $CuCr_2O_4$

In contrast with CuO, XPS shows that there is only a partial reduction of copper in $CuCr_2O_4$ due to CO prereduction (Fig. 1). However, the 2p signals of Cr (not shown) were not affected, indicating that CO prereduction of copper chromite mainly affects copper. Two types of oxygen could also be detected (Fig. 2), one of which disappeared after treatment with CO. It can be suggested that an oxygen species associated to copper is removed by CO pretreatment, though chromium prevents excessive reduction of copper.

As in the case of CuO, the $CuCr_2O_4$ samples were contaminated with carbon, and this was removed by heating. Calculations from AES for fresh $CuCr_2O_4$ samples, showed that there is an atomic ratio Cu/Cr smaller than the stoichiometric one. However, an increment in this ratio after CO reduction is found and, remarkably, this increase is almost identical to the increase in catalytic activity attained by the catalyst after CO prereduction (Table 1).

Table 1. Effects of pretreatment with CO in $CuCr_2O_4$

Variable	Calcined	Pretreated	Increase factor
Activity*	0.38	0.95	2.5
χ_{Cu}**	0.11	0.27	2.4

*Activity, as CO conversion at 200°C and atmospheric pressure

**χ_{Cu}, atomic ratio = Cu/(Cu + Cr)

These results suggest that activation of $CuCr_2O_4$ is intimately related to surface copper enrichment produced by CO prereduction. This might be accounted for by chemical segregation, similar to that observed in alloys /6/. Alternatively, it could be also explained by selective take-up of oxygen from surface cupric ions by CO, because these ions tend to adsorb oxygen in excess, that oxygen being easily eliminated by CO /7,8/.

4. LITERATURE

1. M.Shelef, K. Otto, H. Gandhi: J. Catal. 12, 361 (1968)
2. Y.F. Yu Yao, S.T. Kummer: J. Catal. 46, 388 (1977)
3. J.T. Kummer: Advan. Chem. Ser. 143, 178 (1975)
4. F. Severino, J. Laine: Ind. Eng. Chem. Prod. Res. Dev. 22, 396 (1983)
5. C.D. Wagner, W.M. Riggs, L.E. Davis, J.F. Mourder, G.E. Muillenberg (eds.): Handbook of X-Ray Photoelectron Spectroscopy (Perkin-Elmer Corporation, Minnesota, U.S.A.)
6. R. Bouwman, L.H. Toneman, A.A. Holscher: Surf. Sci. 35, 8 (1973)
7. J. Gundermann, C. Wagner: Z. Phys. Chem. B37, 157 (1937)
8. K.P. De Jong, J.W. Geus, J. Joziasse: J. Catal. 65, 437 (1980)

AES Study of Hydrodesulfurization Catalysts

J.L. Brito[1], *G.R. Castro*[2], *F. Severino*[1], *and J. Laine*[1]

[1]Centro de Química, I.V.I.C. Apdo. 21827, Caracas 1020-A, Venezuela
[2]Escuela de Física, U.C.V. Apdo. 21201, Caracas 1020-A, Venezuela

1. INTRODUCTION

Auger Electron Spectroscopy (AES) has been scarcely used in studies of real catalysts, specially in the case of oxides and sulfides, in spite of the fact that it is a powerful technique for qualitative and quantitative analysis of surfaces. It may also provide information on the chemical state of the elements being determined, but its interpretation is not as straightforward as in XPS; e.g., most chemical information in AES concerns low-energy peaks, where line overlapping in the case of multicomponent systems might obscure interpretation. Nevertheless, its high surface sensitivity and good lateral resolution make AES specially suitable for studying catalysts.

In the present work, AES is used to characterize the surface composition of supported hydrodesulfurization (HDS) catalysts.

2. EXPERIMENTAL

Three catalysts were prepared using the "dry" impregnation method, as described elsewhere /1/. Their compositions were: AlMo (10.4 wt% MoO_3); AlCoMo (10.4 % MoO_3 + 3.55 % CoO); and AlNiMo (10.4 % MoO_3 + 3.55 % NiO), the remainder up to 100% being γ-Al_2O_3. Samples of each catalyst were calcined in air at 500 and 600°C. Composition and firing temperature are reported below as, e.g. AlMo-500, AlNiMo-600, etc. HDS activity experiments were also described in the previous work /1/.

AES spectra were measured by means of a Varian UHV system (base pressure 10^{-10} torr), equipped with a Cylindrical Mirror Analyzer. A special holder allowed to introduce several samples simultaneously and heating them under UHV. Powdered samples were poured in small stainless steel buckets and pressed into flat pellets under 5 Tons. Then they were introduced into the UHV chamber (oxidic samples), or were pretreated ex situ in a flow reactor before introducing them in the AES system. Two types of pretreatment were performed: heating at 400°C for 2 hours under H_2 (reduced samples) or under H_2S (sulfided samples). Afterwards, the samples were purged in He and cooled to room temperature. Contact with air could not be prevented, but it was minimized.

The effects of thermal and chemical treatments on the heights of Al and Mo peaks were studied, in order to get information about the chemical state of the elements. Peaks chosen were the ones at 51, 1378, 186 and 221 eV (Al^1, Al^2, Mo^1 and Mo^2). The remaining elements were studied using their most intense peak in the 100-2000 eV range. For quantitative analysis, sensitivity factors were estimated by standard methods /2/, using pure compounds.

3. RESULTS AND DISCUSSION

3.1 Qualitative Results

Auger peaks of Al, Mo, C, N, O, Co and Ni were observed for the oxidic and reduced samples. In addition, S was detected in sulfided samples. C and N are contaminants which might arise from the materials used in the preparation of catalysts (e.g., N from nitrates or ammonia present in the impregnated solutions; C from organic impurities adsorbed and decomposed on the surfaces). From another AES study on Cu-Cr catalysts /3/, it was anticipated that C contamination from the UHV system should be negligible, as it was observed that an initially high C content was essentially reduced to zero after UHV heating. For the present catalysts, a similar heat treatment produced a drastic increment of surface C and N. The shape of the C peak suggests a graphitic phase. A small signal of Fe (contaminant from the commercial support) was detected in the AlMo-600 sample, for which the absence of Ni and Co and the high calcination temperature may favour Fe migration to the surface.

3.2 Effect of Carbon in the Ratio Al^1/Al^2

The ratio of the heights of those peaks varied significantly with thermal treatment under UHV, as can be seen in Table 1, which also shows the amount of C in each catalyst. The variation in Al^1/Al^2 seems to be independent of the chemical state of the solids. In general, however, an increase in the amount of C is followed by a decrease of the ratio of Al peaks. This may be accounted by a shielding effect of carbon (probably as a monolayer), which affects more the lower energy electrons than the high energy ones.

Table 1. Ratio Al^1/Al^2 and concentration of C of different catalysts

Catalyst	R_o	C_o	R_{oT}	C_{oT}	R_r	C_r	R_{rT}	C_{rT}	R_s	C_s	R_{sT}	C_{sT}
AlMo-500	7.5	2.4	4.1	26.7	6.8	7.0	6.4	16.6	10.5	5.8	9.6	4.1
AlMo-600	9.7	1.3	3.6	25.1	5.7	5.8	5.4	15.8	9.8	0.9	8.4	6.9
AlCoMo-500	7.5	1.5	4.7	25.0	8.3	4.9	5.6	16.3	8.3	0.3	6.8	0.7
AlCoMo-600	8.2	0.8	6.7	15.5	10.5	18.2	5.2	13.9	8.8	0.1	8.3	6.6
AlNiMo-500	9.1	0.9	5.0	20.5	5.9	4.3	7.9	10.6	7.9	0.5	8.3	2.9
AlNiMo-600	9.1	1.5	4.1	24.8	6.0	6.6	5.2	18.9	7.8	0.5	8.4	7.6

Symbols - R: ratio Al^1/Al^2; C: carbon amount (atomic %); o: oxidic sample; r: reduced sample; s: sulfided sample; T: UHV heated

3.3 Effect of the Chemical State of the Catalysts on the Ratio Mo^1/Mo^2

The ratio between the heights of Mo peaks varied with the state of the samples and their thermal history, as can be seen in Table 2. A similar behaviour has been reported in the case of MoO_3 single crystals submitted to UHV heating or other reducing conditions /4/, but no clear trends have been published nor any satisfactory explanation has been offered. Our observations in pellets of polycrystalline, unsupported compounds suggest that the variation of the oxidation number of Mo from +6 to +4 accounts for the diminution of the Mo^1/Mo^2 ratio. The values of this ratio for pure samples are: Mo (VI) (MoO_3) $\simeq$ 1.9; Mo (IV) (MoS_2) $\simeq$ 1.4. Literature data /2/ on Mo (0) (metal) give a ratio of 1.1. There is a clear diminution of the ratio of Mo peaks with decreasing oxidation of the element, which might arise from differences in electronic structure of Mo in each oxidation state.

Table 2. Ratio Mo^1/Mo^2 for different catalysts

Catalyst	ox	ox + T	red	red + T	sulf	sulf + T
AlMo-500	1.8	2.0	2.2	2.0	1.2	1.2
AlMo-600	2.0	2.3	2.6	1.9	1.5	1.6
AlCoMo-500	2.0	2.2	2.2	2.1	1.5	1.3
AlCoMo-600	1.6	1.6	2.3	2.1	1.8	1.6
AlNiMo-500	2.0	2.0	2.2	2.1	1.6	1.7
AlNiMo-600	2.0	2.4	2.2	2.3	1.6	1.7

Symbols - ox: oxidic, red: reduced, sulf: sulfided samples; T: UHV heated

Comparing the values of oxidic and sulfided catalysts (Table 2), a trend similar to bulk compounds is observed. The results for reduced samples suggest that air exposure re-oxidized the surface of these samples, in agreement with the very high sensitivity of the reduced state of molybdena catalysts to oxygen exposition /5/. Sulfided samples also adsorb O_2 at room temperature /6/ but AES suggests that this does not change the oxidation state of Mo, which also agrees with the observation that the sulfided state is easily regenerate after air exposure by treatment under vacuo /6/.

3.4 Relationship Between Surface Composition and HDS Activity

Surface composition of several catalysts (as atomic %) is shown in Table 3 (for C contents, see Table 1). As uncertainties arise due to the shielding effect of C, the different solids are compared by using the atomic ratios of elements which are also shown in Table 3. These data were obtained using the Al^2 and Mo^1 AES peaks, which were more reproducible than Al^1 and Mo^2.

Table 3. Surface composition of catalysts

Catalyst	Chemical state	Atomic % *					Atomic Ratio *		
		Al	S	Mo	O	Pr	Mo/Al	Mo/S	Pr/Mo
AlMo-500	oxidic	42.0	-	3.7	51.9	-	0.09	-	-
AlMo-600	oxidic	39.8	-	4.2	54.7	-	0.11	-	-
AlMo-500	sulfided	32.3	5.6	3.1	53.2	-	0.10	0.55	-
AlMo-600	sulfided	31.8	10.5	3.0	53.8	-	0.09	0.29	-
AlCoMo-500	oxidic	42.3	-	3.4	52.2	0.6	0.08	-	0.18
AlCoMo-600	oxidic	44.5	-	4.7	49.2	0.8	0.11	-	0.17
AlCoMo-500	sulfided	35.5	6.2	1.7	55.9	0.5	0.05	0.29	0.29
AlCoMo-600	sulfided	32.2	8.1	3.3	55.7	0.7	0.10	0.43	0.21
AlNiMo-500	oxidic	41.9	-	3.8	53.0	0.5	0.09	-	0.13
AlNiMo-600	oxidic	38.0	-	3.4	56.5	0.7	0.09	-	0.21
AlNiMo-500	sulfided	35.9	6.8	1.8	54.5	0.5	0.05	0.27	0.28
AlNiMo-600	sulfided	32.8	8.9	2.3	54.9	0.6	0.07	0.27	0.26

* Promoter (Pr = Co or Ni)

Mo and Al concentrations correspond to a molar ratio Al_2O_3:MoO_3 of about 5. Assuming that a monolayer of Mo oxide covers the support surface (which should be a good assumption, see e.g. MASSOTH /5/), this might indicate an excessive depth of analysis (more than five layers). It can be inferred that Mo does not cover all the external surface, in agreement with a report by EDMONDS and MITCHELL /7/, who showed that firing of dry, uncalcined catalysts produced migration of Mo compounds from the external surface to the internal, porous structure of the catalysts.

The amount of promoters (Ni or Co) is also very low, even with regard to Mo concentration; the calculated ratio Pr/Mo should be about 0.68. The results in Table 3 show that a considerable amount of promoter is undetectable by AES, which is in line with the known trend of these type of metals to enter the alumina lattice, thus forming sub-surface spinel-like phases /5/. In Fig. 1-a, it may be seen that, except for the AlNiMo-600 sulfided catalyst (open circle), the ratio Pr/Mo shows acceptable correlations with HDS activity, for both sulfided and oxidic (not pretreated) states. This agrees with recently proposed models of HDS catalysts, which ascribe the activity to Pr-Mo-S sites /8/ or to Pr sites highly dispersed in MoS_2 /9/.

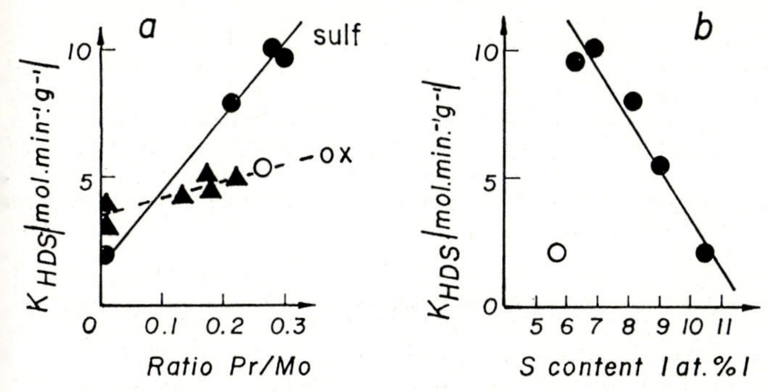

Figure 1. Relationship between HDS activity and surface composition.

Regarding S content, it is shown in Fig. 1-b that an inverse correlation between HDS activity and S content exists for sulfided samples (only AlMo-500 does not fit at all, open circle). HDS activity is frequently assigned to anionic (sulfur) vacancies /5/. Thus, a more defective structure with the same amount of active metals, but less sulfur, should have better catalytic activity.

In conclusion, it has been shown that AES could give chemical information about the state of Mo in HDS catalysts. Also, good correlations between surface composition, as measured by AES, and catalytic activity have been found.

4. LITERATURE

1. F. Severino, J. Laine, C. Cáceres, J.L.G. Fierro, A. López Agudo: To be published
2. G.E. McGuire: Auger Electron Spectroscopy Reference Manual (Plenum, New York, 1979)
3. G.R. Castro, P. Tacconi, S. Yunes, F. Severino, J. Laine: This Congress
4. L.C. Dufour, O. Bertrand, N. Floquet: Surf. Sci. 147, 396 (1984)
5. F.E. Massoth: Advan. Catal. 27, 266 (1978)
6. J. Bachelier, J.C. Duchet, D. Cornet: J. Catal. 87, 283 (1984)
7. T. Edmonds, P.C.H. Mitchell: J. Catal. 64, 491 (1980)
8. H. Topsøe, B.S. Clausen: Catal. Rev.-Sci. Eng. 26, 395 (1984)
9. J.P.R. Vissers: Ph.D. Thesis. Eindhoven University (1985)

Mössbauer Study of Small Iron Particles of the Iron Phases in a Clay Catalyst Employed in the Hydrotreatment of Heavy Oils*

H. Constant[1], *F. González-Jiménez*[2], *R. Iraldi*[2], *and M. Rosa-Brussin*[3]

[1]Departamento de Física Aplicada, Facultad de Ingeniería,
[2]Departamento de Física, [3]Departamento de Química, Facultad de Ciencia
Universidad Central de Venezuela, A.P. 21201, Caracas 1020-A, Venezuela

1. INTRODUCTION

Mössbauer spectroscopy has been widely employed in the characterization of the electronic states of iron in natural clays /1/, as well as in clay-catalysts used in hydrodesulfurization (HDS) and hydrodemetallization (HDM) processes of heavy oils /2/.

One of the specific contributions of Mössbauer spectroscopy i these studies, compared, for example, with X ray diffraction, is that very small particles containing the Mössbauer isotope can be identified, when the other techniques are no longer useful. In the scope of the Mössbauer effect with ^{57}Fe, when a magnetically ordered compound is studied in large particles (e.g. for iron compounds diameter $\geq$ 400 A) a typical Mössbauer six line Zeeman spectrum can be observed, allowing its identific tion. For the same compound studied in very small particles, the magnetic array should fluctuate, due to thermal excitation, at higher frequencies than the characteristic one of the magneti hyperfine interaction, leading to a typical paramagnetic spectru (1 line or 2 lines if a quadrupole splitting is present). This behaviour is called superparamagnetism. In this case, if the temperature is lowered, the fluctuation frequencies should be slowed down enough to observe again the characteristic magnetic spectrum of the compound. Another feature to be pointed out is that these small particles have a large number of superficial atoms, the conventional transmission Mössbauer technique is then useful for studying iron electronic states in the surface.

In a previous work, we reported the existence of a relationship between the HDM catalytic activity and the appearance and abundance of pyrrhotites ($Fe_{1-x}S$) in the clays employed in the HDS-HDM of heavy oils /3/. The same behavior was observed employ ing a system modelling the real process, by sulfiding previously the clay catalyst until pyrrhotites appear and later making a demetallization of a vanadyl porphyrin (TFP-VO) at atmospheric pressure.

In the present work we try to clarify possible mechanisms of formation of the pyrrhotites by decomposing the HDM process

* Work partially supported by CDCH (Consejo de Desarrollo Científico y Humanístico de la Universidad Central de Venezuela).

on heavy crudes in a simplified model process separated in different steps: Reducing the catalyst by H_2, sulfurating the catalyst and demetallizing the oils.

2. EXPERIMENTAL PROCEDURE

A natural clay catalyst, previously treated with H_2SO_4 and later calcined, was submitted to a treatment with H_2 at 20 psi pressure, following the experimental procedure: I) the sample is heated at 425C for two hours II) the catalytic bed is impregnated with gasoil at T=150C III) step 1 is repeated for 1 hour IV) step 2 is repeated. After this treatment the clay was used for the demetallization of vanadyl-tetraphenyl porphyrin (TFP-VO) diluted in gasoil, in three different ways: 1) Employing nitrogen in the process 2) Employing hydrogen in the process 3) Before the demetallization process, as in 2, the catalyst was sulfurated with CS_2 (2%) diluted in gasoil (T=400C, pressure H_2=20 psi.).

Afterwards the catalysts were studied by Mössbauer spectroscopy. The spectra were recorded at room temperature in a spectrometer running in the triangular symmetric mode. The source was ^{57}Co in Pd. The Mössbauer results were fitted supposing lorentzian shaped absorption lines. The adjusted parameters were IS (Isomer Shift), GA (HWHM), H1 (sum of the depths of all the lines of each subspectrum), QS (Quadrupole nuclear splitting) and finally CH (magnetic hyperfine field). From the areas of each subspectrum, the percentage abundance of each iron compound was determined (supposing that the Mössbauer f factors are equal for all the compounds).

3. ANALYSIS OF THE RESULTS

The Mössbauer spectra recorded and hyperfine parameters used to fit these spectra are shown in fig. 1 and table I. The results obtained for these spectra present the following common features:

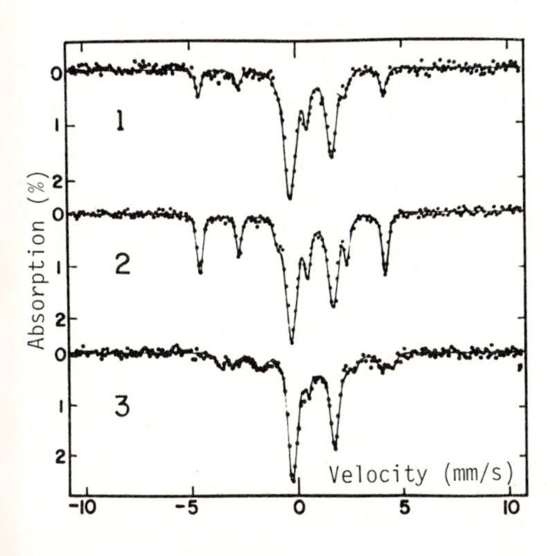

Fig. 1 Mössbauer spectra of the clays used in the three different processes described in the text.

Table I. Hyperfine parameters found to fit Mössbauer spectra of Fig. 1. Isomer shifts are given with respect to α-F

Nº	GA	DI/mm/s/	QS/mm/s/	CH/KG/	%
1	0.20	-0.16			6.6
	0.30	1.10	2.42		56.4
	0.20	0.54	0.69		16.7
	0.20	0.01	-0.02	330	20.3
2	0.20	-0.13			3.5
	0.30	1.08	2.43		47.9
	0.20	0.49	0.67		10.9
	0.20	0.01	0.00	331	37.7
3	0.30	-0.12			3.5
	0.30	1.13	2.44		58.8
	0.22	0.49	0.67		12.5
	0.20	0.71	0.13	311	6.8
	0.20	0.71	-0.09	300	7.0
	0.20	0.71	0.13	266	11.4

a) An Fe^{2+} doublet characterized by an electric quadrupole interaction QS=2.44 mm/s and an isomer shift IS=1.12 mm/s, these values are in agreement with those obtained for Fe^{2+} in the crystalline parts of the clay in the octahedric positions cis and trans, of the clays /4/. The linewidth of this spectrum indicates that it is a superposition of spectra due to the contributions of the two indicated sites, that have similar parameters.

b) An Fe^{3+} doublet with DI=0.54 mm/s and QS=0.69 mm/s which should correspond to iron in oxides in the superparamagnetic states or in the structure of the clay.

In the spectra of the non-sulfurated catalysts (1 and 2), contributions attributable to iron in the metallic (Fe°) are present in two forms:

i) The characteristic magnetic spectrum of metallic iron α-phase with particle diameters greater or equal to 350 A.

ii) A single line with IS=-0.15 mm/s corresponding to superpa magnetic particles with size lower than 350 A /5/.

In the spectrum of the sulfurated catalyst the magnetic spectrum of metallic iron was no longer present, however the following could be observed:

i) Three magnetic subspectra with parameters corresponding to those of iron sulfides, actually pyrrhotites, $Fe_{1-x}S$, coming from the sulfuration of iron.

ii) A single line attributable to small superparamagnetic particles with IS=-0.12 mm/s.

4. DISCUSSION

The fact that in samples 1 and 2, magnetic and superparamagnetic spectra of iron are simultaneously present, is evidence for the presence of particles of iron with sizes between 15 and 350 A/5/. These particles should arise from two different processes: The reduction of the small particles of iron oxides which were observed in the spectrum of the calcined clay/2/, and, observing that metallic iron Fe^{0} has a volume eight times greater than those of Fe^{2+} and Fe^{3+}, the eclosion of iron coming from the clay.

In the sulfuration process the pyrrhotites are formed from the metallic iron. This is an indication that in the real hydroprocessing of heavy oils with these clays, metallic iron should be an intermediate phase in the formation of the pyrrhotites observed in it/3/.

5. REFERENCES

1. S. Mϕrup, J. Frank, J. van Wonterghem, H. Roy-Poulsen and L. Larsen. FUEL, 64, 527 (1981).
2. M. Rosa-Brussin, H. Constant, R. Iraldi, E. Jaimes and F. González-Jiménez. Acta Científica Venezolana. 35, Nº 5-6, 332 (1984).
3. F. González-Jiménez, H.Constant, R. Iraldi, E. Jaimes and M. Rosa-Brussin. HYPERFINE INTERACTION. 28, 927-930 (1986).
4. P.A. Montano in Mossbauer spectroscopy and its chemical applications, Ed. J.G. Stevens and G.K. Shenoy. Advances in Chemistry Series 194, ACS, 1981.
5. J.A. Dumesic, H. Topsϕe, S. Khammounma and J. Boudart. J. CATAL. 37, 503 (1975).

Effect of Grain Boundary Structure on Segregation

J. Reyes-Gasga and L.D. Romeu

Instituto de Física, Universidad Nacional Autónoma de México,
A.P. 20-364, 01000 México, D.F. México

The purpose of this paper is to show that grain boundary structure does influence the amount of solute segregated to it. We try to establish a relation between segregation and boundary type as defined by orientation relationship and boundary geometry in an attempt to identify the "speciality" criterion for grain boundaries.

1. Introduction

The solute concentration in grain boundaries is generally different from that of the crystal. This phenomenon is called Grain Boundary Segregation, and has a conspicuous influence on many metal properties. For instance, it can augment the intergranular fracture tendency (1).

The thermodynamic treatment of segregation has been developed by many authors both for binary and multicomponent systems (2,3,4). Nevertheless, no theory is yet capable of explaining how boundary properties control the amount of segregation, nor how these properties are modified by the solute content.

Since solute atoms deform the matrix lattice, they will interact with the strain fields near a grain boundary. In accordance with existing grain boundary models (5), certain grain boundaries are "special" in the sense that they produce small strain fields, and therefore have a small free energy.

The best known example of these are the coincidence boundaries (6), which are supposed to be characterized by a very good atomic fit. Yet, when the orientation relationship has a small deviation from perfect coincidence, a strain field will develop which may be periodic and will interact with solute atoms nearby. Therefore, special boundaries are not expected to be susceptible to impurity segregation because this will increase their elastic energy. It has been shown that coincidence boundaries are indeed reticent to impurity segregation, but they are not the only ones (7,8). This suggests that there are other boundaries besides coincidence which are also special (7,8). In general, for almost every physical property, there is a family of boundaries which behave in a "special" manner. Although most of these "special" boundaries are generally coincidence boundaries, they are not the only ones and at present there exists no grain boundary model capable of explaining what distinguishes these "special" boundaries from others.

2. Experiment

Samples of Cu-0.5% Sb alloy were used in the present study. This alloy has been experimentally shown to exhibit segregation of Sb to the grain boundaries (9). Bulk samples were annealed in high vaccum at 1173°K for 1 hour in order to homogenize the sample. The temperature was then decreased to 773°K to allow Sb to segregate to grain boundaries without forming large precipitates. At this temperature, the diffusion coefficient of Sb in Cu is larger than that of Cu self-diffusion (10). The samples were then quenched in liquid nitrogen (77°K) and thin foils prepared for observation under the electron analytical microscope.

The analysis consisted of the following steps: a) microanalysis at several points on a line perpendicular to the grain boundary plane. This measures the antimony concentration as a function of the distance from the boundary. The separation between consecutive test points was one tenth of a micron with an error 0.05 microns; b) microanalysis at different points along the boundary to measure concentration fluctuations; c) selected area diffraction patterns (SADIFF) from the grains enclosing the boundary in order to determine the orientation relationship and d) obtaining micrographs of the boundaries under study.

3. Results and discussion

40 boundaries were studied and classified in accordance with the coincidence model (see table 1) totaling 6 coincidence boundaries (C), 30 non-coincidence boundaries (NC) and 4 low-angle boundaries (LAB). The coincidence, low-angle and 23 non-coincidence boundaries showed no Sb.

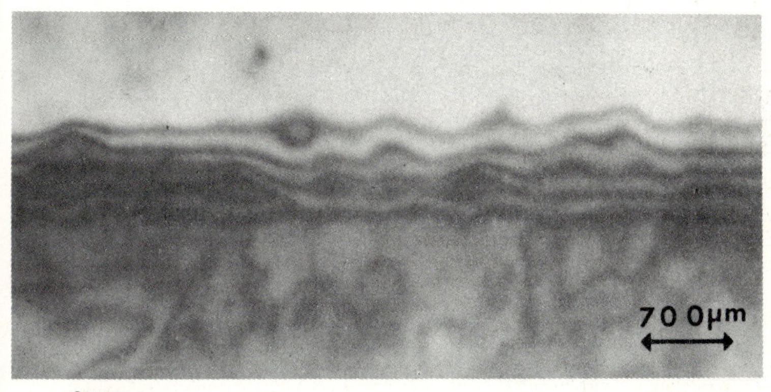

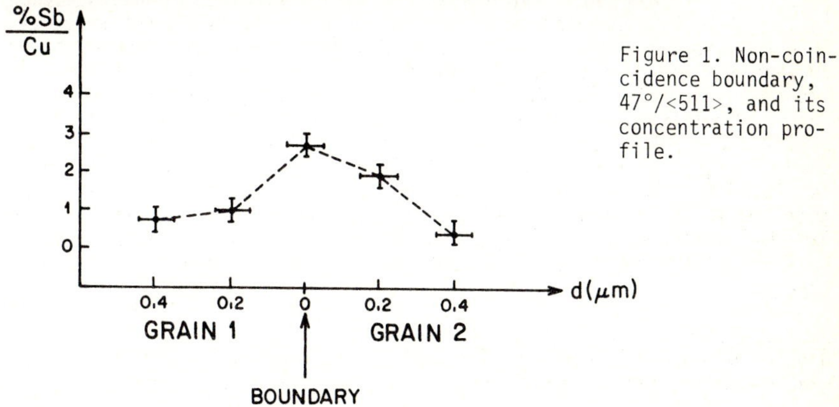

Figure 1. Non-coincidence boundary, 47°/<511>, and its concentration profile.

Table 1

Type group	I	II	III
C	0	3	3
NC	8	9	13
LAB		4	
Total	8	16	16

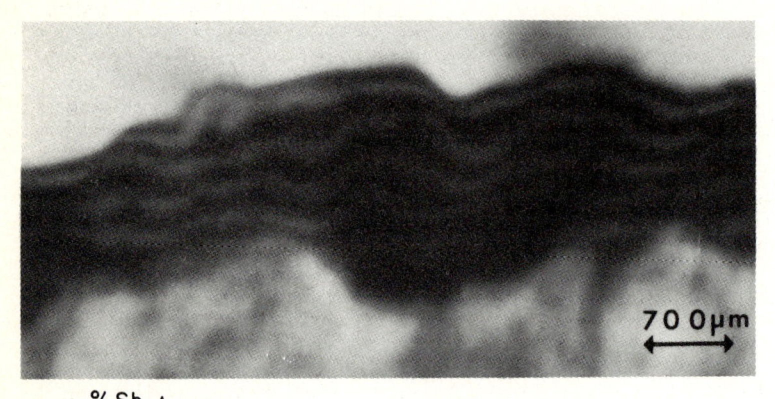

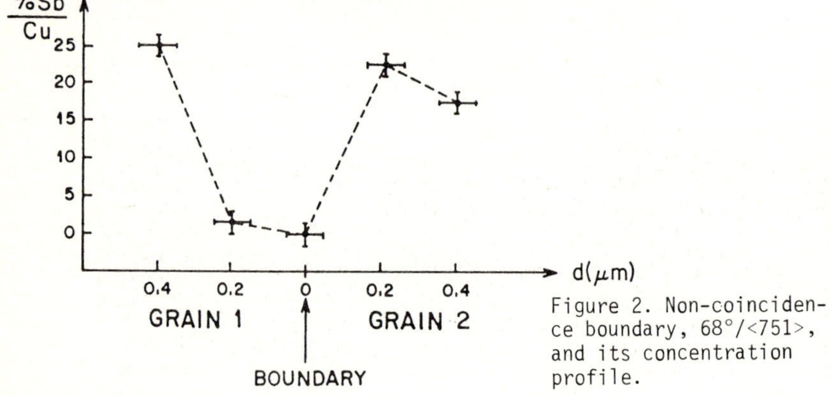

Figure 2. Non-coincidence boundary, 68°/<751>, and its concentration profile.

The boundaries were classified in accordance with their Sb content as follows: Group I: higher Sb concentration in the grain boundary than the in the grains (fig. 1). Group II: same antimony concentration in the boundary as in the grains. Group III: higher Sb concentration in the grains than in the grain boundary (fig. 2).

Coincidence grain boundaries had no impurity content. This result is in agreement with the coincidence model. However, our results show that also some non-coincidence boundaries contain no antimony. This result has no explanation within the coincidence model.

The microanalysis showed that the antimony concentration along grain boundaries does not change appreciably. Yet, when the grain boundary was faceted, the amount of Sb in some facets was different from others. This indicates that the orientation relationship is not a sufficient criterion to determine the boundary type. According to our results, the boundary plane orientation is important for impurity segregation and thus for the boundary structure and should be considered in the definition of a boundary.

Grain boundary migration was seen to be related with the solute concentration in the boundary. Micrographs showed that those boundaries which migrated, as evidenced by gross profile irregularities, had no antimony (fig. 2). On the other hand, those boundaries which showed a measurable Sb content did not migrate appreciably (fig. 1).

Finally, some boundaries showed what we call antisegregation, which is evidenced by Sb depletion in the boundary (boundaries in group III, fig. 2). This effect was present in 16 of the 40 boundaries studied: 3 coincidence and 13 non-coincidence boundaries. No evident relation among these boundaries was found. Apparently, these boundaries produce a repulsive strain field for the Sb atoms. This is an open field for future research.

4. Conclusions

1. The quantity of Sb segregated to grain boundaries depends both on their orientation relationship and on the orientation of the boundary plane itself.

2. There is not a clear distinction between boundaries which show segregation and those which do not. Although coincidence boundaries show no segregation, the converse is not true.

3. Grain boundary migration is limited by segregation. Migration is higher in the boundaries without Sb.

4. Some grain boundaries showed "antisegregation" where the solute concentration in the boundary is less than in the adjacent grains.

5. References

1. E. D. Hondros, M. Seah: Inst. Met. Rev. Review 222 (1977).
2. W. Losch: Scripta Met. 13,661 (1979).
3. M. P. Seah: Acta Met. 28,955 (1980).
4. C. L. Briant, R. P. Messmer: Phil. Mag. B42,569 (1980).
5. G. A. Chadwick, D. A. Smith: Grain Boundary Structure and Properties, Materials Science and Technology (Academic Press., 1976).
6. M. L. Kronberg, F. H. Wilson: Trans. AIME. 185,501 (1949).
7. H. Gleiter, G. Hermann, G. Baro: Acta Met. 24,353 (1976).
8. H. Gleiter, H. Sautter, G. Baro: Acta Met. 25,467 (1977).
9. M. C. Inman, D. McLean, H. R. Tipler: Proc. Roy. Soc. A,538 (1963).
10. R. E. Hoffman, D. Turnbull, E. W. Hart: Acta Met. 3,417 (1955).

Surface Science and Analysis for Materials Technology Within IBM

C.R. Brundle

IBM Almaden Research Center, 650 Harry Road, San Jose, CA 95120, USA

Abstract: An overview of surface and interface analysis within IBM is presented. The types of materials concerned; the areas in which surface analysis is having major impact; and the relative importance of different analytical techniques are discussed, together with perspectives for the future. Some recent examples of surface analysis studies, for both basic science reasons and applied technology reasons, are discussed. They are taken from IBM Research Division work and are chosen to illustrate the range of materials involved.

1. INTRODUCTION

IBM employs a large number of people who would classify themselves as "Surface Scientists" throughout its Research Division, development laboratories, and manufacturing divisions. The range of materials on which they work is enormous. It is often assumed by the academic world that most of the work is on semiconductors. This is a misconception, probably arising from the fact that the semiconductor part of the industry was the first to make effective use of modern surface analytical techniques starting some 15 to 20 years ago. In fact, the fraction of "Surface Science" personnel engaged primarily on semiconductor technology related work is probably no more than a third, the other two major areas being Packaging and Interconnect Technology and Storage Technology, both of which involve work on a wide variety of materials. Of course, part of the work done on both semiconductor and other material surfaces in the Research Division is done for basic research reasons, unconnected in any direct way with technology other than that the class of materials being studied may be chosen because of their technological relevance. Again, even in this basic research area certainly no more this 50% of the work is on semiconductors.

In this review, I give a summary of the types of materials being worked on in the different technological areas. The different ways in which surface science/analysis can impact the technological areas are then briefly reviewed and an assessment of the relative importance of the major available surface analytical tools is given. A perspective of how this may change and develop in the future is then given.

The second section of this review gives some selected examples of the types of surface analysis work done, all from the Research Division. The selection is intended to illustrate the range of materials and techniques involved and to show the connections to IBM technology. Readers should be aware that the work in the Research Division represents only a small fraction of the total work being done in surfaces, interfaces, and films. Each IBM manufacturing division has its own analytical and development laboratories.

2. MATERIALS

The different technological areas involved can be divided into three major groups. There are other smaller areas, such as electrophotography, for instance, but for this review, I cover only the major groupings.

2.1 Semiconductor Technology

This area relates primarily to logic and memory requirements and is the "classical" area of high technology within the computer area with which the outside world is well-acquainted. It is the world of semiconductor devices comprising integrated circuit chips, chip modules, and semiconductor memories. The vast majority of this area still involves primarily Si as the semiconductor and SiO_2 as the dielectric. In addition to the obvious analytical area of Si and SiO_2 surfaces and Si/SiO_2 interfaces, the processing and construction of semiconductor devices requires knowledge of such areas as Si and SiO_2/metal interfaces (metals used for electrical connections and diffusion barriers), Si and SiO_2/polymer interfaces (polymer films used in lithography to produce integrated circuitry and sometimes as a dielectric), and the low concentration doping of Si thin layers with such elements as P, As, and B.

2.2 Packaging

The area of packaging really includes everything other than the semiconductor devices themselves in the finished "packages" that are connected by conventional cables within the computer. The most readily recognized area to the outsider would be a circuitboard on which individual semiconductor devices (diodes, transistors), chips, or chip modules (packages of chips) are placed for interconnection. It also includes: (a) the technology of making chip modules, that is, packaging together many chips in a module which attaches to the board; (b) providing personalized connection distribution "boards" on which chips are mounted (first-level packaging), which then allows attachment to a main circuitboard (second-level packaging). The basic high technology in all these things relates to the art of getting an enormous number of interconnections and insulations made reliably (that is, no shorts, no opens, no violations of electrical specifications) in the minimum of space. It has become a high technology area because of the increasing miniaturization and complexity of the semiconductor devices themselves; that is, to take advantage (speed, size, materials cost) of these devices requires an equal sophistication in how they are connected to each other. As an example, consider a typical computer circuitboard 15 years ago compared to now. Fifteen years ago, such a board consisted of a piece of thick epoxy insulation with a few Cu connection lines mounted on each side. Transistors were soldered to this, and connections between boards were made by cables. Today, a board may be 2 ft. by 2 ft. in size and have 30 individual layers (total thickness ca. one quarter inch), comprising insulation, circuitry, and power cores with interconnections between layers made by mechanical and laser drilling plus a variety of methods for Cu plating the drilled holes. The Cu circuitry may have linewidths and heights of only 50μ and there may be a total of a third of a km. of wiring in each layer. The IBM 3081 main frame computer requires only four boards of this type as its total complement for all the chip modules used. Without these boards, an enormous number of connecting cables would have to be used. One can see, then, that though the lithographic dimensions of the packaging area ($\sim 50\mu$) have not decreased to those used in the semiconductor devices themselves ($\sim 1\mu$), the technological requirements have become quite sophisticated. In addition, the range of materials and processes involved is much wider than for the semiconductor technologies. Interfaces between metals and metals, metals and epoxies, metals or metal oxides and polymers, epoxy and polymers, polymers and polymer, etc., are involved. The processing steps for production of a modern circuitboard may include solid/gas interactions (dry etching), solid/liquid (etching, cleaning corrosions protection, adhesion promotion, plating) and solid/solid interfaces (lamination). A large amount of organic, polymer chemistry and electrochemistry is involved.

2.3 Storage Technology

The third major area of technology is storage, that is reading and writing from and to magnetic tapes or disks or optical storage media. The storage business is equally important to the semiconductor logic and memory side with conventional magnetic storage making up the major part. Over the years, storage densities on disks have gone up and up but the *same basic magnetic oxide particulate media and inductive soft magnetic heads* still predominate. The improvements have come about by a continued push to the engineering limits, for instance, closer and closer flying heights of head to disk which require more carefully controlled treatment of the disk surfaces and better control of the ambient environment in a disk file. This has lead to a greater demand for more basic and detailed understanding of the surfaces and interfaces involved. In the future, thin film metal alloy magnetic storage and optical storage materials may become more important. This introduces the possibility of a whole range of complicated alloy materials, layered structures, protective overcoats, etc. The study of the surfaces and interfaces of these materials having special magnetic or optical properties will become more common as they find their way into the storage industry.

2.4 General Comments

Surfaces and, therefore, surface analysis are generally important during processing whereas interfacial properties, and, therefore, analysis, becomes more important after production when one is concerned with the long-term integrity of the product.

Some surface analytical techniques are very important even when it is not really a surface property that is at stake. For instance, the doping depth profile concentration of thin Si films can only be effectively measured by SIMS depth profiling when the films become too thin for electrical methods. SIMS then becomes a product quality control technique.

In many cases, fine lateral resolution is, of course, important as well as depth resolution. For semiconductor technology and packaging technology, I quoted figures of 1μ and 50μ, respectively, earlier. In magnetic recording head technology, similar dimensional requirements are common. In many cases, however, there are calls for surface analysis where no particular lateral resolution is required.

In terms of the areas where surface and interface analysis is having an impact currently on the manufacture of the computer materials discussed above, we can consider three situations.

2.4.1 Intrinsic Needs for Surface Analysis. This would include situations where the dimensions involved are so small that one requires the surface/interface analysis technique to monitor the tolerances of the manufactured product. The case of using SIMS to check dopant concentrations in Si films mentioned above fits this category. The use of SEM to monitor 1μ lithography dimensions is another. It is generally in the semiconductor device area where sophistication is so great that these intrinsic cases are found regularly today.

2.4.2 Reliability Considerations. This means time greater than zero instabilities of surfaces or interfaces which may affect reliability. These could be mechanical (strain), chemical (reactivity, diffusion, electrochemistry), or tribological (e.g., the effect of flying a head over a disk almost in contact). These types of concerns are found throughout all the technology areas, not just semiconductor technology.

2.4.3 Yield. Manufacturing a product often involves many steps. For example, the large circuitboard example discussed earlier involves over 200. Though it is possible to produce product by having only functional tests as controls at each step, the functional control often only tells when there is a problem, not what the reason is for the problem. Keeping the overall yield up for the manufacture of this board has, therefore, required extensive surface and interface analysis at many of the individual steps in order to identify the nature of the problem and therefore rectify it.

3. RELATIVE IMPORTANCE OF DIFFERENT ANALYTICAL TECHNIQUES

In this brief survey, I address the current situation with some comments at the end for future perspectives

3.1 Most Common Surface Instrument

The most common surface instrument is undoubtedly the SEM, with the number within IBM in the US probably exceeding a hundred. Most dedicated SEM's are used for topological information only, with relatively little EDX analysis being done. SEM attachments are sometimes added to other surface analytical instruments to help locate the spatial region to be studied by the particular analytical technique in question.

3.2 Most General Purpose Technique

X-ray photoelectron spectroscopy (XPS or ESCA) is the technique which has the widest range of uses. It is used for the surface analysis of virtually all materials. It fares particularly well against electron beam techniques for organics and other sensitive materials and for insulating surfaces because of its lower propensity for surface damage and surface charging. Its big disadvantage, compared to electron beam techniques, is its lack of good spatial resolution. The best that commercial instrumentation can do at present is around 150μ spot size. If this were to be improved to 1μ, orders for twice the number of instruments that already exist in IBM would probably be generated in a very short time. The reason would be its chemical state analysis capability, which is more often exploited than its quantitative aspects, even though XPS is generally capable of better quantification than AES, for instance. It is a technique used by a wide range of people from organic chemists through to electrical engineers and for subjects ranging from the chemical identification of organic contaminates to interdiffusion at metal-metal interfaces.

3.3 Techniques Most Used For Small Area Analysis

The electron beam microprobe and Auger spectroscopy just about tie here, though they are used for very different purposes. Wavelength Dispersive X-ray Fluorescence (WDX), as provided in a modern commercial e-microprobe, can give elemental analysis using spot sizes down to ~100Å and in the less quantitatively accurate energy dispersive mode (EDX), as used in SEM's, down to ~30Å. The technique is not particularly surface sensitive, however, with signal originating from up to several μ depth. This means that the general usefulness of the technique decreases when working with films that are much thinner than this. Also, the electron beam spot size is not at all representative of the spatial extent from which the X-ray signal comes owing to the spreading out of the primary electron beam from inelastic collision within the penetration depth. In addition, the e-microprobe has no possibility for any chemical state information. AES can, in principle, operate at a spatial resolution of down to about 250Å in commercially available instruments. It is, of course, sensitive to only the top few monolayers and can provide chemical information from chemical shifts. In practice, however, it is almost always used in a semiquantitative elemental analysis mode and *never* at 250Å spatial resolution. The practically useful ranges of spatial resolutions are in the 1,000Å-several μ range. Though the electron beam can be made much smaller, the amount of current required in that beam to get a usable signal often causes complete havoc and also the sample mounting stages are not sufficiently stable (thermal drift) to allow data collection over the long time needed for decent statistics at 250Å resolution. In addition, most practical surfaces are quite rough on this scale which seriously degrades spatial resolution at the high-resolution end. The advantage of having an AES capable of a 250Å e-beam spot size is in its use in the SEM mode, where the much lower current requirement than for AES means that pictures can be taken quickly. In this way, very small features can be identified and the

AES done with the beam on the feature even if the spatial resolution is much worse than the feature size. If we cannot identify the topological feature in the first place, it is very had to get an Auger Spectrum of it!

3.4 Most Accurate Analysis Techniques

The most used quantitative technique, albeit over a very restricted range of materials, is SIMS. It is used routinely and extensively in the dynamic mode to depth profile dopants (B,P,As) in Si over a wide dynamic range to absolute detection limits that are orders of magnitude below the detections limits of the other techniques. Quantification, the problem with much of the other uses of SIMS, is overcome by calibration standards using ion implantation where total concentrations and implant distributions can be accurately predicted. The thicknesses of films profiled in this manner are usually in the 1000'sÅ range. SIMS usage for quantitative analysis other than dopants is much more limited at present and there is, as yet, little use of static SIMS for chemical information of practical surfaces. Owing to its great absolute sensitivity, SIMS is used for the detection of trace elements in a variety of situations, but quantification without standards is always a problem. Surprisingly, the imaging mode of SIMS is, as yet, not exploited extensively.

Another technique capable of accurate quantitative analysis is Rutherford Back-Scattering (RBS). This is a technique not usually associated with surface analysis, and indeed one usually uses it to provide elemental profiles over the few hundred Å to 10μ range with a depth resolutions of only around 100Å. The great advantage of RBS is that the backscattering cross sections are accurately known and so an elemental analysis can be made accurately quantitative in a straight-forward manner *without standards*. A disadvantage is its limited elemental resolution and lack of any chemical state analysis capabilities. The billiard-ball momentum transfer process of the technique results in poor ability to separate elements of similar masses for the higher mass elements. In addition, RBS normally has no spatial resolution capabilities, though specialized microbeam systems do exist. Despite these two drawbacks, RBS is extensively used for quantitative thin film analysis of both metal and semiconductor films over the thickness ranges mentioned above. In the selected examples section, examples will be given of how the technique can be made to have a far greater surface sensitivity and to yield much more than just quantitative elemental analysis.

The third technique that can provide very accurate elemental analysis is Wavelength Dispersive X-ray Fluorescence (WDX) using the e-microprobe, though, of course, for thin films (μ's) it is a bulk technique for all practical purposes. In additions, it is not effective for elements of $Z<11$ owing to low X-ray yields and the requirements for thin window or windowless X-ray detectors.

XPS and AES are, of course, also used for quantitative analysis but are really not capable of the same accuracy as the above techniques. They are used where good depth resolution is required but, like SIMS, can generally only depth profile by utilizing destructive sputtering. An exception to this is by using variable angle XPS (or AES). Here, non-destructive profiling over the top ~20Å is possible. An example is given later.

4. FUTURE PERSPECTIVES FOR SURFACE ANALYSIS

In general, it seems that over the foreseeable future all areas of surface and interface analysis will become more important for materials technology in the computer industry owing to the increasing sophistication and miniaturization of the materials, particularly in the non-semiconductor areas. Some guides as to which techniques will become more important, and for what, can be given.

Chemical state analysis will continue to gain importance leading to increased demand for XPS with the additional requirements of better spatial resolution. Vibrational spectroscopy in the form of FT IR, surface Raman spectroscopy, and High Resolution Electron Energy Loss (HREELS) is gaining in importance in providing functional group identification for surface and thin films of polymers and organics. There is an increasing demand for complete molecular identification of polymers and large organics, not just functional group analysis. In addition, one would like to be able to do this with good spatial resolution. Only a mass spectroscopy based technique, such as SIMS, currently has any possibility of this. Recent work using TOF mass analysis and pulsed sources has detected masses up to 9,000 amu and shown that under the right conditions large parent molecular ions can be produced by the SIMS process with relatively little fragmentation [1].

Improved methods of small area analysis will be important. Analytical SEMS with combinations of EDX, WDX, and AES attachments are beginning to be more common instead of just being used for topography information. Though small-spot XPS is currently only capable of $\sim 150\mu$ resolution, all the major XPS instrument companies are now working in this area and improvements are to be expected. One company [2] has a prototype instrument that uses electron imaging in very strong magnetic fields (7 tesla) as a means of photoelectron microscopy which is capable, for lower photoelectron energies, of a few μ resolution. The focussed high brightness and pulsed properties of synchrotron radiation present possibilities for photoelectron microscopy also.

The advent of UHV electron microscopes has narrowed the discipline gap between microscopy and surface analysis. Microstructural information on the few 10's Å scale can be complemented in the modern STEM by EDX and transmission core level EELS. The latter contains not only elemental identification but also chemical information from core level shifts (like XPS). The spatial extent of the analysis depends on the sample thickness because of the inelastic collision spreading of the beam. For 100 KeV instruments, samples should be only a few hundred Å thick for optimum results, and at this thickness, 50Å spatial resolution is possible. UHV STEMs with sample preparation and treatment chambers and Auger attachment are also beginning to appear [3]. This leads to the possibility of detailed microstructural and surface and bulk elemental and chemical analysis within one microscope.

At this point, it is appropriate to mention that X-ray diffraction has always been a work-horse for providing structural information on thin film. The more routine laboratory diffraction arrangements are capable of providing structural information on metal and oxide films of down to a few 100Å thickness, though are more typically used for μ films. They have no depth resolution capability, however. Therefore, for instance, for an oxide film which had a different phase in the top 20Å, normal XRD would not detect the phase. For ultra-thin films (<100Å) and for the sort of problem addressed above, one needs to go to *grazing incidence* XRD, either using Seeman-Bohlin geometries in the laboratory, or by using synchrotron radiation. In the latter case, structural information on *monolayer* thicknesses for metals, and 20Å thickness for light elements (*e.g.*, SiO_2 or Si) are achievable.

Owing to the movement to thinner and thinner films, or multilayers of films, the e-microprobe is becoming less suitable for quantitative very thin-film analysis. RBS, on the other hand, is finding more and more uses particularly when a range of ion beam energies and high-resolution electrostatic energy analysis are used [4,5]. This allows trade-off between depth resolution and maximum depths sampled. Also, the materials being studied by RBS are changing. Traditionally, RBS has been used for metals and semiconductors but is now finding use for polymer and polymer-metal interfaces and for hydrogen analysis in the Recoil Forward Scattering (RFS) mode [6].

One aspect common to all the different techniques which must be further developed and utilized is the use of computer control for data taking and particularly for data analysis. Too often data is taken and *not* analyzed because of the time and expertise requirement. Sophisticated software including "expert systems" for the spectroscopy, plus image processing for the microscopy are required.

5. SELECTED EXAMPLES

Four examples are given as illustrations of the range of materials and surface/interface analysis involved.

5.1 Oxidation of Permalloy Surfaces

Ni/Fe alloy in about a 80/20 ratio (permalloy) is used extensively in magnetic recording head technology within IBM and other companies. The thickness of the permalloy films used to be in the μ and greater range. As the technology advanced, films became thinner. Films in the 1000'sÅ range are common now and eventually films as thin as a few hundred Å will become important. The thinner the films, the more significant surface corrosion becomes to performance, both from the magnetic property point of view and from the greatly increased mechanical tolerance requirements. Two types of corrosion must be considered: corrosion due to environmental conditions in the disk file and corrosion during the manufacture itself. Since couples between different metals are present in a real head, wet electrochemical activity is likely to be the dominant problem in the often aggressive manufacturing steps. The environment of a working head can at least be controlled and, though electrochemistry may still play a role through humidity, the long term gaseous environment now becomes important in controlling corrosion. Here, we consider two related studies at opposite ends of the basic/applied spectrum: the initial chemisorption of oxygen and oxidation of a Ni(75)/Fe(25) (100) single-crystal surface and the corrosion behavior of 1000Å permalloy films in an accelerated corrosion testing chamber.

In the single crystal study, the O(1s), Fe($2p_{3/2}$), and Ni($2p_{3/2}$) XPS signals were monitored as a function of oxygen exposure under UHV conditions [7]. The O(1s): clean surface Ni($2p_{3/2}$) ratios were used to establish the oxygen coverage via the known photoionization cross section. In addition, the measurements were taken as a function of electron emission angle to vary the surface sensitivity and therefore provide a very near-surface depth profiling. For example, for the Ni($2p_{3/2}$) signal, the electron mean free path length is such that at an emission angle of 45° 90% of the intensity originates from about the top 13Å of the crystal, whereas at 15° the figure is about 4Å. In addition to these quantitative elemental aspects, one can use the chemical shift and shake-up satellite details of the spectra to distinguish O_2 from O_{ads} from OH_{ads}, Ni° from Ni^{II}, and Fe° from Fe^{II} from Fe^{III}. Of course, to do this reliably requires a bank of XPS data on Ni and Fe oxides and hydroxides. This "standards" approach to identifying chemical species in XPS is the normal and most effective way the technique is used.

An example, some of the results are given in Fig. 1 and Table I. From the figure, it can be seen that at room temperature within the top 4Å (the 13° angle data) the oxygen oxidizes the Fe in preference to the Ni, that Fe^{III} is formed preferentially, and that the preferential oxidation leads to significant segregation of Fe to the surface (see Table I for quantification). Heating an approximately 3 monolayer oxide surface to 300°C emphasizes all the above effects (Fig. 1) leading to a surface layer with no Ni^{II} in it and massive amounts of Fe^{III} (Table I). This demonstrates that the situation achieved at room temperature was kinetically limited by the ability of Fe to diffuse to the surface through the thickening oxide layer.

In the practical corrosion study on permalloy films, which was performed in the same laboratory by a different group [8], the knowledge obtained from the single crystal XPS work on the surface segregation and oxidation states of the products was used as analytical input to the study. In this work, the rate of corrosion of the film was estimated from the weight gain

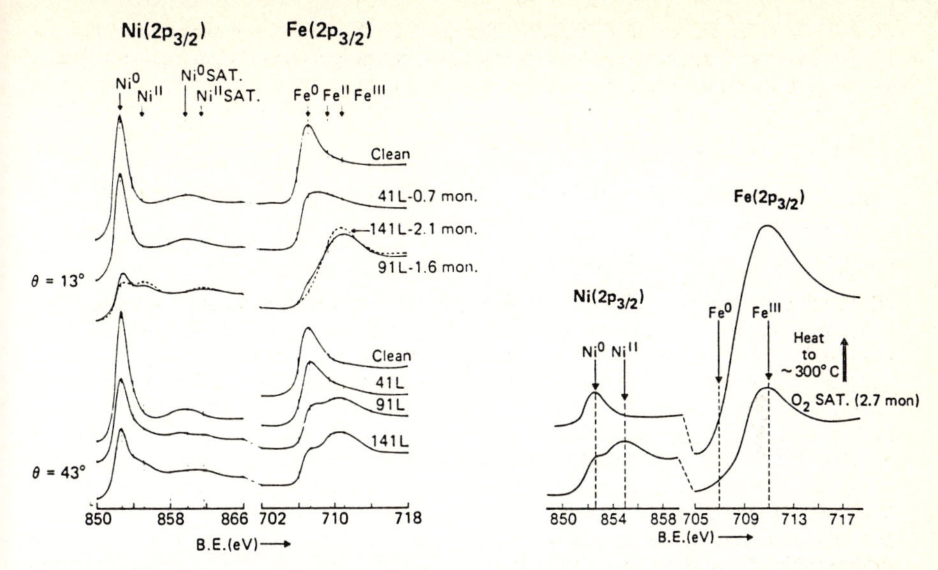

Fig. 1. (Left) $Ni2p_{3/2}$ and $Fe2p_{3/2}$ XPS spectra for oxygen adsorption on Ni(75)/Fe(25) (100) at 300K [7]. Exposures are given in Langmuir (L) and oxygen atom coverage, as determined from the O(1s) intensities, in monolayers. Data from two emission angles, 13° and 43°, are given. The dotted vertical lines indicate the expected B.E.'s for the metal and various oxide states, as determined from standards. Sat. refers to shake-up satellite positions. (Right) Spectra for saturation oxygen coverage at 300K heated to 300°C for a few minutes [7].

Table I. XPS Determined Surface Composition of Ni_{75} Fe_{25} (100) during Oxygen Adsorption and Oxidation [7]

Oxygen Atom Coverage (monolayer)	% Fe(a) In Top 4Å	% Fe(a) In Top 13Å
0	28	25
1.0	32	25
1.7	32	25
1.7	40	28
2.7	46	35
2.7(b)	85	56

(a) Determined from the $Ni2p_{3/2}/Fe2p_{3/2}$ ratios at two different angles. The % age given assumes uniform distributions through the sampling depths.
(b) The 2.7 monolayer case for adsorption at 300K was heated to 600K for a few minutes.

as a function of time in the accelerated corrosion testing chamber, which is designed to have an acceleration factor of about 100 over a "typical" environmental condition. The results for a 1000Å film with three different pretreatments are shown in Figure 2. Clearly, the corrosion resistance of the film is very dependent on the way the film is deposited and pre-treated with deliberate oxidation/passivation of the film deposited at elevated temperature being most effective. AES depth profiling of the three as-prepared films revealed that the passivation process resulted in a fairly thick (200Å) layer of iron oxide being formed at the surface with

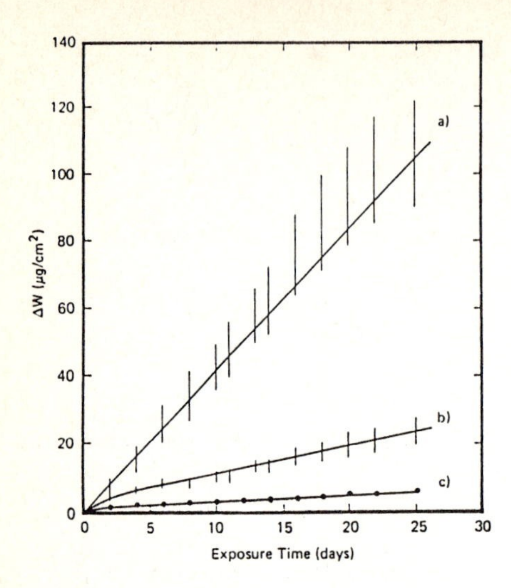

Fig. 2. Plot of weight gain of 1000Å permalloy films versus exposure time in an accelerated corrosion testing chamber. Three different types of initial films were monitored; (a) deposited on a substrate at room temperature; (b) deposited at 200°C; and (c) deposited at 200°C and subsequently heated in air at 160°C for 125 hours [8].

nickel being completely absent; thus, the process is an extension of what was found under UHV conditions on the single crystal surface. Detailed XPS spectra confirmed that the iron present is in the Fe^{III} state and showed that under these non-UHV conditions about a quarter of the oxygen was formed as hydroxide rather than oxide. Finally, optical microscopy and SEM showed that in the protective layers the oxide film was smoother and had fewer defects and lower porosity than room-temperature oxidized films. These are presumably the characteristics necessary for the oxide film to inhibit the formation of corrosion pits.

Oxidation and corrosion studies are of importance in IBM over a wide variety of materials. The Ni/Fe case was picked as an example simply because of familiarity. Other examples which could equally well have been given include niobium oxidation (Josephson Junction Barriers), Co/Cr thin film oxidation (thin films storage disks), Cu/A1 oxidation (Cu/A1 alloys are widely used in electrical connection metallurgy), and Sn/Pb oxidation (solder pads extensively used in joining chips to boards).

5.2 Si Technology; Metal Silicide Formation

When certain metals, such as Pt, Pd, Ni, are deposited on Si and annealed, reactions take place to form a new surface phase, a metal silicide. A range of stoichiometries and thicknesses can often be formed and the products can have a variety of electrical properties which make them candidates for thin film device materials. Much UPS, XPS, and Auger work has been done on these systems, but here I want to discuss some RBS work which demonstrates the tremendous power of the technique for very thin films rather than for films of the conventional thicknesses studied by RBS. This work [4,5] was not performed in IBM Laboratories, but one of the co-authors is from IBM Research and similar other studies have been done.

In conventional RBS, MeV energy He^+ ions are backscattered by collisions with the nuclei of atoms in the sample material (Fig. 3a). They undergo a billiard-ball momentum transfer and so the mass, M_2, of the sample atom can be determined from the ratio of the energy of the backscattered ion mass, M_1, to its impinging energy, E_1/E_0. The high energy edge of the backscattered He^+ energy, Fig. 3b, therefore identifies the mass, M_2. This edge is actually for those scattering events at the surface. Most of the He^+ ions penetrate the

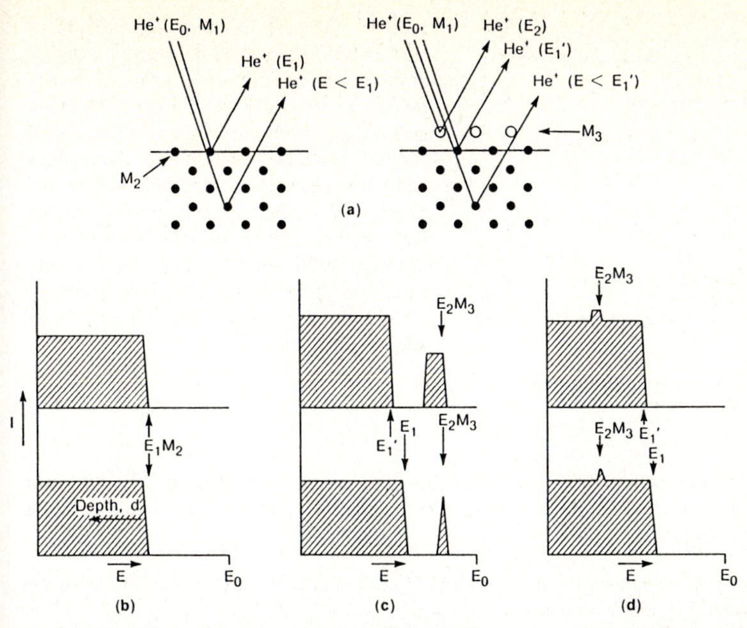

Fig. 3: (a) Schematic representation of the RBS process. Solid circles represent substrate atoms, M_2; open circles represent adsorbate atoms, M_3. (b) Energy distributions for He^+ ions backscattered from a clean, thick, substrate. (c) Energy distribution for He^+ ions backscattered from a thick substrate, M_2, with a heavier atom, M_3, overlayer that is either thick compared to the energy resolving capability of the detector (upper trace) or thin compared to the resolving capability of the detector (lower trace). (d) As (c) except M_3 is a lighter atom than M_2.

surface without backscattering collision and lose energy in small angle scattering events where energy is exchanged with the electron cloud of the atoms. Eventually at some depth, a collision with an M_2 nucleus occurs and the He^+ is backscattered out with the large energy loss characteristic of M_2. It has, of course, lower energy than those ions scattered at the surface which did not have energy loss by electronic collisions during depth penetration. The distance from the high-energy edge in the RBS spectrum, Fig. 3b, therefore, is a measure of depth below the surface. The beauty of RBS is that all the scattering cross sections are accurately known so that thicknesses and quantities of atoms present can be accurately determined. A sample consisting of a layer of M_3 on M_2, where $M_3 > M_2$, would give an RBS spectrum as in Fig. 3c, upper trace, whereas if $M_3 < M_2$ it would look like Fig. 3d, upper curve. If the overlayer were thin compared to the energy resolving capability (and therefore the depth-resolving capability) of the solid-state detector usually used in RBS, then the spectra would look like Fig. 3c and 3d, lower curves. For MeV energy ions and solid-state detectors the lower curve spectra correspond to overlayer thicknesses of around 50 to a few hundred Å, depending on the material and, in particular, the incidence and backscattering angles used. Grazing angle emission enhances the surface sensitivity. If one goes to medium energy ions (100-400 KeV) and high-resolution electrostatic energy analyzers then the depth resolution is greatly improved (to a few Å). In addition, if for single crystal material the scattering is done in a direction where atoms are aligned in strings along specific crystallographic directions (Fig. 4), backscattering can be confined essentially to those atoms in open view of the beam. Those atoms in the "shadow cone" (Fig. 4) cannot receive a direct

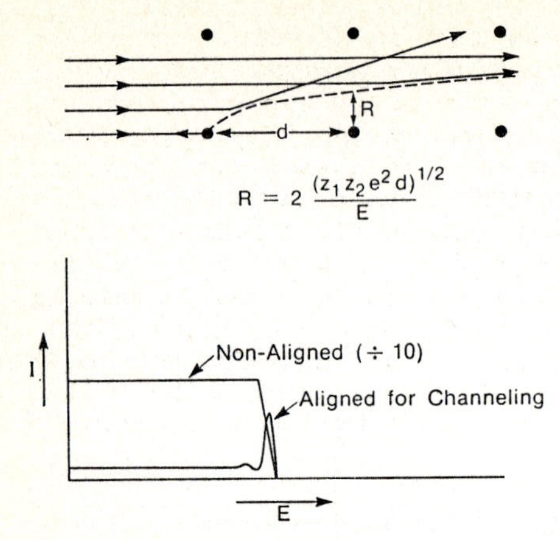

Fig. 4. (Upper) "Channeling" alignment of the He^+ beam in RBS along a specific crystallographic direction. Those ions not being backscattered by a nuclear collision suffer small angle forward scatterings from the electron cloud surrounding the nucleus and are channeled between strings of atoms. A shadow cone is created, radius R, from which no backscattering can occur. (Lower) RBS spectra of a thick, clean, substrate showing the difference in energy distributions between aligned and non-aligned directions.

nuclear backscattering collision. The effect of this is to convert the spectrum in the lower section of Fig. 4 to one in which most of the backscattering beyond the visible atomic layers is suppressed, as shown.

RBS results for the evaporation of Pd on Si(111) taken in this manner [4,5] are shown in Fig. 5. The scattering geometry is shown in the figure and actually both "channeling" of ions in, as described above, and "blocking" of backscattered ions out are used. Examination of the scattering geometry drawn shows that in this mode only the top two planes of Si(111) atoms for a clean surface are visible to the beam, resulting in the RBS spectrum shown in panel. From the area under the peak and the known scattering cross section for 176 KeV He^+, one can determine that the visible Si atom concentration is 2.6×10^{15} atoms/sq cm. This determination requires no standards. In the bottom panel of Fig. 5, a sequence of RBS spectra corresponding to the build up of Pd on the Si surface during evaporation is shown. There is a large amount of information in this sequence. First, one can determine the amount of Pd deposited from the area under the Pd RBS peak and establish the thickness of the Pd layers

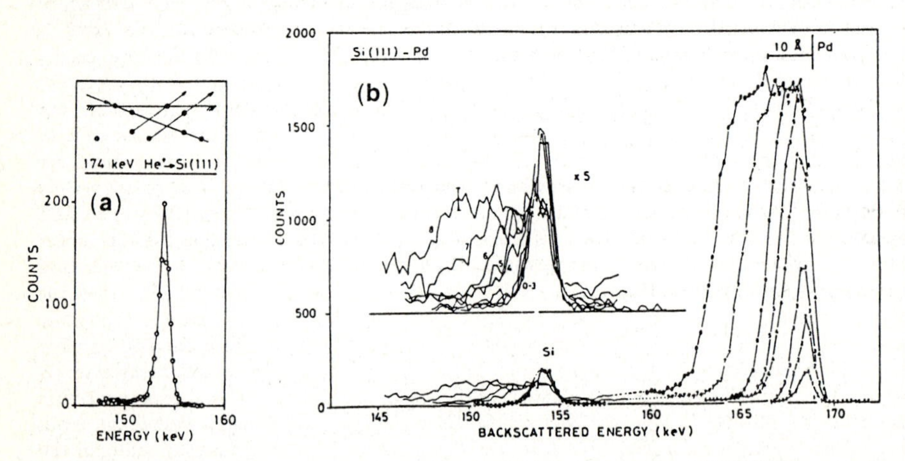

Fig. 5. (Left) Energy spectrum of backscattered He^+ ions in the [111]-[001] scattering geometry of a clean Si(111) surface [4,5]. (Right) Energy spectra of He^+ ions backscattered in the geometry described in the top panel for increasing Pd coverage (traces 1 through 8) at 300K [4,5].

from the width of the Pd peak, and therefore give the Pd concentration at which the first monolayer is complete and multilayers start to grow. The accompanying Si RBS peak does not show a decrease in energy of its leading edge, which means that the Si is never completely buried under the Pd layer. As the Pd layer thickens beyond the monolayer point, the Si RBS peak starts to broaden to lower energy and the intensity under the peak increases. This means that beyond one monolayer of Pd, the Si and Pd interdiffuse moving Si atoms off their original lattice sites and, therefore, increasing the number visible to the He^+ beam in the aligned orientation. The final traces show: (a) that the depth of the displaced Si is the same thickness as the Pd deposited (from peak widths); (b) the stoichiometry of this Pd/Si intermixed phase is Pd_2Si (from the relative intensities under the two RBS peaks); (c) the approx. 20Å of Pd_2Si is very uniform in composition (the trailing edges of the RBS peaks are sharp and the tops are flat); (d) there is no epitaxial relationship of the Pd_2Si to the Si^{III} substrate (variation of the outgoing direction of the He^+ detection does not give a specific direction in which dips in intensity are observed owing to blocking).

Subsequent annealing to 400°C produced large changes in the RBS spectra (not shown) which were interpreted as coalescence into approximately 40Å Pd_2Si islands which now *do* have an epitaxial relationship to the substrate, plus the segregation of some elemental Si on top of these islands. Altogether, these data illustrate the tremendous power of modern RBS analysis for determining thickness, stoichiometries, and epitaxial relationship of ultra-thin films.

5.3 Thin Film Disks: New Magnetic Media

IBM hard disks still use conventional particulate γ-Fe_2O_3 as their magnetic media. It is spin-coated on in an "ink." Candidates for future thin film disks are magnetic alloys or perhaps sputtered γ-iron oxide thin films [9]. The issues are the intrinsic magnetic properties, corrosion resistance, flyability and wear (tribology), and ease of manufacture. For thin iron oxide films prepared by post-oxidizing "Fe_3O_4" films made by reactive sputtering, an interesting phenomenon is observed. A magnetic "dead-layer" at the surface of some of the films has been reported [10,11]. This lack of magnetic moment has been established by Neutron Reflection [10] and Kerr Rotatation [11]. This, of course, is detrimental to the magnetic performance of the films, but it is also an interesting scientific question as to why the dead-layer is there when standard X-ray diffraction would indicate that the films are γ-Fe_2O_3 with no detection of other phases.

We examined such films by depth profiling Auger and by grazing angle 2.4 MeV He^+ RBS [12]. One point to note before discussing the results is that the films are doped with either Co or Os to control grain size and, therefore, coercivity.

The Auger profiles showed that, whereas the dopant was present up to the surface of the "Fe_3O_4" films prior to post-oxidation, it was missing in the surface region for the "γ-Fe_2O_3" produced by post-oxidizing that "Fe_3O_4" film. This is shown for the Co-doped films (2600Å thick) actually used in the earlier Neutron Reflection study in Fig 6. Similar results were found for Os-doped films.

The credibility of Auger depth profiles are affected by the fact that they are (a) destructive and therefore cannot be repeated on exactly the same sample region, and (b) there may be complications from preferential sputtering effects. However, the RBS measurements at grazing angle, which were actually performed with the object of establishing the Fe/O stoichiometrics of the films, confirm the Os depletion results very nicely. Figure 7 shows RBS spectra for a 500Å "Fe_3O_4" film and its post-oxidized partner, using 6° incident angle. One can clearly see that the leading edge of the Os peak is recessed to lower energy for the post-oxidized film, corresponding to Os depletion from the surface region (~70Å). In fact, because of the good separation of Os from Fe in RBS and the high cross section of Os, the

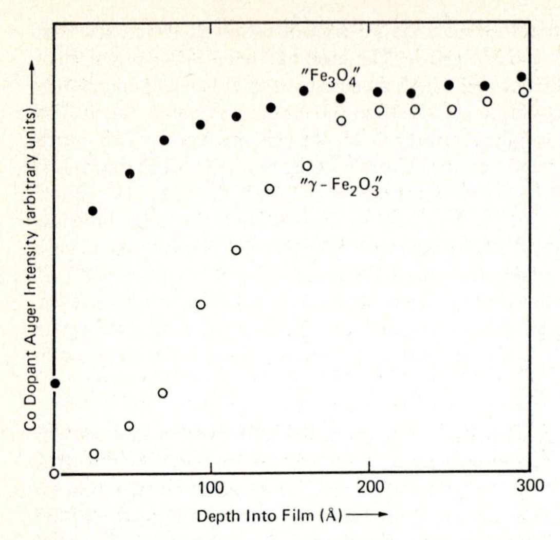

Fig. 6. Co depth profile, by Auger Spectroscopy, for a ~2600Å film of Co-doped "Fe_3O_4" and for a sister sample which has been post-oxidized to "γ-Fe_2O_3" [12].

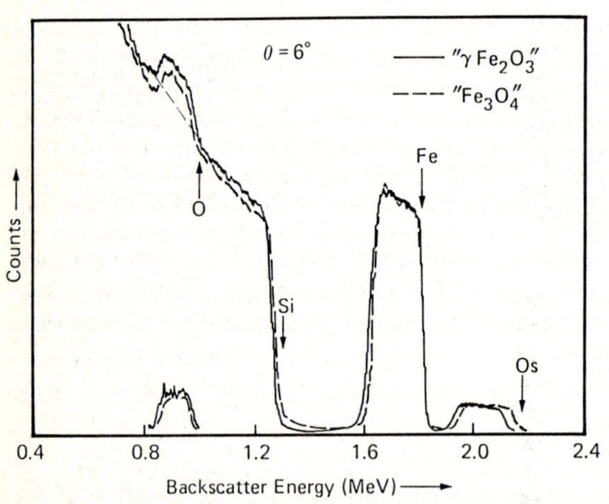

Fig. 7. RBS spectra for an Os-doped ~500Å film of "Fe_3O_4" and for a sister sample which has been post-oxidized to "γ-Fe_2O_3" [12].

absolute concentration of Os and the surface depletion are revealed non-destructively in 10 minutes of data taking. In addition, one can see the expected growth in thickness of the film on post-oxidation, as evidenced by the increased width of the Fe peak and the delayed onset of the Si substrate signal. If the RBS had been done at normal incidence, the depth resolution would have been too poor to observe any of these features.

The intriguing scientific question is how the dopant depletion and the magnetic dead-layer, which occur over the same surface thickness, are related. Is the dead-layer caused by the lack of dopant? To cut a long story short in the limited space available here, the answer is no. We believe that both phenomena are caused by the growth mechanism on post-oxidation of the "Fe_3O_4" film. In this oxidation process, Fe ions diffuse out of the lattice to react with the oxygen and produce a fresh oxide layer on the surface of the original film. For a fully oxidized film, the thickness should increase by about 7%. Apparently, the dopant ions do *not* move out into the growing lattice. In addition, the extra ~7% that grows (the dead-layer) is not γ-Fe_2O_3 at all but α-Fe_2O_3, which is non-magnetic. This latter point – the final solution of the problem – was subsequently established by grazing angle X-ray diffraction which showed *only* α-Fe_2O_3 at the surface, where conventional bulk X-ray diffraction had shown only γ-Fe_2O_3. Thus, we have a nice example of three different techniques, Auger profiling, RBS, and XRD being combined to form the full picture.

5.4 Packaging Technology: Surface Adhesion and Corrosion Problems

Figure 8 shows the cross section of a circuitry level under manufacture in a typical modern IBM circuitboard. There is a thin Cu layer on the epoxy substrate to which a resist has been laminated. Lithography is performed on the resist, opening line channels of approximately 100μ width and depth. This structure is then immersed in an electroless Cu plating tank and Cu lines plate up on the exposed Cu substrate. The plating process for this thickness of line (100μ) takes many hours and the plating bath is hot and quite basic. The bath can attack the interface between the resist and the Cu during the plating, causing adhesion loss. When the lines have been plated the resist is dissolved and the thin Cu substrate etched away, leaving the isolated Cu circuitry. The only purpose of the thin Cu substrate is to initiate the plating in the exposed channels. Since the rest of the thin Cu is eventually removed it might be thought that the integrity of the resist/Cu interface was not important. However, corrosion under the resist can be serious. At its worst, this causes the resist to lift (or even fall off) and extensive plating to occur under the resist, leading to shorts in the circuitry between lines after the Cu etch step (see Fig. 8). Alternatively, and more subtle, is simply the presence of corrosion pits on the Cu which cause local potential changes and debris formation adjacent to the lines while plating progresses. This results in defects in the lines, such as linewidth narrowing (see Fig. 7), which violate electrical specifications and are a reliability concern. To inhibit this corrosion, Benztriazole (BTA) is used to treat the Cu surface. BTA is a well-known corrosion inhibitor for Cu surfaces. It forms a 20-30Å Cu^+ BTA^- complex under the right pH conditions during aqueous application, which is protective. Under the wrong conditions, layers 1000's Å thick are produced, but they afford no protection against corrosion.

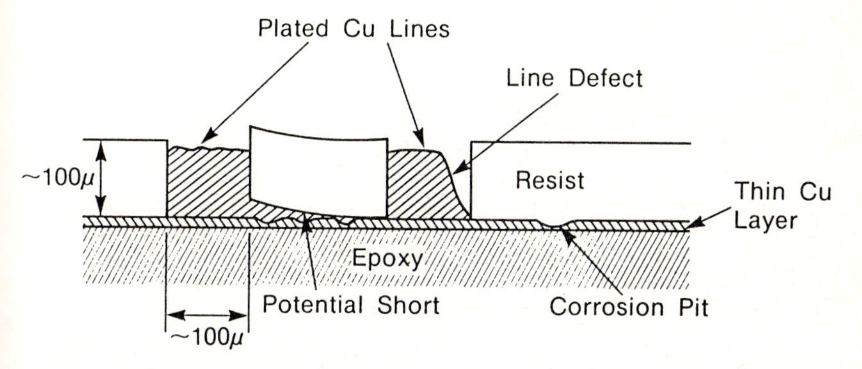

Fig. 8. Cross section through a circuitry layer of a typical multilayer circuitboard during the electroless plating step that deposits the Cu circuitry [14].

In an extended series of unpublished studies on the circuitboard manufacturing processes, we [14] used AES in a depth profiling mode and XPS to determine the thickness of the BTA layer under varying treatment conditions of the Cu (the nitrogen in BTA is used as the signature of its presence). We also correlated the thicknesses with other tests of the corrosion effectiveness, thereby establishing AES and XPS "standards" for properly protected Cu surfaces on real boards. Some of the corrosion problems in the manufacturing process were then tracked down to environmental attack on the BTA layer after application but prior to resist lamination. Proper handling eliminates this problem, but others remain. Line scan N Auger traces across the Cu surface (resist removed simply by physical peeling) after the rinse treatment, which takes place immediately prior to entering the plating bath showed the following. Whereas the correct BTA thickness was present over most of the Cu surface, immediately adjacent to the open line channels it was missing (Fig. 9). Thus, the rinse step which is designed to clean out the BTA on the exposed Cu at the bottom of the line channels also penetrates under the resist and removes it *right in the region where it is required for corrosion protection.* So far, then, the work described has used truly surface sensitive techniques, AES and XPS, to provide control information on layers that, though only 20-30Å thick, are critical in processing steps involving aggressive wet chemistry and producing structures (the 100μ Cu lines) which are not traditionally thought of as requiring detailed surface and interface information. The same techniques, plus SEM, were then used to study the corrosion behavior on the Cu itself. In fact, originally it was not known that the problem was corrosion, but was considered to be simply resist delamination due to "adhesion loss." For badly affected product, we peeled the resist after plating and SEM was used to locate the damaged spots on the Cu adjacent to the plated lines (Fig. 10a). AES was then used to depth profile in these spots. Two things were obvious. First, no BTA was present in these regions and second they were deep corrosion pits with heavily oxidized Cu (CuO by XPS) being present to up to 2μ depth within the pits (Fig. 10b). AES and XPS studies away from the pits, but still near the plated lines, showed only the normal ~20Å surface layer of Cu_2O and a very low level of BTA present. Far away from the plated lines the normal 30Å of BTA was found. Finally, the

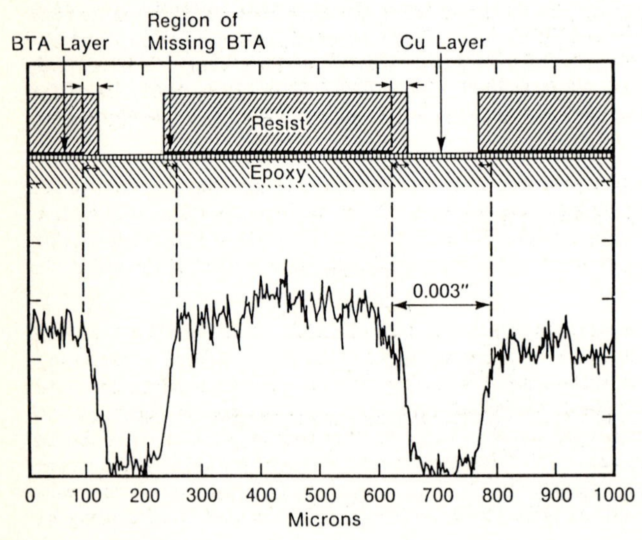

Fig. 9. N Auger line scan across the Cu surface prior to the electroless plating step. To obtain this scan the resist was removed by flexing the board and physically peeling the resist from the Cu surface [14].

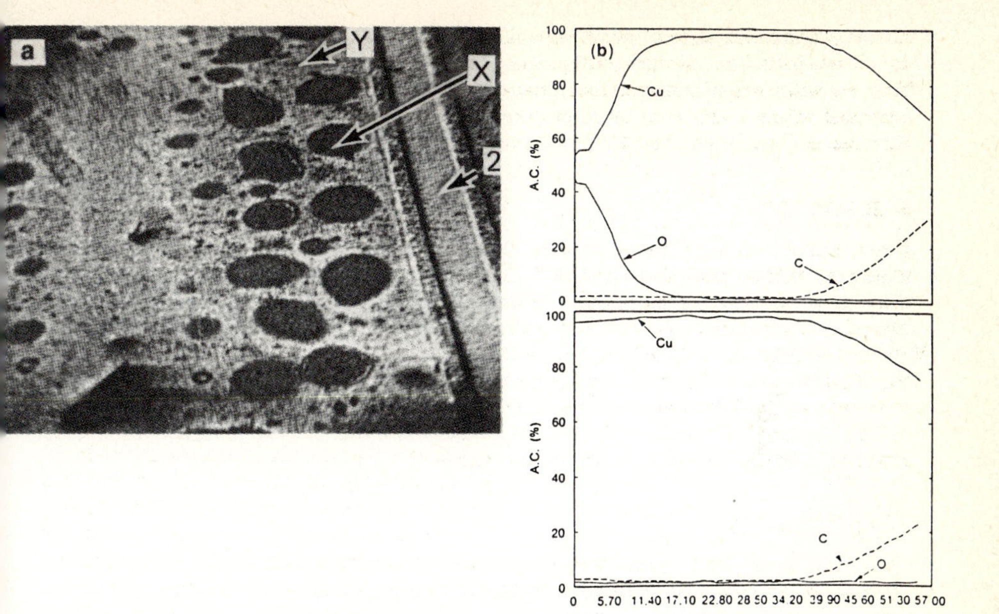

Fig. 10. (a) SEM of a section of the circuitboard after Cu plating and physical removal of the resist by peeling to reveal the Cu surface. The corrosion pits show as black spots [14]. (b) AES depth profiles in region X of the SEM micrograph (upper panel) and regions Y (lower panel) [14].

plating bath could be definitively blamed for all these problems because Na from the bath (contains NaOH) was found on the Cu surfaces adjacent to the line channels, even after only 5 minutes in the bath. No Na is found in region far away from lines where the BTA levels are normal. Large amounts are found in the corrosion pits. Thus, the bath penetrates the resist/Cu interface almost instantly and causes corrosion on the Cu which is already lacking its protective BTA coating.

The surface analysis "detective work" so far had thus established what the problem was and how it occurred. The remaining questions were (a) how to fix the problem and (b) how to monitor for it more efficiently on-line; that is, is there some single way one can establish whether the problem is going to occur other than waiting for bad product at the end of the plating stage?

For (a) a number of changes taken together proved effective. First, the BTA treatment process was optimized (temperature, time, concentration of aqueous application treatment). Second, the time the treated surface spent in an uncontrolled environment before resist application was minimized. Third, the pre-cleaning step prior to plating was made less aggressive. All the above steps were aimed at improving the corrosion resistance of the Cu resist interface to the plating bath. The final step was to have tighter control on the expose and develop steps of the resist lithography. If this is not properly under control, the resist adhesion to the Cu is poor and the plating solution has an easier time penetrating along the interface.

For (b), we knew from Auger measurement that, for "bad" boards, the plating bath penetrated quickly along the interface from a line channel. We, therefore, looked for some simple, quicker, and cheaper way of establishing this than destructive analysis by Auger. In fact, it was found that any penetration along the interface could be followed by a color change observed with an optical microscope through the resist to the interface *provided* the board was examined within a half-hour of immersion in the bath. After this, the resist progressively darkened and changes at the interface could not be followed optically.

6. SUMMARY

I have tried to present in this paper some of the flavor of both the practical and basic research aspects of surface analysis within IBM. IBM has its Research Division spread over three laboratories and has a large range of manufacturing sites where surface analysis is part of everyday technical business. Of necessity, then, my view is a personal one and I apologize to all my IBM surface analysis colleagues whose work I have been unable to mention here. If there is any overall message, it is probably that one should strive to use the right combination of techniques for the problem at hand, rather than the one you are most familiar with or have easiest access to. In addition, I believe great progress has been made in recent years by not dividing work, and people, into "basic" and "applied."

6. REFERENCES

1. D. Hercules, M. Houalla: Fresenius Z. Anal. Chem. 324, 589 (1986).
2. Surface Science Laboratories, Palo Alto, California, U.S.A.
3. E.g., V. G. Scientific Ltd., East Grinstead, United Kingdom.
4. R. Tromp, E. J. Van Loenen, M. Iwami, R. Smeenk, F. W. Saris: Thin Solid Films 93, 151 (1982).
5. R. M. Tromp, E. J. Van Loenen, M. Iwami, R. G. Smeenk, F. W. Saris, F. Nava, G. Ottaviani: Surface Science 124, 1 (1983).
6. P. J. Mills, P. F. Green, C. J. Palmstrom, J. W. Mayer, E. J. Kramer: Appl. Phys. Lett. 45, 9571 (1984).
7. C. R. Brundle, E. Silverman, R. J. Madix: J. Vac. Sci. Tech. 16, 474 (1979).
8. W-Y Lee, G. Scherer, C. R. Guarnieri: J. Electrochem. Soc. 126, 1533 (1979).
9. A. Terada, O. Ishii, S. Ohta: IEEE Trans. Magn. MAG-21, 521 (1985).
10. E. Kay, R. A. Sigsbee, G. L. Bona, M. Taborell, H. C. Siegman: Appl. Phys. Lett. 47, 533 (1985).
11. S. S. Parkin, R. A. Sigsbee, R. Felici, G. P. Felcher: Appl. Phys. Lett. 48, 604 (1986).
12. B. Hermsmeier, C. R. Brundle, J. Baglin, to be published.
13. T. Huang, unpublished data.
14. C. R. Brundle, D. J. Auerbach, D. C. Miller, unpublished results.

Dynamical Charge Transfer Process. SIE from Si and Si:O

M.C.G. Passeggi, E.C. Goldberg, and J. Ferrón

Instituto de Desarrollo Tecnológico para la Industria Química, Universidad Nacional del Litoral, Consejo Nacional de Investigaciones Científicas y Técnicas, Guemes 3450, 3000 Santa Fe, Argentina

1. Introduction

Since the introduction of Secondary Ion Mass Spectrometry (SIMS) and Ion Scattering Spectroscopy (ISS) there has been a growing interest in the basic mechanisms responsible for the final charge states of particles either leaving or approaching a solid surface, as well as the effects on the ionization probability caused by surface impurities.

The goal of the different existing models is mainly related to the velocity functional dependence of the ionization probability and, even in this case, the agreement with the experimental evidence is only partially fulfilled. In particular, deviations from the exponential behaviour for low sputtered ion velocities [1,2] has been accounted for in only two cases [3,4]. On the other hand, the most important models which have intended to rationalize the matrix effects, the Bond Breaking (BB) and the Surface Polarization Model [5], have been never discussed in other than a qualitative way.

In this work we examine the ionization probability for Si atoms ejected from a Si (100) surface, and present quantitative calculations for the oxygen effect on the Si positive secondary ion emission. We adopted a description of the atom-substrate system based on a cluster of 80 atoms, using a Tight Binding (TB) approach in the first neighbor approximation with a sp^3s^* basis set per atom, with parameters taken as those which allow for a good description of the band structure of a Si crystal. The dynamical process is simulated by assuming an exponentially time decaying interaction between the ejected atom and the substrate.

2. The Time Dependent Formalism

We have used a cluster of 80 atoms having a free 100 surface, where dangling bonds belonging to other boundary surfaces were eliminated from the basis set. Oxygen atoms were included in a substitutional site, either incorporated at the subsurface or adsorbed at the surface. In the TB approximation the Si p-electron energies are located above the Fermi level. To correct this unrealistic asymptotic behaviour, we allowed one p-state parameter to vary along the trajectory from its value in the solid to the correct one in the free atom.

We describe the electronic part of the system through the time dependent hamiltonian:

$$H(t)=\sum_{i,\sigma} \epsilon_{ai}(t)\, a^{+}_{i\sigma} a_{i\sigma} + \sum_{n,\sigma} \epsilon_n \psi^{+}_{n\sigma} \psi_{n\sigma} + \sum_{i,m,\sigma} (V_{im}(t)\, a^{+}_{i\sigma} \psi_{m\sigma} + h.c.)\,. \qquad (1)$$

The first term describes the sputtered atom. We assume for one p-state diagonal parameter: $\epsilon_{ap}(t) = (\epsilon_p - V_o) + V_o \exp(-\lambda t)$,

V_0 and λ characterizing the time decaying energy. The second term in (1) describes the cluster substrate hamiltonian; and the last term gives the atom substrate interaction, where the matrix elements are assumed to decay in time as: $V_{im}(t) = V_{im}(0)\exp(-\lambda t)$.

The dynamical evolution of the system can be made in terms of the time evolution operator $U(t,t0)$, with the initial condition $U(t0,t0) = 1$. As $H(t)$ is a one particle Hamiltonian, $\psi(0)$ is a Slater determinant constructed by filling the molecular state Φ_α eigenfunctions of $H(0)$, by increasing order of energies up to the Fermi level. Therefore $\psi(t)$ is obtained from $\psi(0)$ as:

$$\psi(t) = \det\{|U(t,0)|\Phi_\alpha>\}. \tag{2}$$

With five orbitals per site, the selection of all different electronic configurations of the adsorbate contained in $\psi(t)$, leading to a specified charge state, becomes more complex than in the usual case where only one orbital is involved. Thus the probability of having a singly ionized Si atom is given by

$$p^+=2\{\sum_i<n_{pi}>-\sum_{j>i}<n_{pi}n_{pj}>+<n_{p1}n_{p2}n_{p3}>\}*\{1-\sum_i<n_{pi}>+\sum_{j>i}<n_{pi}n_{pj}>-<n_{p1}n_{p2}n_{p3}>\}, \tag{3}$$

where

$$<n_{pi}> = \sum_{\alpha=occ}|<a_i|U(t,0)|\phi_\alpha>|^2,$$

$$<n_{pi}n_{pj}>=<n_{pi}><n_{pj}>-|\sum_{\alpha=occ}<a_i|U(t,0)|\Phi_\alpha><\phi_\alpha|U(0,t)|a_j>|^2,$$

$$<n_{p1}n_{p2}n_{p3}>=<n_{p1}><n_{p2}><n_{p3}>-\sum_{i(j,l\neq i)}<n_{pi}>|\sum_{\alpha=occ}<a_j|U(t,0)|\phi_\alpha><\phi_\alpha|U(0,t)|a_l> \tag{4}$$

We have ignored the configurations s^1p^2 and p^3, and those involving the occupancy of the s* state, as we verified they contribute negligibly to p^+.

According to (3) we are requested to know the matrix elements of the evolution operator between the asymptotic states centered at the sputtered atom and the MO's occupied at t=0. In order to write the motion equation in matrix form, we expanded it in terms of the MO eigenstates of $H(\infty)$ as:

$$i\frac{\partial}{\partial t}<\phi_i(\infty)|\tilde{U}(t,0)|\phi_\alpha> = \sum_k \exp\{i[E_i(\infty)-E_k(\infty)]t\}* <\phi_i(\infty)|H_1(t)|\phi_k(\infty)><\phi_k(\infty)|\tilde{U}(t,0)|\phi_\alpha>. \tag{5}$$

Because of the exponential form assumed for the decaying interaction, an integration by parts can be performed with the result that an iterative scheme to compute the required matrix elements is obtained.

3. Results and Discussion

Figure 1 shows schematically the different configurations selected to study the emission of silicon ions. The ejected atoms, 2 and 3 are respectively first and third neighbors of the oxygen atom in the case of fig. 1a, and second and fourth neighbors in the case of fig. 1b.

The plot of p^+ as a function of the inverse of the ion velocity is shown in fig. 2. These results show, at high ion velocity the typical exponential behaviour: $p^+ \simeq \exp(-vo/v)$ with $vo \simeq 1.10\ 7$ cm/s. This value of vo is in reasonable agreement with experimental results [1,2], as well as with previous the-

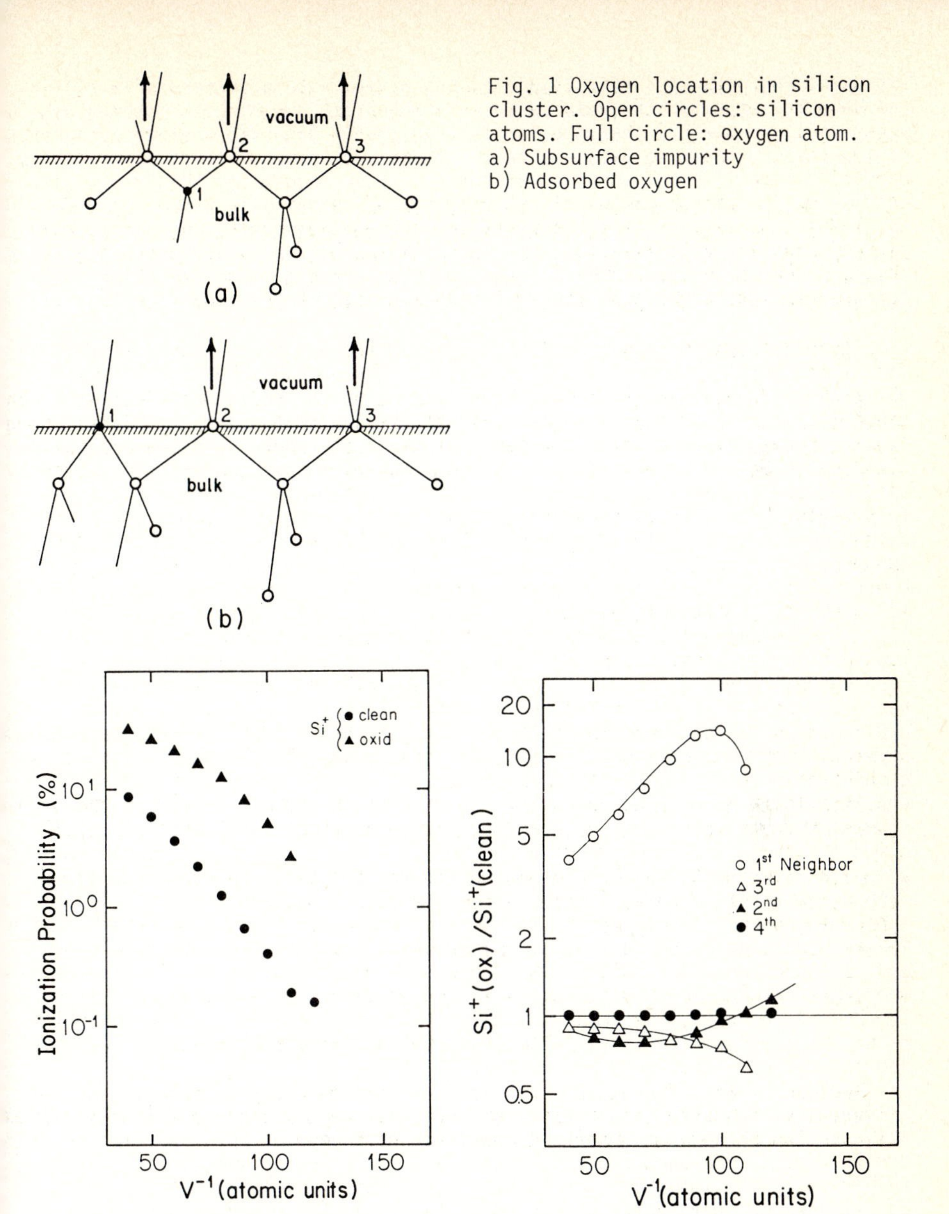

Fig. 1 Oxygen location in silicon cluster. Open circles: silicon atoms. Full circle: oxygen atom.
a) Subsurface impurity
b) Adsorbed oxygen

Figure 2 Ionization probability vs inverse of ion velocity.

Figure 3 Oxidized to clean ionization probability ratio vs inverse of ion velocity

oretical calculations [6,3]. For low velocities ($v \simeq 2.10^6$ cm/s), we found a behaviour of P^+ vs v^{-1} which is weaker than an exponential one. This behaviour has been observed experimentally for O^- [1], and recently for positive ion emission [2], in the same velocity range. The deviation from the exponential dependence has been attributed to the variation of the ion velocity along the ion path [4]. VASILE [2] by using an image force term adjusts the low velocity experimental results to an exponential dependence. However, it was recently shown [7] that the forces required in order to achieve agreement between experimental results and calculations, would be at least one order of magnitude larger than typical interatomic forces, suggesting that the exponential deviations can be hardly attributed to ion velocity variations. In our model calculations, these deviations occur independently of our assumption of constant velocity along the ion path. We have already pointed out that this exponential deviation may be due to interferences among the elementary contributions to the transition amplitude along the trajectory [3].

Figure 2 shows p^+ as a function of v-1 for the ejection of atom 2 (fig. 1a). These results show a strong increase in the probability of Si^+ emission which goes up by a factor of 5 at high velocities to about 10 at low velocities. This is expected on chemical grounds because of the large electron affinity of oxygen, this being the basis of the B.B. model. In order to show clearly the effect of oxygen, Fig. 3 plots the ratio between Si^+ ion yields obtained for an oxidized surface against those corresponding to a clean one. This ratio shows several interesting features. Firstly, the large effect of oxygen on first to other neighbors, up to one order of magnitude, supports our previous assumption on the strongly localized character of the oxygen enhancement. In the case of the 4th neighbor, the silicon atom does not feel the presence of the oxygen impurity. Another important feature appearing in Fig. 3 is the velocity dependence of the oxygen effect, being stronger for low velocities. Finally, an important qualitative result appearing in Fig. 3 is the possibility of oxygen partially lowering the positive ion emission: see the curves for the 2nd and 3rd neighbors. This effect, which can not be explained on the grounds of the high electronic affinity of oxygen, is not relevant from the experimental point of view, since it is impossible to discriminate among silicon ions coming from sites far or near to an oxygen atom, and this lowering will be masked by the strong enhancement of ions coming from sites near the impurity. However, this effect opens the possibility for oxygen to be capable of enhancing the negative ion emission, a phenomenon which is experimentally well known. The lowering of Si^+ ion yield for atoms located far from the oxygen atom may be understood considering that for a Si atom not bounded to the oxygen, the main change occurs through a global change in the band states. In this case, the arguments turn to be similar to those of the tunneling model. Within this model, we may expect a Si^+ yield decrease whenever the work function of the Si-O system decreases (incorporated oxygen). In the present case, our results for Si^+ tend to agree with such a prediction, mainly at low ejection velocities, although the changes in the band states, and in the work function are very small.

These results also support previous suggestions about the coexistence of at least two mechanisms governing the effect of oxygen on the ion emission [5]: one related to the bond breaking model, and the other to models like the tunneling one. The importance of each mechanism being dependent on the distance of the ejected atom to the oxygen impurity. Since the stronger effect is produced by oxygen over its near neighbors, and it is originated in its high electron affinity, it appears reasonable that no correlation will be found between Si^+ yield enhancement and the global changes in the electronic structure of silicon due to oxygen, as was recently measured by YU and coworkers [8]

4. References

1. M.L. Yu: Phys. Rev. Lett. 47, 1325 (1981)
2. M. Vasile: Phys. Rev. B29, 3735 (1984)

3. E.C. Goldberg, J. Ferrón and M.C.G. Passeggi: Phys. Rev. B30, 2448 (1984)
4. N.D. Lang: Phys. Rev. B27, 2019 (1983)
5. P. Williams: Appl. Surface Sci. 13, 241 (1982)
6. G. Blaise and A. Nourtier: Surface Sci. 90, 496 (1979)
7. W.L. Clinton and S. Pal: Phys. Rev. B33, 2817 (1986)
8. M.L. Yu, J. Clabes, and D.J. Vitkavage: J. Vac. Sci. A3(3)1316 (1985)

Quantitative Auger Analysis of FeSi(100) Using Noble Gas Ion Sputtering

A. Ballesteros, C.E. Rojas, and G.R. Castro

Laboratorio de Física de Superficies, Departamento de Física,
Facultad de Ciencias, Universidad Central de Venezuela, A.P. 21201,
Caracas 1020-A, Venezuela

1 Introduction

The surface composition of a binary alloy or compound A_xB_{1-x} may be different from its bulk composition because of a number of different effects like the formation of a surface oxide layer, surface segregation of one of the components or preferential sputtering if the sample is bombarded by noble gas ions. This precludes the direct use of binary alloys of known bulk composition as reference standards for the calibration of Auger signals. In quantitative Auger Electron Spectroscopy (AES) the sensitivity factor of element A relative to element B is commonly taken as the ratio of the Auger peak height measured on a standard sample of pure A to the Auger peak height measured on a standard sample of pure B. This elemental relative sensitivity is then modified by a matrix correction factor since the Auger peak height is known to be affected by matrix dependent parameters like atomic density, escape depth and backscattered electron current. HALL and MORABITO /1/ have shown how to evaluate numerically these effects and hence a considerable improvement has been achieved in the quantification of AES.

For a proper determination of the relative elemental sensitivity factor the standards reference data must be recorded on the same instrument used for the analysis of the unknown sample and under the same experimental conditions. It is usually assumed that sample and standards all have the same surface roughness. The numerical evaluation of the matrix correction factor is easy to carry out only in the dilute limit case, in which the matrix consists of almost entirely one element. This evaluation is usually based on the densities of the reference standards, on an universal curve for the electron inelastic mean free path and on general semiempirical expressions for the electron backscattering factor. It would then be desirable to have a quantification method for AES based on measurements only of the sample under analysis, whose bulk composition is known, and without the need to resort to numerical estimates of the matrix correction factor.

In this work we use a quantification method for AES of a $Fe_{0.5}Si_{0.5}$(100) sample based on its known bulk composition and on the preferential sputtering phenomenon of alloy surfaces.

2 Experimental Results

Ion sputtering experiments were performed with a $Fe_{0.5}Si_{0.5}$ single crystal sample cut and polished to expose the (100) face. The sample was mounted on a holder with provision for direct heating. By rotating a precision manipulator the sample could be positioned to face either the ion gun for sputtering or the cylindrical mirror analyser for the subsequent Auger analysis. Ar^+ with a kinetic energy of 2250 eV and Ar^+, Ne^+ and Kr^+ with a kinetic energy of 1000 eV were used as sputtering ions. Auger analysis was performed under UHV conditions using a 2000 eV, 0.8 μA primary electron beam at normal incidence and a Varian CMA with 1 V peak-to-peak modulation. Before the sputtering experiments the sample was annealed at 500°C for 12 hours. Ion sputtering was carried out with the sample at room temperature and using ion intensities of about 10^{14} ions /cm^2 sec.

The low energy, surface sensitive, Si 91 eV and Fe 47 eV Auger signals were used to monitor the changes in surface composition induced by sputtering. The Auger signals corresponding to the bombarding species were always very small, so ion implantation can be considered as negligible.

The ratio R = I(Si 91 eV)/I(Fe 47 eV) of the Si 91 eV to the Fe 47 eV Auger peak heights was obtained as a function of the accumulated sputtering time t for Ar^+(2250 eV), Ar^+(1000 eV), Ne^+(1000 eV) and Kr^+(1000 eV). The ratio of signals corresponding to the sample at thermal equilibrium was R_{eq} = 1.18. Figure 1 shows the plot R vs t for the case of sputtering with 2250 eV Ar^+ions. Initially, for $t < 20$ sec., there is a rise in the values of R, followed by a monotonic fall that finally attains a steady state value R_∞.

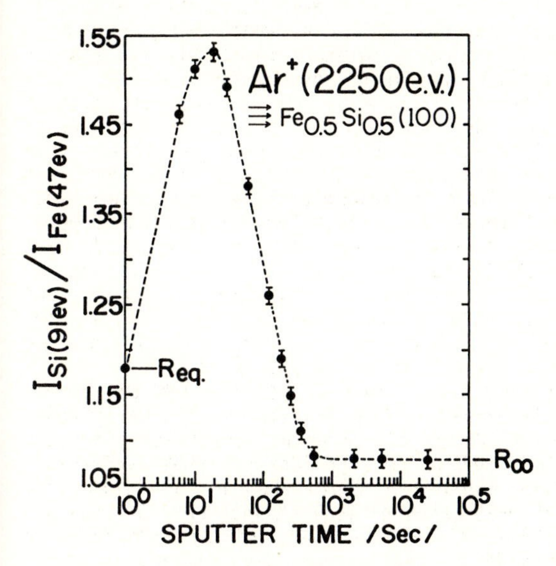

Fig. 1 Time variation of I(Si 91 eV)/I(Fe 47 eV) observed for sputtering of $Fe_{0.5}Si_{0.5}$(100) using 2250 eV Ar^+ions.

Similar R vs t plots were obtained for the sputtering experiments using 1000 eV Ar^+, Ne^+ and Kr^+ ions. In these cases the initial rise occurs for $t < 30$ sec., the final steady state value R_∞ being different for each type of projectile.

3 Quantification Method

The quantification method for AES to be used is based on the phenomenological model of preferential sputtering formulated by HO et al. /2/ and by CHOU and SHAFER /3/. According to this model, the process of sputtering of an homogeneous multicomponent sample is seen as the formation and development of an altered layer which recedes into the target with ion bombardment. After a sputtering time t_0 the thickness of the altered layer becomes constant and the surface composition of the alloy varies with time till a steady state value is achieved. If we express the surface composition of a binary alloy AB in terms of the atomic fraction C of, let us say, component A ($C = C_A$, $C_A + C_B = 1$), then its time variation $C(t)$ is related to the steady state final surface composition $C(\infty)$ and to the initial surface composition $C(0)$ by the expression

$$\ln H(t) = - t/\tau \;, \quad (t>t_0, \tau \text{ constant}) \tag{1}$$

where the function $H(t)$ is defined as the ratio

$$H(t) = \frac{C(t) - C(\infty)}{C(0) - C(\infty)} \tag{2}$$

$C(\infty)$ and τ being dependent on the bulk composition C_0.

In a homogeneous binary sample $C(0) = C_0$, i.e. the initial surface composition is the same as in the bulk. This kind of sample could be used as a reference standard; the ratio $R_0 = I_A(C_0)/I_B(C_0)$ of their Auger I_A and I_B signals could be used to calculate any other surface composition C, yielding a ratio of signals $R = I_A(C)/I_B(C)$, using the expression

$$C = \frac{C_0 R}{C_0 R + (1 - C_0) R_0} \;. \tag{3}$$

The problem is that R_0 is difficult to determine directly because the surface composition is usually different from the bulk composition. However, we can make use of the linear relationship between ln H and t. By replacing (3) into (2) one can write (1) as

$$\ln \frac{R_0\{R - R_\infty\}}{\{R_0 - R_\infty\}\{R_0(1 - C_0) + C_0 R\}} = - t/\tau \tag{4}$$

and use this expression to determine R_0. The method consists then in choosing a value of R_0 such that a least-squares fit of the latter part of the sputtering data should pass, according to (4), through the origin when extrapolated to $t = 0$. Initial departures from linearity can occur if $C(0) \neq C_0$. This procedure can be carried out iteratively starting with a suitable value of R_0 that then can be corrected according to the value of the intercept of $\ln H(t)$ at $t = 0$. The value of R_0 so obtained can then be used in (3) to calculate surface compositions.

From the sputtering data the values R_∞ corresponding to the steady state final conditions were obtained. Figure 2 shows the plot of ln H vs t for Ar^+ (2250 eV) and for Ne^+(1000 eV). Initially, there is a rise in the values of $\ln H(t)$, followed by a linear fall as predicted by (1). The straight line extrapolated to the origin corresponds to the least-squares fit to the latter part of the data, used to obtain the R_0 values. Similar plots were obtained for the sputtering with 1000 eV Ar^+ and Kr^+ ions.

In Table 1 we collect the values of R_∞, R_0 and τ obtained for the four sets of sputtering conditions.

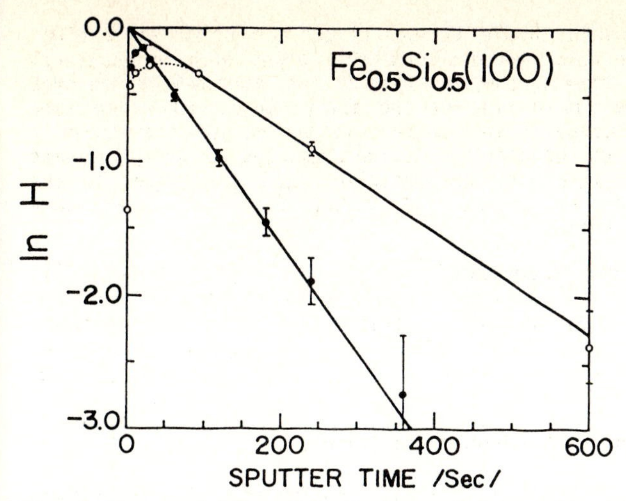

Fig. 2 Time variation of ln H observed for sputtering of $Fe_{0.5}Si_{0.5}(100)$.
: Ar^+(2250 eV), o : Ne^+(1000 eV).

Table 1 Values of R_∞, R_0 and τ for $Fe_{0.5}Si_{0.5}(100)$.

Sputtering condition	R_∞	R_0	τ/sec/
Ar^+(2250 eV)	1.08	1.62	137
Ar^+(1000 eV)	1.10	1.61	340
Ne^+(1000 eV)	1.01	1.66	260
Kr^+(1000 eV)	1.06	1.63	320

4 Discussion

There is an initial rise of the ln H vs t curves for $t < 20$ sec in the case of sputtering with 2250 eV Ar^+ ions and for $t < 30$ sec in the case of sputtering with 1000 eV Ar^+, Ne^+ and Kr^+ ions. These times are related to the initial surface composition and must not be confused with the much shorter time t_0 after which the thickness of the altered layer becomes constant. This initial deviation from linearity indicates that the original equilibrium surface composition was different from the bulk composition. The surface composition C(t) that appears in (2) corresponds to the mean composition within the altered layer, while the surface composition C given by (3) is basically an effective average composition within the escape depth of the Auger electrons used for the analysis. We then expect the model to be valid if the width of the altered layer is comparable to the escape depth of the Auger electrons, which we estimate to be between 2 and 4 atomic layers. As the sputtering proceeds, ln H vs t follows a straight line related to the bulk composition C_0 and independent of the original surface conditions. This linearity brings support to the sputtering model and allows us to determine R_0 by the extrapolation procedure described above. The fact that sputtering with different projectiles and at different energies gives the same value of R_0 confirms the reliability of the present quantification method for AES of FeSi. As can be seen in Table 1, the values of R_0 obtained for the different sputtering conditions are very close to each other so their average value $\overline{R}_0 = 1.63$ can be used in (3) in order to calculate the surface composition of the $Fe_{0.5}Si_{0.5}$(100) sample. In this way we obtain that at thermal equilibrium the surface is depleted in Si to 42%; upon ion sputtering the Si composition initially rises to 47-49%, and finally attains a steady state composition of 38-40%.

The same method was used successfully in reference /2/, with 500 eV Ar^+ ions, for the calibration of the Auger signals in a polycrystalline CuNi alloy presenting a surface oxide layer. The fact that this quantification procedure can be applied for different sputtering conditions and in systems so different as CuNi and FeSi, allows us to recommend its use in other binary systems.

5 Literature

1. P.M. Hall and J.M. Morabito: Surf. Sci. 83, 391 (1979)
2. P.S. Ho, J.E. Lewis, H.S. Wildman and J.K. Howard: Surf. Sci. 57, 393 (1976)
3. N.J. Chou and M.W. Shafer: Surf. Sci. 92, 601 (1980)

Index of Contributors